Methods for Enhanced Oil Recovery

Methods for Enhanced Oil Recovery

Fundamentals and Practice

Baghir A. Suleimanov
Elchin F. Veliyev

WILEY-VCH

Authors

Prof. Baghir A. Suleimanov
Deputy Director
Oil Gas Scientific Research Project Institute,
SOCAR
Baku, Azerbaijan

Dr. Elchin F. Veliyev
Manager of Analytical Researches Laboratory
Oil Gas Scientific Research Project Institute,
SOCAR
Baku, Azerbaijan

Cover Design: Wiley
Cover Image: © Han maomin/Shutterstock

Library of Congress Card No.: applied for

British Library Cataloguing-in-Publication Data
A catalogue record for this book is available from the British Library.

Bibliographic information published by the Deutsche Nationalbibliothek
The Deutsche Nationalbibliothek lists this publication in the Deutsche Nationalbibliografie; detailed bibliographic data are available on the Internet at http://dnb.d-nb.de.

Print ISBN: 9783527354160
ePDF ISBN: 9783527848300
ePub ISBN: 9783527848294
oBook ISBN: 9783527848317

Typesetting: Lumina Datamatics
Printing and Binding: CPI Group (UK) Ltd, Croydon, CR0 4YY

C9783527354160_300625

Contents

Preface

The science and practice of Enhanced Oil Recovery (EOR) have evolved significantly in recent decades, becoming an essential component in extending the productive life of oil fields and maximizing hydrocarbon extraction. This book, *Methods for Enhanced Oil Recovery—Fundamentals and Practice*, provides a comprehensive exploration of both theoretical and practical aspects of modern EOR techniques, making it a valuable resource for students, researchers, and engineers in the petroleum industry.

The early chapters cover fundamental reservoir engineering concepts, including the geology of oil fields, the classification of hydrocarbons and reserves, and the principles behind primary and secondary recovery methods. These foundational topics are essential for understanding the dynamics of oil reservoirs and their behavior under different recovery techniques.

A substantial portion of this book is dedicated to advanced methods of enhancing oil recovery, with in-depth analyses of various modern techniques. Readers will find detailed explanations of thermal, gas, chemical, and microbial EOR methods, as well as emerging innovations. Each method is presented not only with a theoretical background but also with practical guidelines for field implementation. Special attention is given to the criteria for selecting an appropriate EOR method for specific reservoirs and the technological processes necessary to implement these methods in operational settings.

A dedicated chapter focuses on the laboratory evaluation of oil-bearing rock properties. This section covers essential testing methods, such as porosity and permeability measurements, wettability tests, and interfacial tension evaluations. The detailed procedures for laboratory analysis provide readers with the tools to better understand the behavior of reservoir rocks, enabling more accurate predictions of EOR efficiency and outcomes.

Real-world applications play a prominent role throughout the book, with case studies from oil fields in various regions illustrating both the successes and challenges faced in implementing EOR technologies. Furthermore, the book explores innovative EOR techniques that are currently under development or in pilot testing, highlighting the industry's future direction.

In addition to the technical aspects, the book also addresses the economic evaluation of EOR methods, emphasizing the importance of balancing technological efficiency with economic feasibility to guide decision-making in practical scenarios.

Whether you are a student looking to deepen your understanding of EOR or an engineer seeking practical guidance for field applications, this book provides both a solid theoretical foundation and practical insights. We hope it will inspire innovation and contribute to the successful deployment of EOR techniques in oil fields worldwide.

Introduction

The average oil recovery across the world is about 30%. At the same time, proven geological oil reserves in the world, as of the end of 2019, amount to 1.73 trillion barrels (237 billion tons). It is obvious that the current efficiency of oil field development, including enhanced oil recovery (EOR) methods, is insufficient. In connection with this, the relevance of research in the field of EOR increases.

The book brought to your attention is devoted to a detailed study of existing methods for increasing oil recovery, as well as the prospects for increasing their efficiency. EOR methods are part of improved oil recovery (IOR) methods, which include, in addition to EOR, new technologies for drilling and completing wells, control and regulation of development, modern monitoring methods, as well as primary and secondary field development methods.

The current contribution of EOR to total oil production is of great importance for analyzing the prospects for increasing their efficiency. Figure 1 shows the distribution by year of the share of oil production due to EOR in relation to the total production of the USA and China, which are the largest oil producers using EOR.

As can be seen from Figure 1, oil produced as a result of EOR in China significantly exceeds the same figure in the USA, while the indicated figure for China continues to grow (from 14.88 in 2010 to 18.32 in 2016), and for the USA—decrease (from 8.67 in 2010 to 5.83 in 2016). The share of oil produced through EOR in China is approaching 20%. This is the best indicator for countries in the world (the world average is 3.3%, according to 2014 data). It is clear that there is significant potential for improving the efficiency of EOR.

Figure 2 shows the share of each country using EOR in the total oil production from EOR by country of the world, according to data from 2010 to 2014.

As can be seen from Figure 2, four countries—the USA, China, Venezuela and Canada—produce almost 80% of the additional oil obtained using EOR. The distribution of oil according to the applied methods of EOR is also of interest. The specified distribution is shown in Figure 3.

As can be seen from Figure 3, a number of EOR methods give up to 98% additional oil. These are steam injection, chemical methods, carbon dioxide, and hydrocarbon gas injection. World indicators for generalized methods of EOR (Figure 4) allow us to conclude that thermal methods are the main method of increasing oil recovery, although their share is decreasing from 58.2% in 2010 to 48.2% in 2014. At the same time, the share of chemical methods with 14.7% in 2010 to 26.2% in 2014. The share of gas EOR methods remains virtually unchanged.

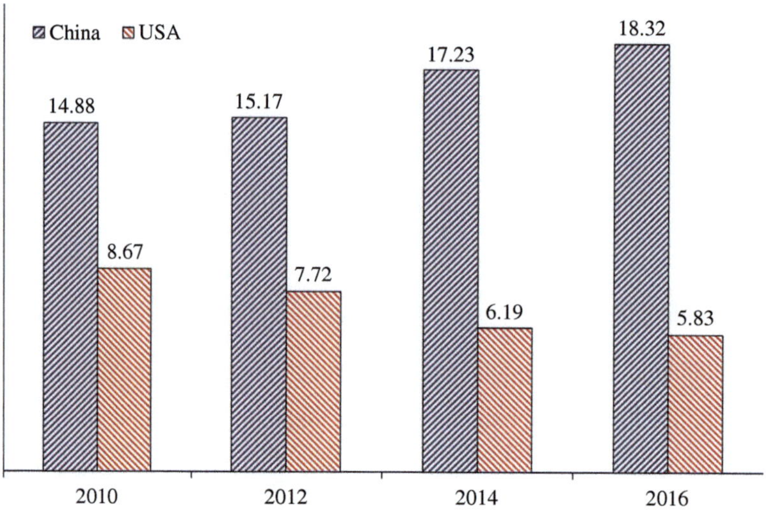

Figure 1 Distribution by year of the ratio of oil production due to EOR (%) in the USA and China.

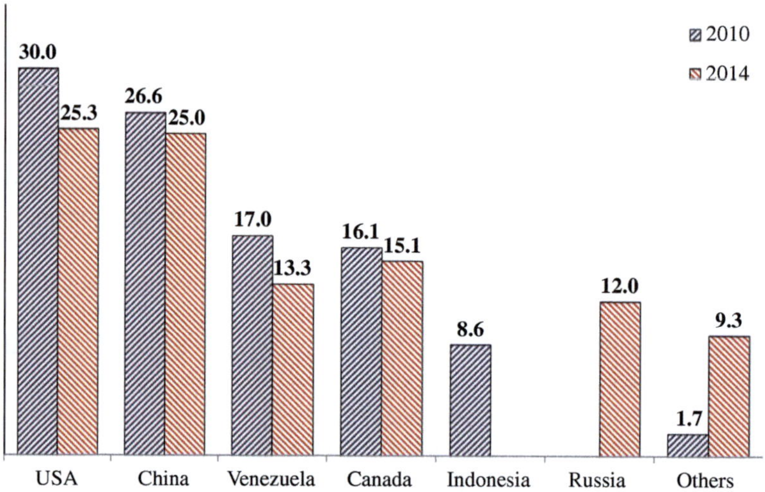

Figure 2 Share of each country in world oil production using EOR (%).

Three generalized EOR methods—thermal, chemical, and gas—continue to dominate. The forecast for the market value of EOR in the USA until 2025 (Figure 5) shows that a similar trend will continue in coming years.

Let us now consider the basic principles of creating new technologies for enhancing oil recovery. When creating a new method for increasing oil recovery, it should be taken into account that it will be effective only when the technical contradiction inherent in the technological process is overcome. For example, a technical contradiction when flooding an oil field is that maintaining reservoir pressure requires

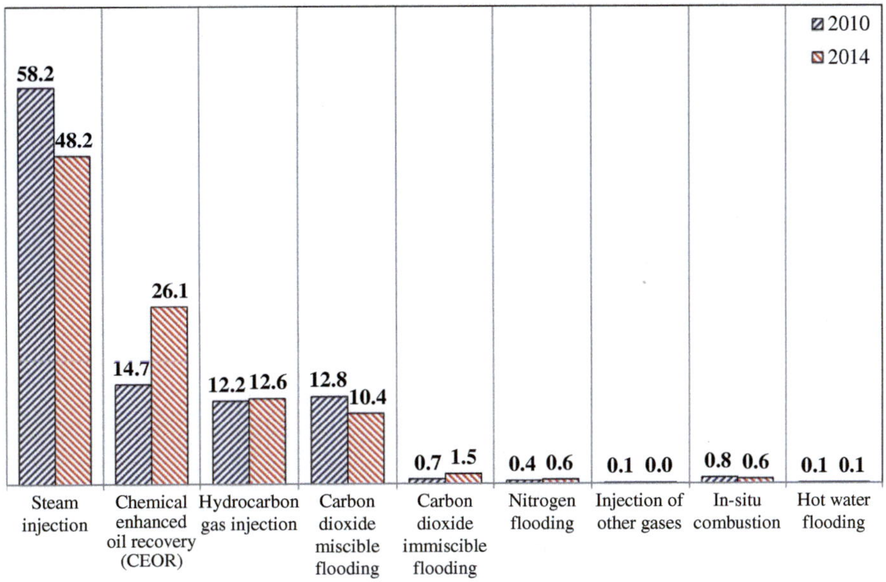

Figure 3 Distribution of additionally produced oil by enhanced oil recovery methods (%).

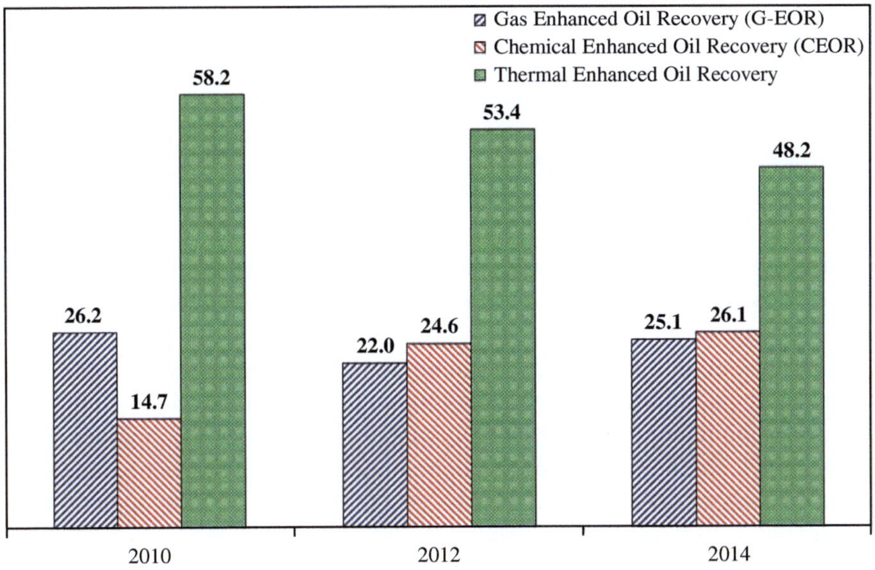

Figure 4 Distribution of additionally produced oil by general methods of enhanced oil recovery (%).

injecting large volumes of water, along with Moreover, injection of a large volume of water leads to disruption of the stability of the displacement front and early breakthrough of water to production wells. Overcoming this contradiction is possible by using non-Newtonian systems that ensure both the maintenance of reservoir pressure and the stability of the displacement front.

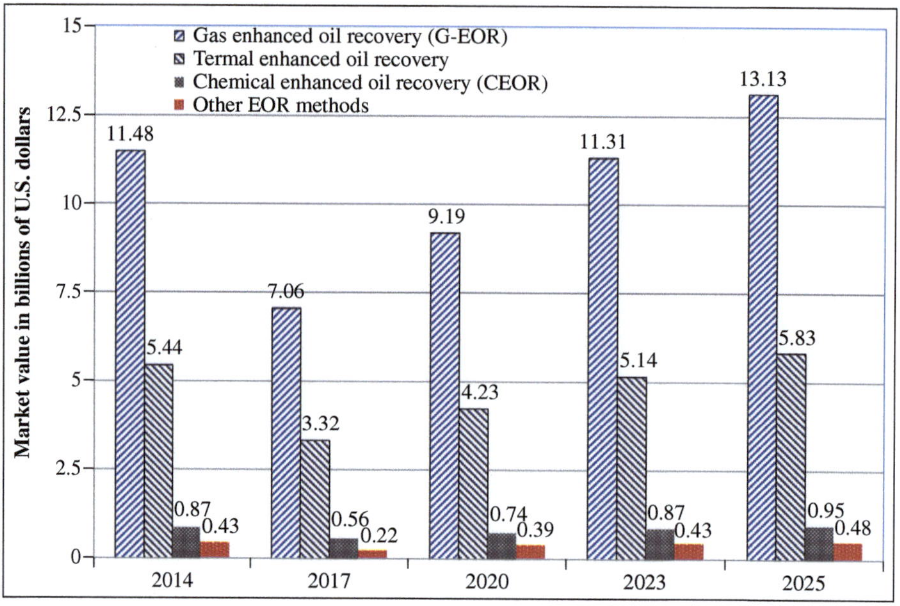

Figure 5 Forecast of the market value of EOR in the USA until 2025.

One of the main requirements for the technological process of influencing the bottomhole zone is selectivity. Therefore, targeted delivery of the working fluid to the least permeable or contaminated zones of the formation is necessary. At the same time, the working fluid, if necessary, should be easily washed out of the well so as not to interfere with subsequent oil production. It is known that high-impact coverage is achieved when using high-viscosity fluids, while low-viscosity fluids are removed from the formation more quickly. Thus, mutually exclusive requirements for the working fluid determine a technical contradiction when influencing the bottomhole zone of wells.

The technical contradiction in oil production is that, with on the one hand, it is necessary to maintain high oil extraction rates, and with on the other hand, high oil production leads to early watering of wells and a sharp decrease in phase permeability for oil. This technical contradiction can be overcome by effective isolation of water inflows, which will ensure the maintenance of a sufficiently high oil recovery and will help reduce field development time. Obviously, the main requirement for an effective method of isolating water inflows is the selectivity of the effect, in which the phase permeability to water decreases (technical success), and the permeability to oil increases (economic success) or remains at the same level.

This approach to solving technical problems was developed by Genrich Saulovich Altshuller and called it «*The Theory of Inventive Problem Solving*» (TRIZ). It is easy to see that Altshuller's approach is based on Hegelian dialectics. The (Hegelian) triad "thesis – antithesis – synthesis" allows solving many technical problems. The technical contradiction, born of the confrontation between thesis and antithesis, gives birth to new knowledge—synthesis.

In the book we will try to consider the technical contradictions of technological processes that lead to the creation of new, more promising methods for increasing oil recovery.

The book consists of 12 chapters and provides a comprehensive overview of enhanced oil recovery (EOR) methods. It begins with core concepts in reservoir engineering and fluid flow in porous media, followed by the classification of hydrocarbons, oil reserves, and recovery factors.

Chapters 4 through 8 cover major EOR techniques, including thermal, gas, chemical, and microbial methods. Chapter 9 discusses recent advances in EOR technologies.

Chapter 10 focuses on the practical steps of EOR implementation, while Chapter 11 addresses presents laboratory methods for evaluating the properties of oil-bearing rocks. The final chapter discusses the economic assessment of EOR projects.

Together, these chapters offer a balanced mix of theory, practical application, and current industry practices.

Bibliography

Altshuller, G. S. (2023). *Find an idea: An introduction to TRIZ – The theory of inventive problem solving: Annotated.* Kindle edition. https://www.amazon.com/s?i=digital-text&rh=p_27%3AAltshuller%2BG.S.&s=relevancerank&text=Altshuller+G.S.

Alvarado, V., & Manrik, E. (2011). *Methods for Enhancing Oil Recovery. Planning and Application Strategies.* Premium Engineering LLC.

BP. (2020). *Statistical Review of World Energy, 2020.* https://www.bp.com/content/dam/bp/business-sites/en/global/corporate/pdfs/energy-economics/statistical-review/bp-stats-review-2020-full-report.pdf

Fragoso, A., Selvan, K., & Aguilera, R. (2018, April). Breaking a paradigm: Can oil recovery from shales be larger than oil recovery from conventional reservoirs? The answer is yes! *Paper SPE-190284-MS Presented at the SPE Improved Oil Recovery Conference. Tulsa, Oklahoma, USA.*

Guo, H., Dong, J., Wang, Z., et al. (2018, April). 2018 EOR survey in China. Part 1. *Paper SPE-190286-MS Presented at the SPE Improved Oil Recovery Conference. Tulsa, Oklahoma, USA.*

Liu, Z.-X., Liang, Y., Wang, Q., et al. (2020). Status and progress of worldwide EOR field applications. *Journal of Petroleum Science and Engineering, 193*, 107449.

Muggeridge, A., Cockin, A., Webb, K., et al. (2014). Recovery rates, enhanced oil recovery and technological limits. *Philosophical Transactions of the Royal Society A: Mathematical, Physical, and Engineering Sciences, 372*, 20120320.

Statista. (n.d.). Market value of enhanced oil recovery in the United States from 2014 to 2025, by technology. https://www.statista.com/statistics/1060491/us-enhanced-oil-recovery-market-value-by-technology/

The Hunt for Advantaged Oil. (2021, June 4). Oil & companies news. *Hellenic Shipping News.* https://www.hellenicshippingnews.com/the-hunt-for-advantaged-oil/#:~:text=Global-ly%2C%20the%20average%20oil%20recovery,secondary%20recovery%20are%20around%2032%25

1

Basic Concepts in Reservoir Engineering

1.1 Rocks and Their Types

A rock is a consolidated mixture of minerals. By “consolidated,” we mean hard and strong; real rocks don’t fall apart in your hands! A mixture of minerals implies the presence of more than one mineral grain, but not necessarily more than one type of mineral. For example, a rock can be composed of only one type of mineral, such as limestone, which is commonly made up of only calcite. However, most rocks are composed of several different minerals. Rocks can also include nonminerals, such as fossils or organic matter within a coal bed or in some types of mudstone.

A critical point to remember is the difference between a mineral and a rock. A mineral is a naturally occurring, inorganic solid with a specific chemical composition and a crystalline structure. Minerals are the building blocks of rocks and are defined by their unique physical and chemical properties. Examples of minerals include feldspar, quartz, mica, halite, calcite, and amphibole. These minerals vary in hardness, color, luster, and crystal form, and they are often used to identify and classify different types of rocks.

On the other hand, a rock is typically a mixture of several different minerals. For instance, granite is a common rock composed of quartz, feldspar, and mica. The proportions and types of minerals present in a rock determine its characteristics and classification. Rocks are categorized into three main types based on their formation processes: igneous, sedimentary, and metamorphic.

1. *Igneous rocks*: These rocks form from the cooling and crystallization of magma, which is molten rock beneath the Earth’s surface. When magma cools slowly beneath the Earth’s crust, it forms intrusive igneous rocks such as granite, which have large, visible crystals. Conversely, when magma erupts onto the surface and cools quickly, it forms extrusive igneous rocks such as basalt, which have smaller crystals.
2. *Sedimentary rocks*: These rocks are formed from the accumulation and lithification of sediment. The sediments produced by weathering and erosion are eventually deposited in various environments, such as rivers, lakes, deserts, and oceans.

Methods for Enhanced Oil Recovery: Fundamentals and Practice, First Edition.
Baghir A. Suleimanov and Elchin F. Veliyev.

Over time, these sediments accumulate in layers and undergo lithification—a process of compaction and cementation—to form sedimentary rocks. Sediments can be fragments of other rocks, mineral grains, or biological materials. Sedimentary rocks are often layered and can contain fossils. Common types include sandstone, formed from compacted sand grains, and limestone, formed primarily from the remains of marine organisms. Sedimentary rocks provide valuable information about Earth's history and past environments.

More than 70% of the area of all continents is covered with sedimentary rocks, and most of the mineral deposits are directly associated with them.

Depending on the forming processes, sedimentary rocks are divided into the following three groups:

- clastic (crushed stone, sand, pebbles, gravel, clay);
- chemical (various salts, silica); and
- organogenic (limestones, fossil fuels).

Common sedimentary rocks include:

- Sandstone: Formed from compacted sand grains.
- Shale: Formed from compacted clay particles.
- Limestone: Formed primarily from the remains of marine organisms.

3. *Metamorphic Rocks*: These rocks form when preexisting igneous or sedimentary rocks are subjected to high temperatures, pressures, or chemically active fluids, causing them to undergo physical and chemical changes, leading to metamorphism. Metamorphic processes can result in the formation of new minerals and the reorganization of mineral grains into a more compact, crystalline structure. Examples of metamorphic rocks include schist, which has a foliated texture, and marble, which forms from limestone.

1.1.1 The Rock Cycle

The rock cycle is a fundamental concept in geology that describes the dynamic transformations of rocks within the Earth's crust. The rock cycle is a continuous process, and rocks can be recycled through the cycle many times from one type to another through various geological processes (Figure 1.1). For example:

- Igneous to Sedimentary: Igneous rocks exposed at the surface are weathered and eroded into sediments, which are then compacted and cemented to form sedimentary rocks.
- Sedimentary to Metamorphic: Sedimentary rocks buried deep within the crust are subjected to heat and pressure, transforming them into metamorphic rocks.
- Metamorphic to Magma: Metamorphic rocks can melt under extreme conditions to become magma, restarting the cycle.

The rock cycle is a vital part of the Earth's system. It helps to recycle the Earth's materials, and it also plays a role in the formation of mineral deposits. The rock cycle is a complex process, but it is essential for understanding the Earth's geology.

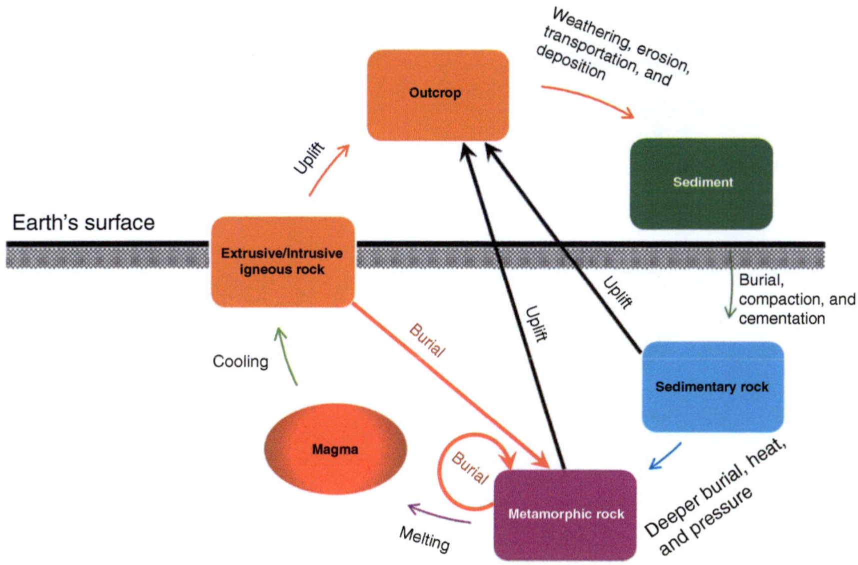

Figure 1.1 Schematics of the rock cycle.

This cycle is driven by two primary forces:

1. *Earth's internal heat engine*: The immense heat from the Earth's interior causes convection currents in the mantle, leading to the movement of tectonic plates. This movement drives processes, such as volcanic activity, mountain building, and the formation of igneous and metamorphic rocks. For example, when tectonic plates collide, they can push rocks deep into the Earth's crust, where they are subjected to high temperatures and pressures, forming metamorphic rocks (Figure 1.2).
2. *The hydrological cycle*: Powered by solar energy, the hydrological cycle involves the continuous movement of water on, above, and below the surface of the Earth. This cycle includes processes, such as weathering, erosion, transportation, deposition, and precipitation (Figure 1.3). For instance, rainwater can cause the weathering of rocks, breaking them down into smaller particles that are transported by rivers and deposited as sediment in lakes or oceans, eventually forming sedimentary rocks.

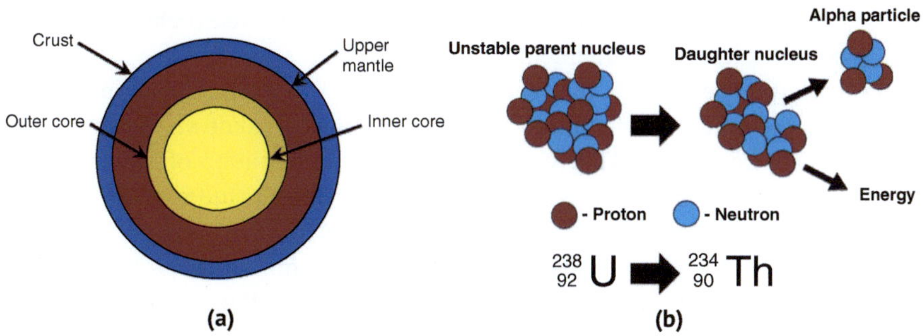

Figure 1.2 Internal structure of Earth (a) and main source of internal heat/energy (b).

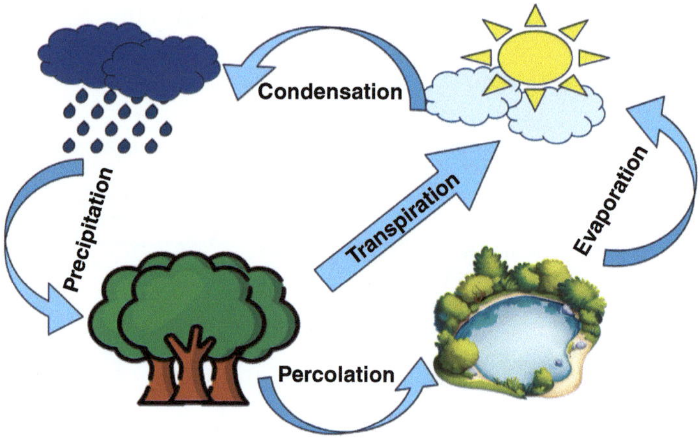

Figure 1.3 Schematics of the hydrological cycle.

To understand the rock cycle, it's convenient to start with magma, which is molten rock beneath the Earth's surface. Magma has temperatures ranging from about 800 to 1,300°C, depending on its composition and the pressure. When magma cools and solidifies, it forms igneous rocks. The location where this cooling occurs determines the type of igneous rock formed:

- *Intrusive igneous rocks*: These form when magma cools slowly beneath the Earth's surface, allowing large crystals to develop. Granite is a common example of intrusive igneous rock.
- *Extrusive igneous rocks*: These form when magma erupts onto the surface (as lava) and cools quickly, resulting in smaller crystals. Basalt is an example of extrusive igneous rock.

Weathering and erosion are critical processes in the rock cycle that break down rocks into smaller particles, which can then be transported and deposited to form sedimentary rocks. Types of weathering:

1. *Physical (mechanical) weathering*: This type of weathering breaks rocks into smaller pieces without changing their chemical composition. It includes:
   - Frost wedging: Water seeps into cracks in rocks, freezes, and expands, causing the rock to break apart.
   - Thermal expansion: Repeated heating and cooling cause rocks to expand and contract, leading to fragmentation.
   - Biological activity: Plant roots grow into cracks in rocks, and animals burrow into the ground, contributing to the mechanical breakdown of rocks.
2. *Chemical weathering*: This involves the chemical alteration of minerals within the rocks, leading to their breakdown. Key processes include:
   - Hydrolysis: Water reacts with minerals to form new minerals and soluble ions. For example, feldspar transforms into clay minerals.
   - Oxidation: Oxygen reacts with minerals, especially those containing iron, to form oxides. This process is responsible for the rusting of iron-rich rocks.
   - Dissolution: Soluble minerals, such as halite and calcite, dissolve in water, especially in acidic conditions.

3. *Biological weathering*: This type of weathering involves the contribution of living organisms. For instance, lichen and moss can produce acids that chemically weather rocks, while tree roots and burrowing animals physically break down rock.

1.1.2 Erosion

Erosion involves the transportation of weathered materials by natural agents. The primary agents of erosion include:

- Water: Rivers and streams carry sediments downstream, where they are deposited in floodplains, deltas, and oceans.
- Wind: In arid regions, wind can transport fine particles over long distances, creating features like sand dunes.
- Glaciers: Moving glaciers pick up and transport large quantities of rock debris, depositing them as glacial till when the ice melts.
- Gravity: Gravity causes rocks and sediments to move downhill through processes like landslides and rockfalls.

Understanding the rock cycle is essential for geologists as it provides insights into Earth's geological history, the formation and distribution of natural resources, and the processes that shape the planet's surface. For example:

- Natural resources: Many natural resources, such as minerals and fossil fuels, are found in specific rock types formed through the rock cycle. For instance, coal forms from buried plant material in sedimentary rock layers, while valuable minerals can be concentrated in igneous and metamorphic rocks.
- Geological hazards: Knowledge of the rock cycle helps in predicting and mitigating geological hazards, such as volcanic eruptions, earthquakes, and landslides.

The rock cycle is not unique to Earth; it can also occur on other planetary bodies, although with significant differences. For instance:

- The moon: The Moon lacks an atmosphere and liquid water, and its tectonic activity is minimal, resulting in a virtually inactive rock cycle.
- Mars: Mars shows evidence of past volcanic activity and ancient river valleys, indicating a once-active rock cycle. However, its current rock cycle is much less active than Earth's due to the absence of liquid water and significant tectonic activity.
- Studying these differences helps scientists understand the unique geological histories of other planets and the factors that influence their evolution.

1.2 Forms of Occurrence of Sedimentary Rocks

Sedimentary rocks are composed mainly of almost parallel layers (strata), differing in physical and chemical properties (Figure 1.4).

The surfaces separating the layers from each other are called the base if it is located below the layer in question and the roof if it is located on top. The line descending perpendicularly from the roof to the sole is called the thickness of the layer (or thickness of the layer); it is also the shortest distance between the roof and the base.

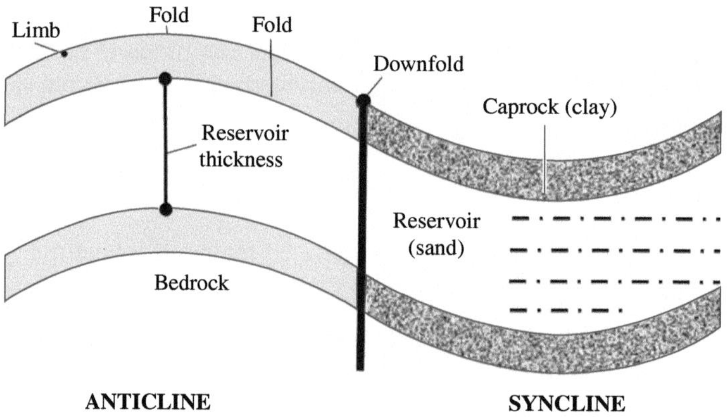

Figure 1.4 Mode of rocks occurrence.

A layer composed of impermeable rocks is called a seal, and one formed from permeable rocks is called a reservoir. In nature, ideally, horizontal occurrence of layers is rarely found; as a rule, they have a wavy (folded) occurrence. This occurrence is due to the fact that the earth's crust is not static, and various oscillatory, tectonic processes constantly occur in it.

The folds formed during these processes are called anticlines and synclines.

Anticline—an anticline is a convex bend in the Earth's crust with a core in the center, with the core composed of older rocks and the outer layers composed of younger rocks.

Syncline—a syncline is the mirror opposite of an anticline, i.e., a concave curve in the Earth's crust with ancient rocks in the outer part and young rocks in the central part. The structural elements of the anticline and syncline are shown in Figure 1.4.

Most of the oil and gas fields are confined to anticlines. Anticlines have an average length of up to 10 km and width up to 3 km. The largest oil field, Ghawar (Saudi Arabia), is located in an anticline 225 km long and 25 km wide.

1.3 Hydrocarbon Reservoirs

For a petroleum engineer, the property of rocks known as permeability is of particular interest, which will be discussed in detail below. For now, let's dwell on the fact that, depending on the permeability of rocks, they are divided into reservoirs and seals. Seals are practically impermeable rocks that act as a dividing surface between reservoir layers.

Reservoirs are rocks that have the ability to contain and release fluids. Mostly reservoir rocks are of sedimentary origin.

The following types of reservoirs are distinguished by pore void structure:

1. *Porous reservoirs*: They consist of voids formed by grain-like debris. For example: sands, sandstones, etc. This type has the best reservoir properties compared to others.

2. *Cavernous reservoirs*: They are mainly associated with carbonate strata and are composed of voids (caverns) formed as a result of leaching or dissolution of salts that make up the rock.
3. *Cracked reservoirs*: This type of reservoir is formed from impermeable rocks; however, numerous fractures of varying sizes allow them to accommodate hydrocarbons. For example, limestones.
4. *Mixed reservoirs*: They are quite common and are a combination of the above types of reservoirs, with the first word in the name indicating the type of predominant rock. For example, cavernous—porous reservoir.

The best reservoir properties. Other types of collectors may also have good abilities to contain and release liquids and gases, as well as pass them through themselves.

It should be noted that there is also a classification of reservoirs by lithological composition:

- terrigenous (silts, sands, etc.),
- carbonate (dolomite, chalk),
- siliceous rocks,
- volcanogenic-sedimentary.

Hydrocarbon deposits are most often found in terrigenous and carbonate reservoirs.

1.4 Oil and Gas Traps

A trap is a part of a reservoir, the conditions of which provide for the accumulation of hydrocarbon reserves. A trap, in essence, is a permeable rock bounded by impermeable layers (seals) in which fluids are in static conditions and distributed according to the law of gravity. This distribution forms the classic structure of an oil field (Figure 1.5).

The fluid interface is named according to the fluids bordering it:

- GWC—gas-water contact,
- WOC—water-oil contact,
- GOC—gas-oil contact.

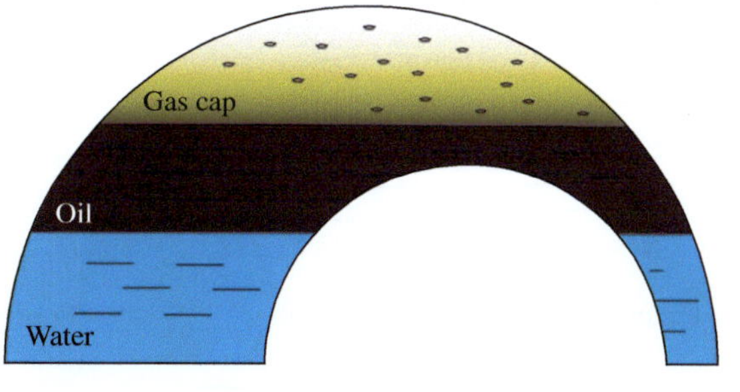

Figure 1.5 Oil field schematic.

There are two main types of traps:

- structural,
- nonstructural.

Nonstructural traps, in turn, are divided into two types:

- stratigraphic,
- lithological.

1.4.1 Structural Traps

This type of trap, formed due to the migration of hydrocarbons into the fold of the anticline, is called *Anticline (fold) traps* (Figure 1.6). Hydrocarbon migration occurs either along the limbs of anticlines or along tectonic faults.

During tectonic movements of the earth's crust, disturbances in the form of rock formations are often formed. If such a disturbance results in an overlap of the reservoir layer with an impermeable screen (seal), such a trap is called a fault trap (Figure 1.7).

The types of structural traps described above generally have the greatest industrial significance in terms of the availability of potential hydrocarbon reserves.

1.4.2 Lithological Traps

This type of traps is formed due to the lithological dynamics of reservoir rocks, that is, the replacement of permeable rocks with impermeable. The reasons for this process are different, but among the most common of them are the following: pinchout, fracturing and changes in the permeability of reservoir rocks (Figure 1.8). Such traps are formed in the form of sand lenses in clayey deposits (Figure 1.8a) or in reef bodies (limestones) covered by poorly permeable rocks (Figure 1.8b).

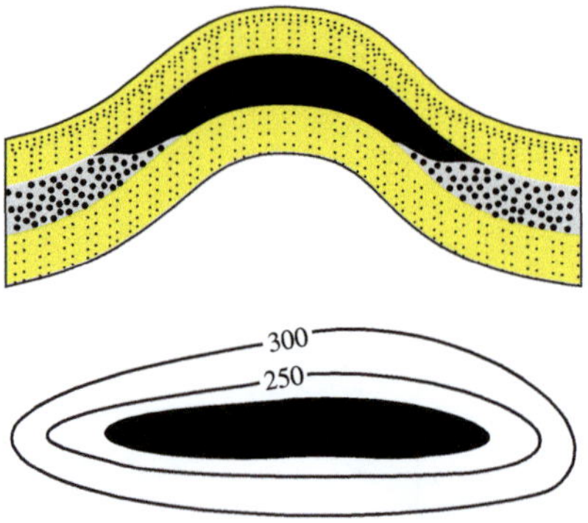

Figure 1.6 Anticline (fold) trap.

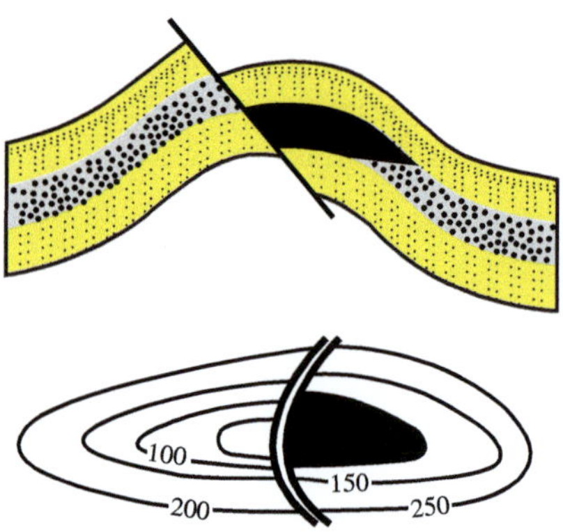

Figure 1.7 Fault trap.

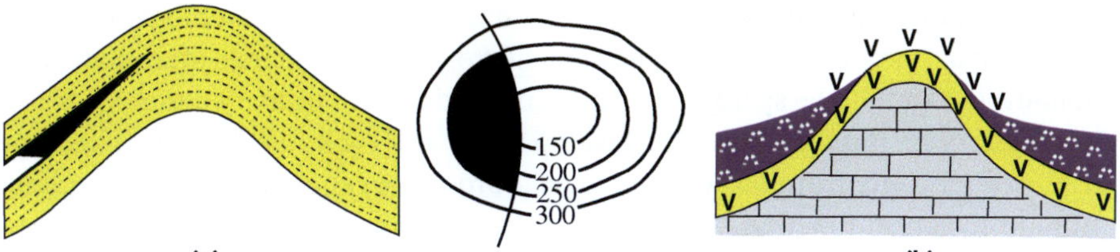

Figure 1.8 (a) Pinch-out trap; (b) a trap confined to a reef massif.

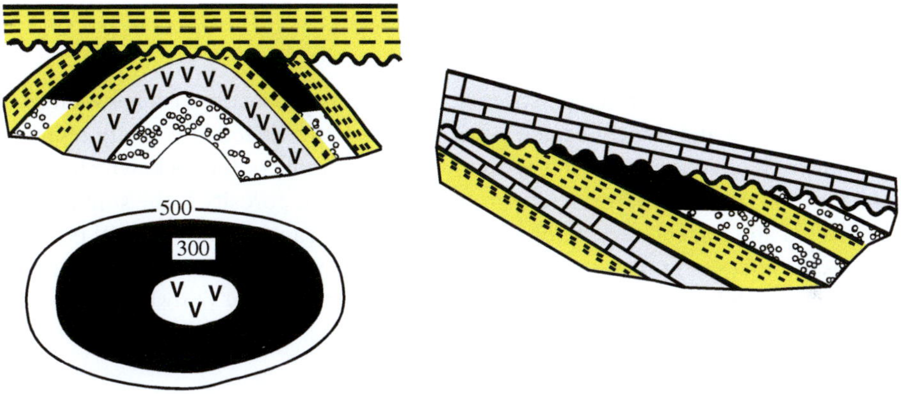

Figure 1.9 Stratigraphic traps.

1.4.3 Stratigraphic Traps

This type of trap is formed by the unconformable overlapping of reservoir rocks with fluid seals (poorly permeable layers). Sometimes a hydrodynamic seal created by formation waters (associated with high-pressure horizons) during filtration along tectonic disturbances can serve as a similar overlapping seal (Figure 1.9).

Quite often, the formation of traps is influenced by several factors, and combined structural and lithological traps are formed.

1.5 Rock Porosity

Porosity is a critical property of reservoir rocks, defined as the ratio of the void space (pores) volume to the bulk volume of the rock, usually expressed as a percentage. This property is fundamental in determining the storage capacity of reservoir rocks for fluids, such as oil, gas, and water. Porosity indicates how much fluid a rock can hold and influences the movement of these fluids through the rock.

1.5.1 Primary and Secondary Porosity

Primary porosity develops during the initial deposition of the rock material. As sediment is deposited and buried, the spaces between the grains form the primary porosity.

Over time, processes like compaction (where grains are pressed closer together under the weight of overlying materials) and cementation (where minerals precipitate from groundwater and fill the spaces between grains) reduce this primary porosity.

Secondary or induced porosity develops after the rock has formed. This type of porosity results from various geological processes, such as fracturing, dissolution, or recrystallization. For example, fractures in shales and limestones or solution cavities known as karsts are forms of secondary porosity. These secondary features can significantly enhance the ability of rocks to store and transmit fluids.

1.5.2 Effective and Total Porosity

From a reservoir engineering perspective, it is essential to distinguish between connected porosity and nonconnected pores (Figure 1.10). Effective porosity refers to the ratio of interconnected void spaces to the bulk volume of the rock, which is crucial for fluid flow within the reservoir. In contrast, total porosity is the sum of both connected and nonconnected porosities. Typically, in sandstones, the total porosity equals the effective porosity, as most pores are interconnected. However, in carbonates such as dolomites and limestones, nonconnected vuggy porosity may occur, where some pores do not contribute to fluid flow.

1.5.3 Diagenesis and Its Impact

Diagenesis refers to the physical and chemical changes that occur in sediment after its deposition, transforming it into sedimentary rock. This process can significantly affect both porosity and permeability. Diagenesis includes various processes, such as mechanical compaction (which reduces pore space by pressing grains closer together), mineralogical changes (altering the mineral composition of the rock), cement precipitation (filling pores with new mineral growth), and mineral dissolution (enlarging pores by dissolving existing minerals) (Figure 1.11). These processes can either enhance or reduce the porosity and permeability of the rock.

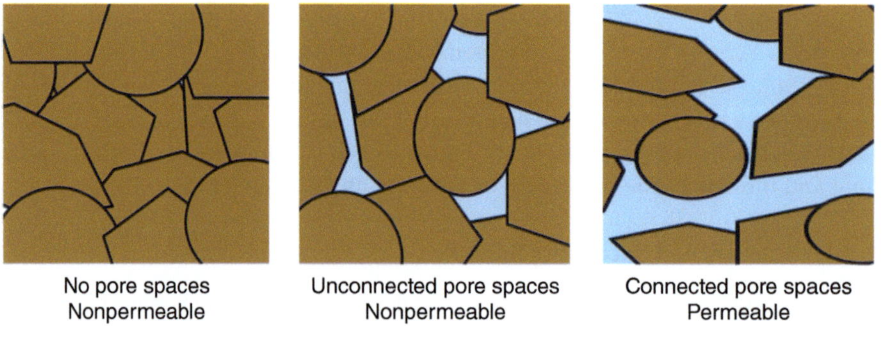

Figure 1.10 Connected/nonconnected pores.

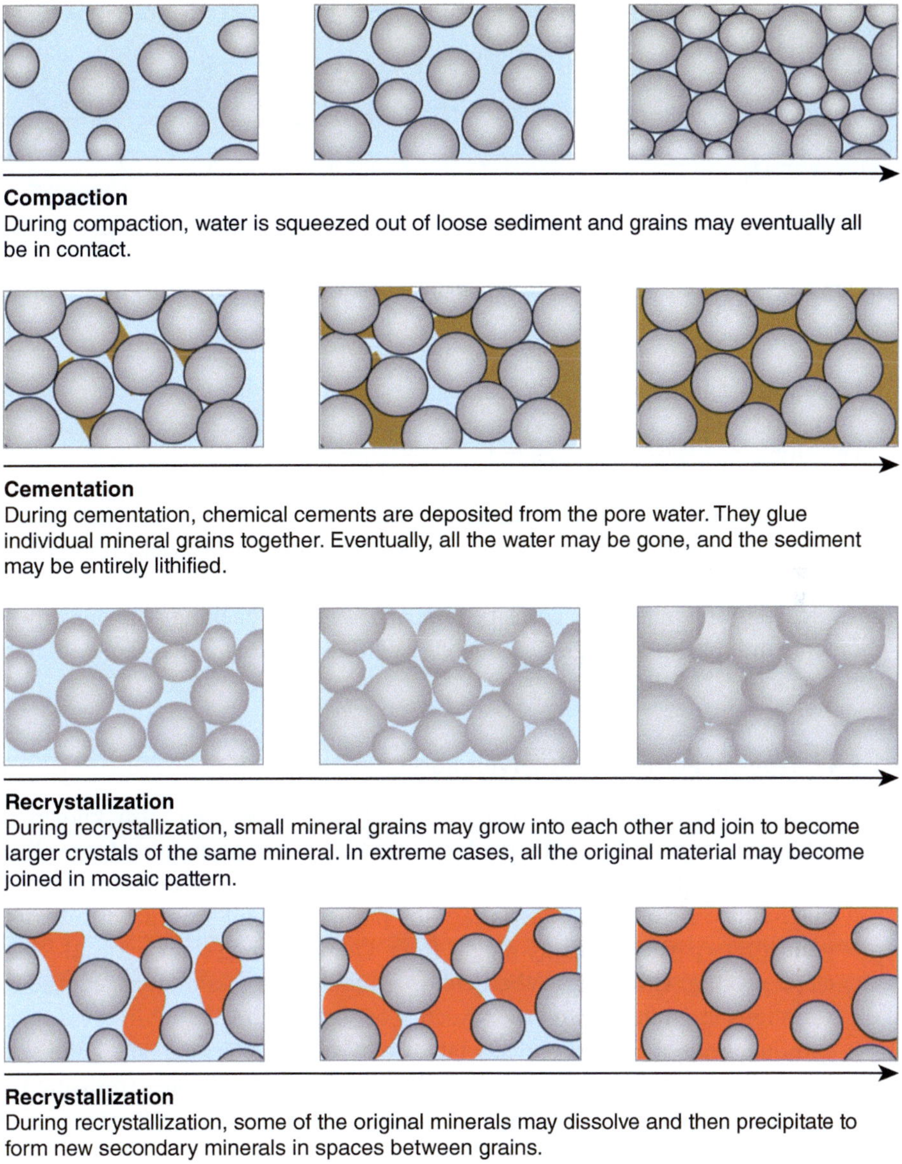

Figure 1.11 Diagenesis.

The porosity of a rock can be influenced by the arrangement of its grains (Figure 1.12). For instance, in a cubic packing arrangement, where grains are arranged in a simple cubic pattern, the porosity can be calculated by considering a cube filled with eight spheres of radius r. The volume of the cube is: $4r^3$ while the total volume of the eight spheres is $8 \times 4/3\ \pi r^3$. The porosity for cubic packing, which is the least compact arrangement, is 47.6%.

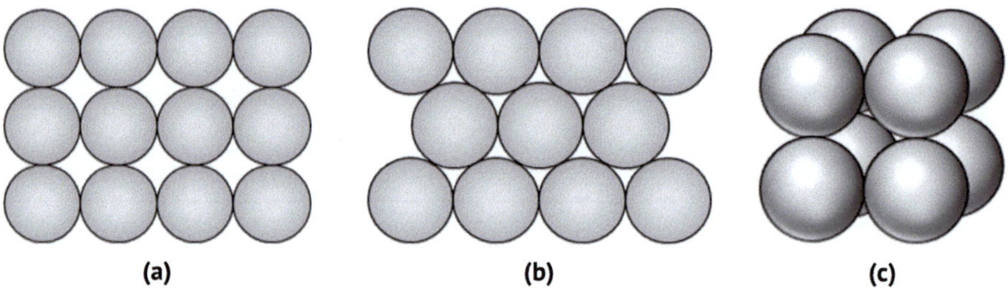

Figure 1.12 (a) Cubic or wide packed; (b) rhombohedral or close-packed; (c) –4 by 4 pack.

Pores space volume is: $(4r)^3 - 8 \times 4/3\pi r^3$.

Porosity: $\varphi = \dfrac{V_B - V_S}{V_B} = 1 - \dfrac{V_S}{V_B} = 1 - \dfrac{32\pi r^3}{3 \times 64 r^3} = 1 - \dfrac{\pi}{6}$,

where V_B, bulk volume and V_S, volume of grains.

In contrast, rhombohedral packing represents the most compact arrangement of grains (Figure 1.12c). In this configuration, the spheres are packed as closely as possible, resulting in a lower porosity compared to cubic packing. The porosity of rhombohedral packing is significantly reduced (25.96%) due to the more efficient use of space.

1.5.4 Types of Porosity in Reservoir Rocks

- *Interparticle (intergranular) porosity*: Interparticle porosity, also known as intergranular porosity, is the predominant type found in sucrosic (sugar-like) rocks (Figure 1.13). In these rocks, pore sizes are typically of the same order of magnitude as, but usually less than, the particle sizes. For uniform spherical grains, interparticle porosity ranges from 47.6% in cubic stacking to 25.9% in close packing. However, in the field, these ideal figures are often reduced by variations in grain size and the presence of shale. Permeability in these rocks is influenced not only by porosity but also by sorting, stacking, and grain size. This can range from high permeability in large-grain sandstones (measured in darcies) to impervious conditions in chalks and siltstones.

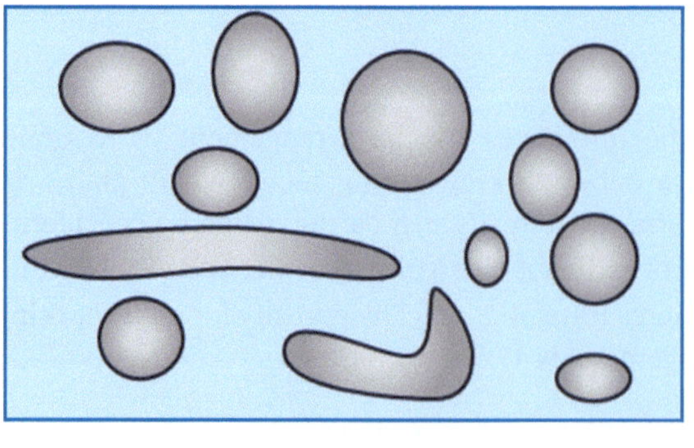

Figure 1.13 Interparticle porosity.

- *Intraparticle porosity*: Intraparticle porosity is the network of pore spaces within individual grains (Figure 1.14). This type of porosity is often revealed by scanning electron microscope (SEM) images. Although intraparticle porosity can contain connate water (water trapped in the rock since its formation), it is usually nearly impervious to fluid flow due to the small size and isolated nature of these pores.
- *Fracture porosity*: Fracture porosity is found in reservoirs that have been fractured, typically in carbonate rocks, such as limestones and dolomites (Figure 1.15). In these reservoirs, the rock is defined as a double porosity medium, consisting of both the matrix (the rock's solid framework) and the fractures. Fractures can significantly enhance permeability by providing additional pathways for fluid flow. Fractured reservoirs are often encountered in crystalline and amorphous rocks, which lack grain size and therefore have minimal interparticle porosity.
- *Vugs and molds*: Vugs and molds, though different in appearance and characteristics, are often classified together (Figure 1.16). Both are formed by the selective dissolution of sedimentary rock by percolating waters. Vugs can vary from the size of a fingernail to large caves and tend to be relatively well connected, allowing for better fluid flow. Molds are typically smaller, often resulting from the dissolution of oolites (small calcite spheres precipitated in seawater). These are known as oolmolds or oolicasts and can exhibit very high porosity, though their connectivity may be poor compared to vugs.

Figure 1.14 Intraparticle porosity.

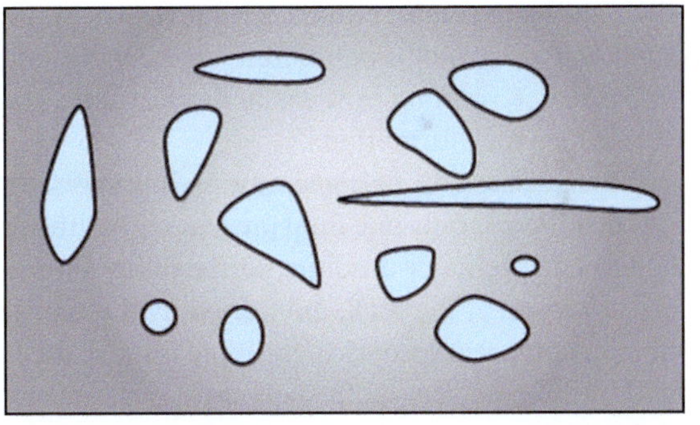

Figure 1.15 Fracture porosity.

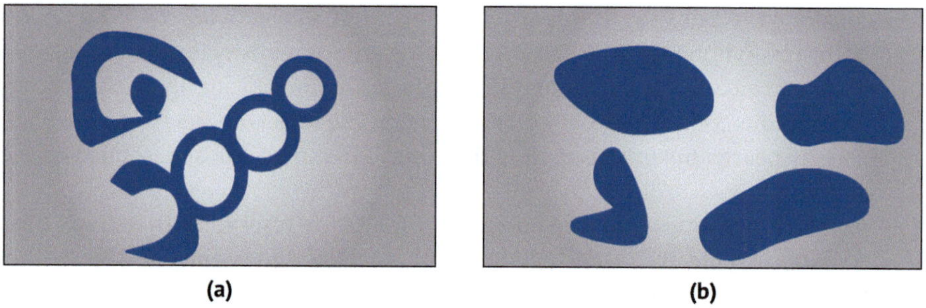

Figure 1.16 (a) Moldic; (b) vugs.

Depending on the diameter, pore channels are divided into three groups:

1. supercapillary (more than 0.5 mm), characterized by free filtration of fluids;
2. capillary (from 0.5 to 0.0002 mm [0.2 μm]), filtration of fluids largely depends on capillary forces; and
3. subcapillary (less than 0.0002 mm [0.2 μm]), filtration of fluids due to large capillary forces is practically impossible, regardless of the porosity value. The most typical examples of such rocks are clays and limestones.

1.6 Rock Permeability

The term "permeability" refers to a crucial property of porous rocks, specifically their ability to conduct fluids. This property plays a significant role in various fields, such as hydrogeology, petroleum engineering, and soil science. Permeability is a key factor in determining how easily fluids, such as water, oil, or gas, can move through subsurface rock formations.

When discussing permeability, it is important to distinguish between different types. If there is only one fluid present within the pore spaces of the rock, this property is termed specific or absolute permeability. Absolute permeability is a fundamental characteristic of the rock, unaffected by the type or properties of the fluid. However, in many practical scenarios, multiple immiscible fluids, such as oil and water, coexist within the pore spaces. In such cases, we refer to the effective permeability of each fluid, which varies depending on the presence and interaction of other fluids in the pores.

To further understand permeability in multi-fluid systems, we use the concept of relative permeability. Relative permeability is the ratio of the effective permeability of a particular fluid to the absolute permeability of the rock. This dimensionless ratio provides insight into how the presence of other fluids affects the flow capacity of the fluid of interest.

Measuring permeability is essentially measuring the fluid conductivity of a material, which is crucial for predicting and managing fluid flow in various applications. The concept of permeability can be analogized to electrical conductivity, where permeability represents the inverse of the resistance that a porous medium offers to fluid flow. Just as electrical conductivity indicates how easily electric current can

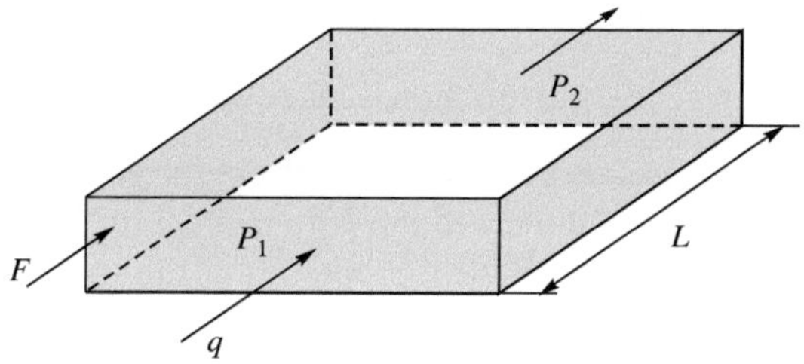

Figure 1.17 Schematic representation of the Darcy equation parameters.

flow through a material, permeability indicates how easily a fluid can flow through a porous rock.

The foundational work on permeability was conducted by Henry Darcy, a French engineer who studied the flow of water through filter sands in the mid-nineteenth century. His pioneering experiments and formulation of what is now known as Darcy's Law laid the groundwork for modern understanding of fluid flow in porous media. As a result, the unit of measurement for permeability is named the "Darcy" in his honor. One Darcy (D) is defined as the permeability that allows a fluid with a viscosity of 1 cP to flow through a porous medium with a pressure gradient of 1 atm/cm at a rate of 1 cm^3/s (Figure 1.17).

Darcy equation:

$$q = \kappa \frac{A}{\mu L}(P_1 - P_2), \tag{1.1}$$

where q, flow rate (cc/s); k, permeability of the rock sample (D); μ, fluid viscosity (cP); A, cross-sample section area (cm^2); L, rock sample length (cm); and P, pressure (atm).

The flow is proportional to:

- Rock permeability
- Cross-sectional area of the sample
- Pressure gradient

And is inversely proportional to:

- Fluid viscosity
- Sample length

Permeability has a dimension of an area. Permeability is usually expressed in milli-Darcy (mD) or μm^2. At the same time, $1\ D = 1.02 \cdot 10^{-3}\ \mu m^2 = 1.02 \cdot 10^{-12}\ m^2 = 1000\ mD$. The mD is commonly used because the D is relatively large and is convenient only for highly permeable rocks. The Darcy, as a unit, is not derived from a homogenous system of units, yet reservoir engineering techniques have been developed based on equations that use the Darcy unit system. Consequently, analytical formulas in reservoir engineering often refer to the Darcy unit system.

Apart from the Darcy unit system, two other unit systems are commonly used in the industry: field units and metric units. It is crucial to exercise great care when working with reservoir engineering formulas to ensure that the appropriate unit system is applied consistently and correctly. Misapplication of units can lead to significant errors in calculations and interpretations, impacting reservoir performance evaluations and decision-making processes.

1.6.1 Types of Permeabilities

In practice, a distinction is made between absolute, phase, and relative permeability.

1. Absolute permeability is the permeability of a porous medium for a homogeneous fluid inert to the porous medium (usually an inert gas is used), characterizing the physical properties of the rock, but independent of the properties of the saturating fluid.
2. Phase permeability is the permeability of a porous medium for a given fluid in the presence of another. Phase permeability is determined by the degree of saturation of the pores with the studied fluid (oil, gas, water).
3. Relative permeability is the ratio of phase permeability to absolute permeability (measured in fractions of unity).

Initially, Darcy's work focused on describing the flow of a single fluid, specifically water, saturating 100% of the porous medium. The permeability to a particular fluid is independent of the fluid's properties, such as viscosity. Consequently, the permeability to a fluid that completely saturates the porous medium is a constant and inherent characteristic of that medium. This constant is referred to as **absolute permeability** or **specific permeability**.

Relative permeability is a crucial concept in reservoir engineering, especially when dealing with multiphase flow in porous media. While absolute permeability refers to the permeability of a rock when it is fully saturated with a single fluid, relative permeability describes the permeability of each fluid in the presence of other immiscible fluids within the pore spaces.

Relative permeability is defined as the ratio of the effective permeability of a particular fluid to the absolute permeability of the rock. Mathematically, this can be expressed as:

$$\text{Relative Permeability} = \frac{\text{Effective Permeability}}{\text{Absolute Permeability}}. \tag{1.2}$$

This ratio varies depending on the saturation levels of the fluids in the reservoir.

In a two-phase system, the fluids typically consist of combinations, such as oil and water or oil and gas. In contrast, a three-phase system involves the simultaneous presence of oil, water, and gas. As these fluids flow through the porous media, their interactions are influenced by capillary forces. These forces act to reduce the flow rate of each individual phase in a nonlinear manner, due to the competition for available pore space and the complex fluid–fluid and fluid–rock interactions.

As a result of these interactions, the **relative permeability** of each fluid phase is affected by the presence of the other phases. Consequently, the sum of the relative permeabilities of all phases in a given porous medium is always less than one. This reflects the fact that the presence of multiple fluids reduces the effective permeability of each individual phase compared to a situation where only one fluid is present.

$$k_{ro} + k_{rw} + k_{rg} < 1. \tag{1.3}$$

Absolute permeability is often approximated as the effective permeability to air.

The relative permeability of a fluid is influenced by several factors, including fluid saturations, wettability of the rock, and the interaction between the fluids.

Relative permeability curves are typically obtained through laboratory experiments and are used to describe the flow characteristics of oil, water, and gas in reservoir rocks. These curves help engineers to better understand and manage the production performance of a reservoir, facilitating more efficient and effective recovery strategies.

1.6.2 Klinkenberg Effect

It has been observed that the measured absolute permeability of a sample varies depending on whether air or liquid is used for the measurement. In the case of low-pressure gas flow, the mean free path of gas molecules is comparable to the pore size of the medium. As a result, the likelihood of collisions between gas molecules and the solid grains of the porous media is similar to the likelihood of collisions between gas molecules themselves. This reduces the relative influence of viscous forces during gas displacement, leading to an overestimation of permeability when applying Darcy's law. Consequently, the apparent permeability measured with gas is higher than that measured with a liquid.

The linear relationship between the observed permeability and the reciprocal of the mean pressure can be expressed as:

$$k_L = \frac{k_g}{1 + \frac{b}{\overline{P}}} = k_g - m\frac{1}{\overline{P}}, \qquad b = \frac{m}{k_L} \tag{1.4}$$

where k_L, absolute permeability measured with liquid; k_g, absolute permeability measured with gas; $\overline{P}$, mean flowing pressure during the displacement; b, constant for a given gas in a given medium; and m, slope of curve.

$$k_{app} = \frac{k_L + b}{P_{mean}}, \tag{1.5}$$

where k_{app} is the apparent permeability measured with gas. k_L is the Klinkenberg permeability, representing the permeability at infinite pressure. b is a constant that depends on the mean free path of the gas and the size of the openings in the porous media. P_{mean} is the mean pressure in the core.

The parameter b is influenced by both the mean free path of the gas and the size of the openings in the porous media, indicating that b is a function of permeability. This correction is essential to accurately determine the permeability of porous media when using gas as the flowing fluid, particularly in tight formations where the pore sizes are small.

Reservoirs are conventionally ranked into five classes based on permeability:

- very well permeable (more than 1 μm^2),
- highly permeable (0.1–1 μm^2),
- medium-impermeable (0.01–0.1 μm^2),
- weakly permeable (0.001–0.01 μm^2), and
- impenetrable (less than 0.001 μm^2).

The first three classes of reservoirs are of industrial interest.

1.7 Geological Heterogeneity of Rocks

Geological heterogeneity is one of the most important characteristics affecting the efficiency of the field development process. Typically, two types of heterogeneity are distinguished: microheterogeneity associated with the heterogeneity of the pore space within one formation, and macroheterogeneity associated with the heterogeneity of the oil-bearing strata that make up the reservoir.

Microheterogeneity of the rock manifests itself when there is a significant difference in the fractional composition. A significant variety of fractional composition leads to the formation of capillary and subcapillary pores. In capillary pores the liquid moves, but in subcapillary pores it is motionless due to the molecular-surface interaction of the liquid and the rock. In this case, the continuity of capillary pores overlapped by subcapillaries may also be disrupted. In addition, other parameters in the formation can vary greatly, such as lithological properties, clay content, carbonate content, and degree of cementation. As a result, small zones with different permeability appear within one formation. At the same time, such formations are characterized by average permeability and are considered in calculations as homogeneous (or quasi-homogeneous), despite the fact that microheterogeneity has a significant impact on the effectiveness of the applied methods of influencing the reservoir.

There are two types of macroheterogeneity:

- along the strike of the formation (zonal heterogeneity) and
- along the dip of the formation (layered heterogeneity).

Macroheterogeneity is manifested in the variability of reservoir properties (primarily permeability) of zones and layers along the strike and dip of the formation.

The macroheterogeneity of a deposit is assessed by several coefficients:

- The compartmentalization coefficient is the ratio of the sum of permeable layers in all wells to the number of wells. In the case when the productive horizon is represented by one layer, the separation coefficient is equal to one. How the more permeable layers in the section, the higher the coefficient of compartmentalization.
- Sandiness coefficient is the ratio of the effective thickness of the formation to its total thickness. The sandiness coefficient allows you to estimate the ratio of reservoirs and nonreservoirs in the section of the object.

The higher the coefficient of dissection and the lower the coefficient of sandiness, the higher the degree of macroheterogeneity of the operational object.

Macroheterogeneity along the dip, or layered heterogeneity, is characterized by another indicator, namely the presence of crossflows between layers. This indicator has a significant impact on hydrodynamic calculations and reservoir development. In the case when there are clay interlayers (tires) between the layers, there are practically no crossflows.

1.8 Saturations

In underground oil and gas reservoirs, the pore spaces in the rocks are always completely saturated with fluids. There is no true "void space" in these pores; they are filled with one or a combination of the following fluids:

1. Oil (and associated impurities in the liquid phase)
2. Natural gas (and associated impurities in the vapor phase)
3. Water (either connate water, present since the rock's formation, or water that entered the reservoir later, possibly through injection)

Initially, when the rock was deposited in an aqueous environment, the pore spaces were 100% water-saturated. Over time, during the processes of burial, compaction, and cementation, water composition might have changed, but the saturation remained full unless hydrocarbons entered the pores and displaced some of the water.

Hydrocarbons can enter the pores if the water-saturated rock is near an active hydrocarbon source, like organic-rich shale, and if the pores are in pressure communication with this source rock. Hydrocarbons, being less dense than water, move upwards through the porous and permeable rock until they either escape at the surface or are trapped by an impermeable layer, forming an oil or gas reservoir if the conditions are suitable.

Despite hydrocarbon entrapment, some water always remains in the pores, leading to the fundamental saturation relationship in reservoirs:

$$S_o + S_w + S_g = 1.0, \tag{1.6}$$

where:

$$S_o = \frac{\text{oil volume}}{\text{pore volume}} = \frac{V_o}{V_p},$$

$$S_w = \frac{\text{water volume}}{\text{pore volume}} = \frac{V_w}{V_p},$$

$$S_g = \frac{\text{gas volume}}{\text{pore volume}} = \frac{V_g}{V_p}.$$

It's possible for oil or gas saturation to be zero in some cases, but water saturation is always greater than zero.

Saturation is a key measure of the fluid content in porous rock and directly influences a reservoir's hydrocarbon storage capacity. It is also useful in identifying gas/oil or oil/water contacts based on changes in residual saturation with depth and serves as a correlation variable for estimating reservoir productivity.

1.8.1 Saturation Distribution in Reservoirs

During hydrocarbon accumulation, water saturation in the reservoir can be reduced to a point where it becomes immobile, typically between 5 and 40%, at which no more water can escape from the pores. This immobile water level is known as irreducible water saturation. It is important to distinguish between the different terms used for water saturation in the petroleum literature. Here are the key definitions:

1. S_{wir}—**Irreducible water saturation**: Below this point, water cannot flow.
2. S_{wc}—**Connate water saturation**: This is the water saturation when the reservoir is discovered. It may or may not be irreducible, so careful interpretation is needed.
3. S_{wi}—**Initial water saturation**: This can refer to irreducible, connate, or interstitial saturation (water present in the pores). It may not signify irreducible and can refer to water saturation at discovery or at any other point in time.

Due to the density differences between gas, oil, and water, typical reservoirs have oil floating on water and, if present, gas floating on oil. Even at the very top of a reservoir, there will always be some level of water saturation, at least at irreducible water saturation.

1.8.2 Fluid Distribution in Reservoirs

In a theoretical case without a porous medium (where porosity = 100%), the fluid layers in a container of oil, gas, and water would have distinct interfaces:

- Gas cap: $S_g = 100\%$
- Oil zone: $S_o = 100\%$
- Aquifer (water zone): $S_w = 100\%$

However, in actual porous rock reservoirs, the fluid interfaces are not so distinct. Water saturation is present throughout the oil and gas zones at a level of at least irreducible water saturation. Additionally, fluid contacts (such as the GOC and oil-water contact [OWC]) are spread over a transition zone that can range from a few feet to tens of feet, depending on fluid density differences and the permeability of the rock.

Figure 1.18 shows a simplified example of how oil, gas, and water would arrange themselves if there was no rock or porous material—essentially like fluids in a jar. In this case:

- *Gas cap (top)*: The gas would float on top because it's the lightest fluid. So, in the gas cap, the **gas saturation (S_g)** would be 100%, meaning that there is only gas.
- *Oil zone (middle)*: Below the gas, you would have oil. Here, the **oil saturation (S_o)** would be 100%, meaning only oil is present.
- *Water zone (bottom)*: At the bottom, you'd have the heaviest fluid, water. The **water saturation (S_w)** would be 100%, meaning it's entirely water.

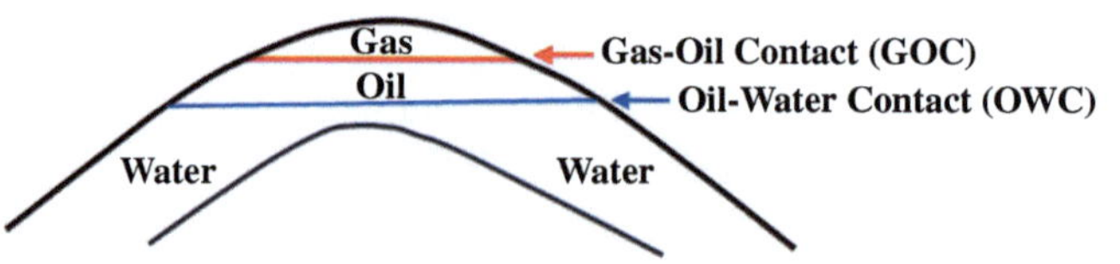

Figure 1.18 Cross section of reservoir showing vertical segregation of fluids.

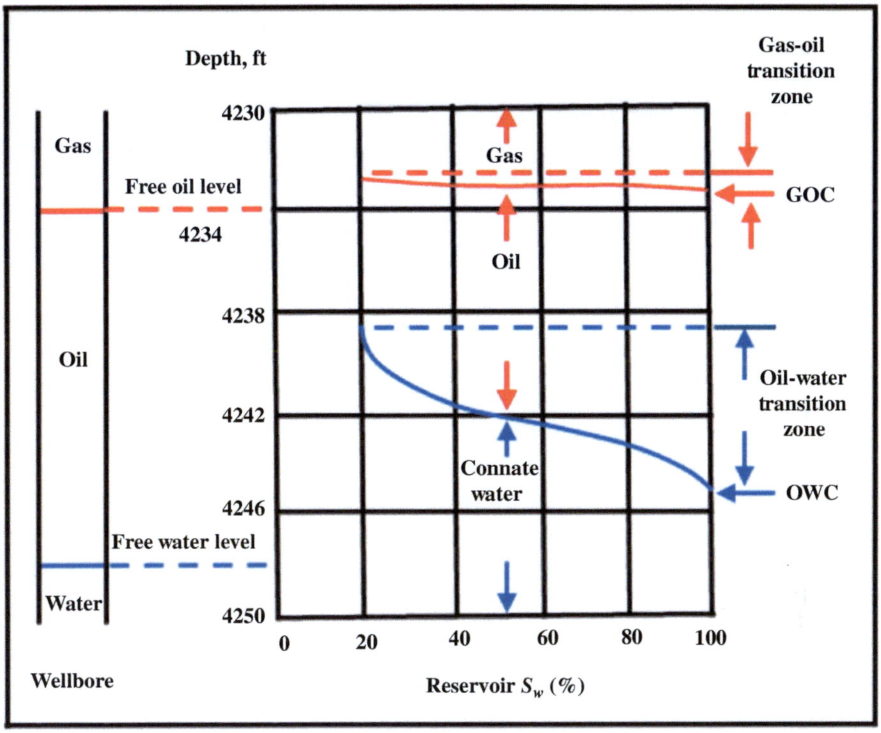

Figure 1.19 Normal initial fluid distribution in a reservoir of uniform permeability and static equilibrium.

This is an **idealized model**, where fluids do not mix, and the interfaces between gas, oil, and water are very clear-cut and distinct.

In the real world, reservoirs are made of porous rock, and things aren't so simple. Figure 1.19 gives a more realistic view of how fluids actually distribute in the pores of a reservoir rock. The transitions between gas, oil, and water aren't sharp, and fluids can be mixed at certain levels.

Here's a breakdown of what Figure 1.19 shows:

1. Gas cap:
   - At the very top of the reservoir, you still have gas, but there is some water mixed in with it. So, even in the gas cap, $\boldsymbol{S_g \neq 100\%}$ because some of the space is taken up by **irreducible water saturation (S_{wir})**—this is water that cannot be displaced, no matter how much gas is present.
   - GOC: This is the point where the gas and oil meet, located at 4233.5 ft.
2. Gas-oil transition zone:
   - This zone is a mix of gas and oil. The gas saturation (S_g) decreases as you go deeper, and oil saturation (S_o) increases.
   - This transition happens over a small depth, about 1 ft (4232.5–4233.5 ft).
3. Oil zone:
   - The bulk of the oil is located below the gas-oil transition. However, even in the pure oil zone, some water is still present due to irreducible water saturation (S_{wir}),

which can't be displaced. So, the oil saturation (S_o) here is close to but not exactly 100%.
- OWC: This is the point where oil and water meet, located at 4245 ft.

4. Oil-water transition zone:
   - As you go further down, the oil decreases, and the **water saturation (S_w)** increases. This transition happens over a few feet (from 4238.5 to 4245 ft).
   - This zone has a gradual change from mostly oil with some water at the top to mostly water with some oil at the bottom.
5. Water zone (aquifer):
   - Below the OWC, there is mostly water. But as always, some oil remains trapped in the pores, even in this zone.
   - The **free water level (FWL)** is at 4248 ft, where the water is the dominant fluid.

In Figure 1.20, the distribution of saturation is shown across the OWC and through the oil-water transition zone, illustrating how saturation levels change across these interfaces in a real-world reservoir.

When measuring fluid saturations from core samples, the core's fluid content is often altered due to two primary processes:

1. Mud invasion: Mud or mud filtrate used in drilling enters the formation and affects the core's original fluid content, especially oil and water saturations.
2. Pressure release: As the core is extracted to the surface, the drop in confining pressure causes gas to expand and fluids to be expelled, especially water.

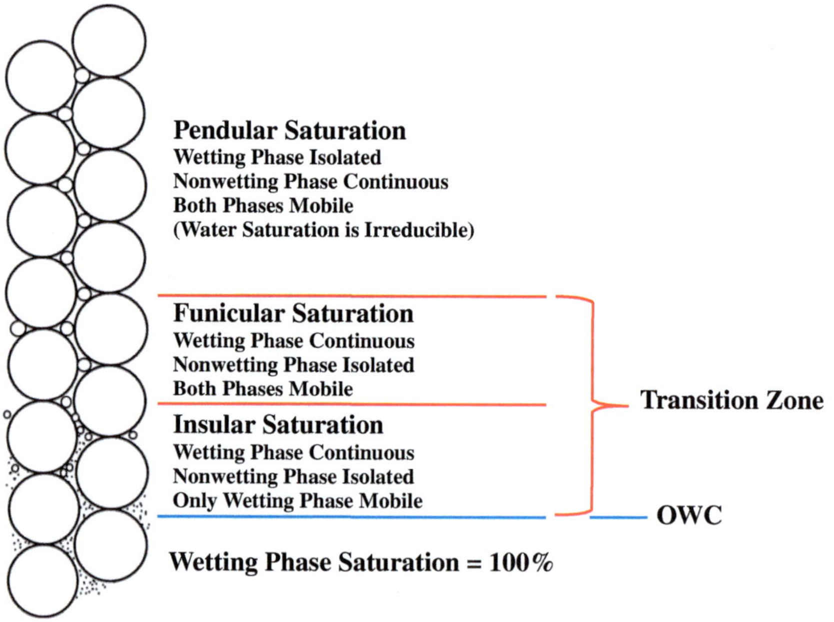

Figure 1.20 Microscopic cross section of OWC and transition zone.

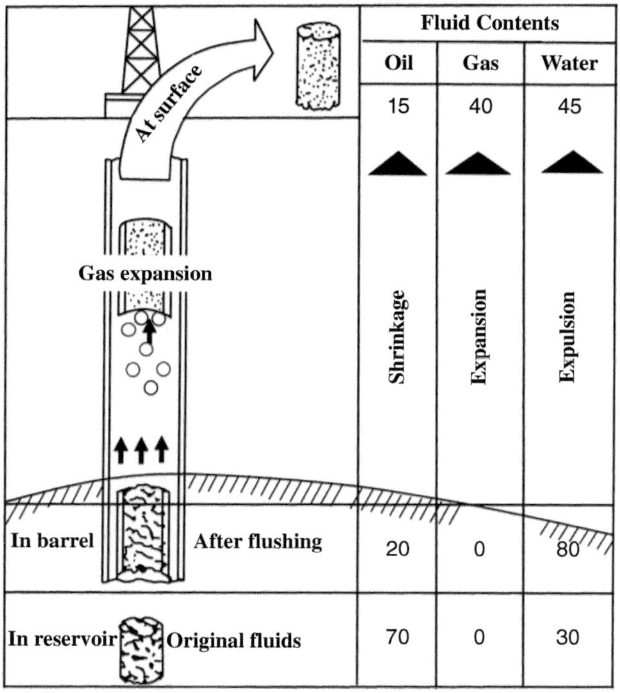

Figure 1.21 Example of saturation changes occurring in the core from *in situ* to surface conditions.

In Figure 1.21, you can observe these changes from *in situ* conditions to the surface. Oil saturation decreases, mainly because of mud invasion, while gas expands as pressure decreases, pushing water out of the core. Additionally, as pressure and temperature drop, oil shrinks in volume, further reducing its saturation.

Several methods have been suggested to minimize these issues. Using oil-based muds (OBM) as the drilling fluid can help reduce mud invasion. Another approach involves using empirical correction factors to adjust measured saturations back to their *in situ* values (Amyx et al., 1960). In some cases, geophysical well logs or capillary pressure measurements are used as alternatives to core analysis for estimating *in situ* saturations.

1.9 Resistivity

The presence of hydrocarbons in a formation can be identified by measuring the electrical resistivity of the rocks. The resistivity depends on several factors, primarily the pore geometry and fluid distribution within the rock. The key variables include the size, type, and interconnection of pore spaces, as well as the type and distribution of fluids, such as oil, gas, and water.

1.9.1 Electrical Properties of Rocks

- The rock matrix itself acts as a perfect insulator and does not conduct any electrical current.
- Conductivity occurs through the fluids present in the pore spaces of the rock, a process known as ionic or electrolytic conduction. This is caused by the movement of ions (such as Na^+ and Cl^-) in the formation water.
- The higher the salinity of the water, the more ions it contains, resulting in greater conductivity. Therefore, highly saline water will exhibit lower resistivity.
- In contrast, oil and gas are insulators, meaning that a hydrocarbon-bearing formation will have lower conductivity (or higher resistivity) than a water-bearing formation of the same porosity.

The electrical properties of rocks and their pore geometry are described by **Archie's Law**, which is fundamental to understanding the volume fraction of fluids in the pore space. Archie's Law is central to the interpretation of well logs, which are used to infer the presence of hydrocarbons.

1.9.2 Basic Concepts: Ohm's Law and Resistivity

To introduce the concept of resistivity in rocks, consider an example using a simple open-top cubic tank with dimensions of one meter in all directions. The tank is electrically nonconductive, except for two metal electrodes on opposite sides (Figure 1.22).

1. The tank is first filled with salt water (simulating formation water), and a voltage (v) is applied across the electrodes. The resulting current (i) is measured.
2. Using Ohm's Law, the resistance of the water can be calculated:

$$r_w = \frac{v}{i}. \tag{1.7}$$

3. The resistivity of the water (R_w) is calculated by scaling the resistance with the aspect ratio of area to length, i.e., the tank's cross-sectional area (A) to its length (L):

$$R_w = r_w \frac{A}{L}. \tag{1.8}$$

   Resistivity is measured in ohm-meters and is inversely proportional to both salinity and temperature—higher salinity or temperature results in lower resistivity.
4. Next, sand is added to the water-filled tank, displacing some of the water. This creates a porous medium that is 100% saturated with water. Since the rock grains (sand) act as insulators, the current must pass through the remaining water, which increases the resistance. The resistance associated with this porous, water-saturated medium is denoted as r_o, and the corresponding resistivity is R_o (Figure 1.23).

The increase in resistance is due to the tortuous path the electrical current must take to circumvent the nonconductive sand grains. This path is longer than the direct path through water, resulting in higher resistance. The tortuous path length is referred to as L_a, and the cross-sectional area through which current flows is A_a.

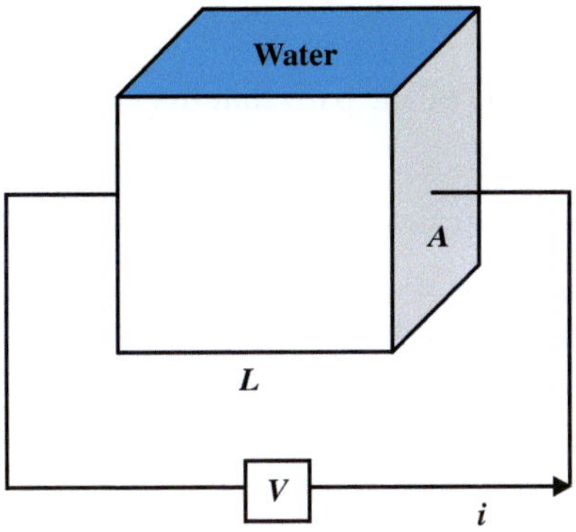

Figure 1.22 Definition of water resistivity.

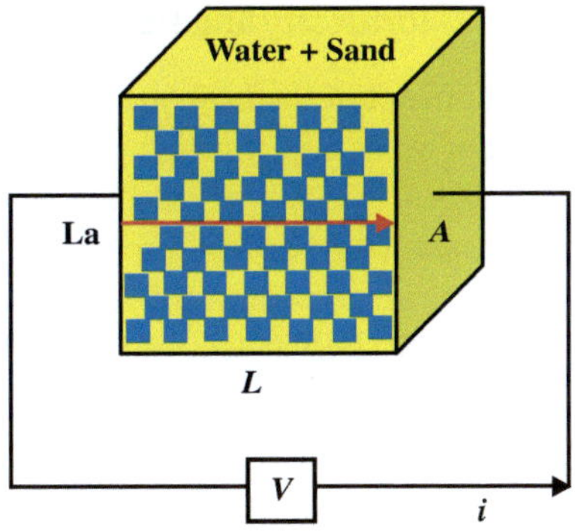

Figure 1.23 Definition of a 100% water-saturated sand.

1.9.3 Formation Resistivity Factor

The resistivity of the rock in this water-saturated state is proportional to the resistivity of the water (R_w) because only the water conducts electricity. This relationship is expressed as:

$$F = \frac{R_o}{R_w}, \tag{1.9}$$

where F is the formation resistivity factor, a proportionality constant that accounts for the increased resistance due to the presence of the rock matrix. Understanding the formation resistivity factor is crucial for accurately interpreting well logs and determining fluid saturation within the formation.

The formation factor (F) is independent of the water resistivity, as shown by the horizontal line in Figure 1.24. The formation factor, defined as the ratio of resistivity in a water-saturated rock to the resistivity of the water, can be described as:

$$\frac{R_o}{R_w} = \frac{A}{A_a}\frac{L_a}{L}, \tag{1.10}$$

where R_o is the resistivity of the water-saturated formation, R_w is the resistivity of the formation water, A is the area of the system, and A_a is the area available to the current, L is the length of the system, and L_a is the tortuous path length the current follows.

Figure 1.24 Relationship between F and water conductivity.

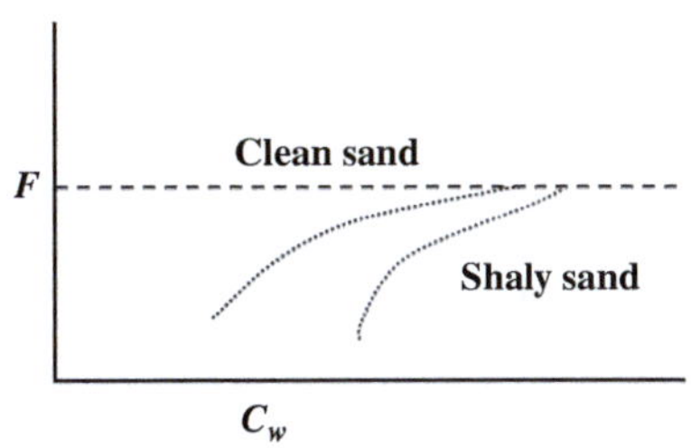

1.9.4 Tortuosity and Porosity

The concept of tortuosity (τ) describes the complexity of the path that the electrical current must take through the pore spaces:

$$\tau = \left(\frac{L_a}{L}\right)^2. \tag{1.11}$$

Porosity (φ) represents the fraction of the rock that is made up of void spaces:

$$\varphi = \frac{A}{A_a}. \tag{1.12}$$

These definitions allow us to express the formation factor (F) as:

$$F = \frac{\sqrt{\tau}}{\varphi}. \tag{1.13}$$

This equation shows that the formation factor is directly related to pore geometry, including the porosity and tortuosity of the formation.

1.9.5 Empirical Relationships and Cementation

Laboratory-measured formation factors provide accurate results for specific rock types. However, lab tests are limited to formations where core samples have been retrieved. In cases where core samples are unavailable, empirical correlations like Archie's Law are often used. Archie established a general relationship between porosity and formation factor:

$$F = a \times \varphi^{-m}, \tag{1.14}$$

where a is the **cementation factor** and m is the **cementation exponent**.

The term **cementation** describes how tightly the grains in a rock are bound together. As the degree of cementation increases, the cementation exponent (m) increases from around 1.3 in unconsolidated rocks to 2.2 in highly cemented rocks. For a formation factor of 20, this change in the cementation exponent translates into a change in porosity from 10 to 26%. The empirical constants in Archie's Law depend on various factors, including cementation, tortuosity, and wettability.

1.9.6 Resistivity Index and Water Saturation

When hydrocarbons are present in the pore spaces, they act as insulators, increasing the formation's resistivity (Figure 1.25). The resistivity of a formation that contains hydrocarbons is denoted as R_t. The resistivity index (I) is defined as the ratio of the resistivity of the hydrocarbon-bearing formation to the resistivity of the formation 100% saturated with water:

$$I = \frac{R_t}{R_o}. \tag{1.15}$$

Archie correlated the Resistivity Index with water saturation (S_w) using the following relationship:

$$I = \frac{R_t}{R_o} = \frac{1}{S_w^n}, \tag{1.16}$$

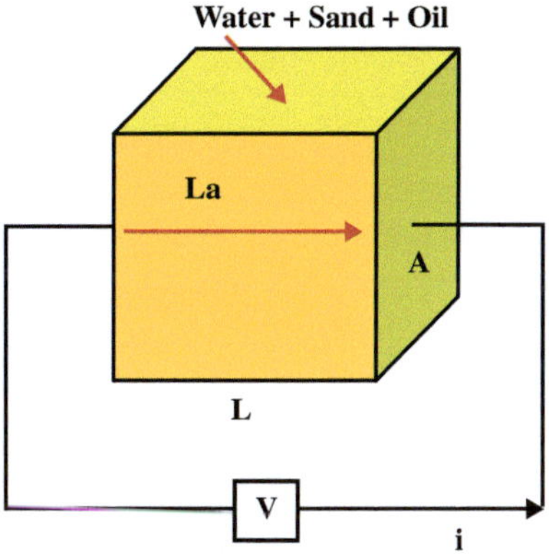

Figure 1.25 Resistance in a hydrocarbon-bearing formation.

where n is the b. When the formation is completely saturated with water ($S_w = 1$), the resistivity of the formation equals R_o (i.e., $R_t = R_o$). Conversely, when the formation contains only hydrocarbons ($S_w = 0$), the resistivity tends toward infinity ($R_t \to \infty$).

By rearranging Eq. (1.16) and assuming ($n = 2$), we can solve for **water saturation**:

$$S_w = \sqrt{\frac{R_o}{R_t}}. \tag{1.17}$$

This equation is applicable when a hydrocarbon-bearing zone overlays a water-bearing zone with similar porosity and water salinity. To generalize this equation for any permeable hydrocarbon zone, Archie extended it by substituting Eq. (1.9) for R_o, resulting in:

$$S_w = \sqrt{\frac{FR_w}{R_t}}. \tag{1.18}$$

This expression is known as **Archie's Relation** or **Archie's Law**, which remains a fundamental tool for determining water saturation and evaluating hydrocarbon reserves in porous formations.

1.10 Capillary Pressure

Reservoir rocks consist of tiny pores, and the small pore sizes, combined with the interfacial tension between the immiscible fluids (such as water and hydrocarbons), lead to **capillary pressure**. This pressure plays a critical role in determining fluid distributions within the reservoir.

Porous reservoir rocks can be thought of as a collection of capillary tubes, where capillary pressure is generated at the interface between immiscible fluids. Capillary pressure is the pressure difference across the interface formed by two immiscible fluids in these small tubes. Figure 1.26 illustrates capillary tubes immersed in both wetting and nonwetting liquids:

- In a **wetting liquid**, the liquid climbs up the capillary tube (capillary rise), as observed with water in a glass tube.
- In a **nonwetting liquid**, the liquid is depressed (capillary depression), as seen with mercury in a glass tube.

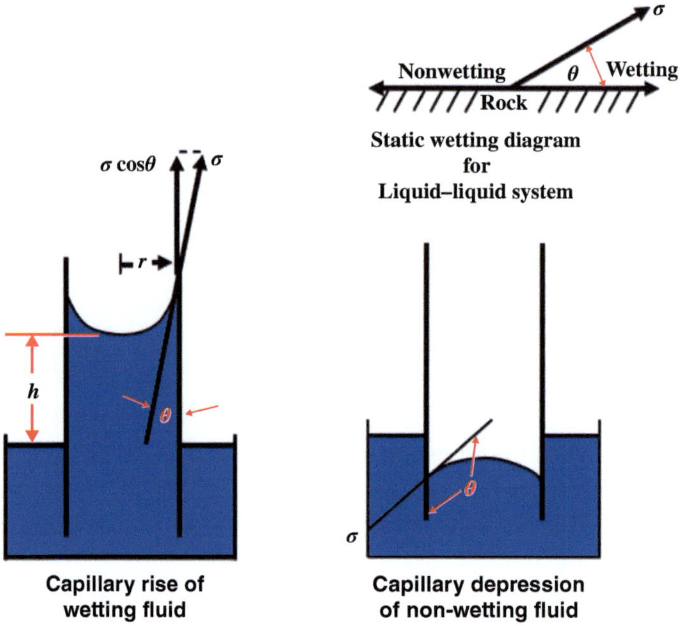

Figure 1.26 Capillary tubes in wetting and nonwetting fluids.

The capillary pressure in a tube is influenced by the interfacial tension between the fluids, the contact angle between the rock and the fluid, and the radius of the tube. Using Figure 1.26, the equilibrium in the capillary tube leads to a force balance equation:

$$\text{force up} = \text{force down},$$

$$\text{force up} = 2\pi r\sigma \cos\theta,$$

$$\text{force down} = \pi r^2 h \Delta\gamma.$$

Since capillary pressure (P_c) is defined as the force per unit area:

$$P_c = \frac{\textit{force up}}{\pi r^2} = \frac{2\sigma \cos\theta}{r}, \tag{1.19}$$

where r is the **pore radius** (cm), σ is the **interfacial tension** (dynes/cm), and θ is the **contact angle** between the fluid and the solid surface (degrees).

Laboratory measurements of capillary pressure often require corrections for several factors:

1. *Closure*: Effects caused by surface irregularities of the core sample.
2. *Microporosity*: Leads to "double curves" due to a different capillary entry point at lower water saturations.
3. *Confining stress*: Affects the magnitude of porosity and permeability.
4. *Presence of clays*: Alters effective water saturation.
5. *Wettability and interfacial tension differences*: These differ between laboratory fluids and reservoir fluids.

Among these, **wettability** and **interfacial tension differences** are particularly important. In practice, capillary pressure is often converted to height above the FWL to develop a **saturation profile**. The height at which water stands above the FWL is proportional to the capillary pressure.

Expanding on Eq. 1.19, the following expression can be derived for capillary pressure:

$$P_c = \frac{\pi r^2 gh(\rho_w - \rho_{air})}{\pi r^2} = \Delta \rho gh, \tag{1.20}$$

where H is the height of capillary rise (ft), P_c is the capillary pressure (psi), ρ_w is the density of water or wetting phase (slugs/ft^3 or kg/m^3), and ρ_{air} is the density of air or nonwetting phase (slugs/ft^3 or kg/m^3).

From Eq. 1.19, we observe that **larger pore radii** result in **lower capillary pressure**. From Eq. 1.20, lower capillary pressure leads to a lower height of water rise in the reservoir.

- Low capillary pressure and low irreducible water saturations are found in reservoir rocks with large pores, such as coarse-grained sandstones, coarse-grained oolitic carbonates, and vuggy carbonates.
- High capillary pressure and high water saturations are associated with fine-grained reservoir rocks.

1.10.1 Capillary Pressure in Reservoirs

In reservoir systems, capillary pressure (P_c) is the pressure difference between the nonwetting-phase pressure (P_{nw}) and the wetting-phase pressure (P_w):

$$P_c = P_{nw} - P_w. \tag{1.21}$$

Capillary pressure is always positive, and in water-wet oil reservoirs, it can be described as the pressure difference between the oil and water phases:

$$P_{cow} = P_o - P_w. \tag{1.22}$$

Similarly, in a water-wet gas reservoir or gas cap:

$$P_{cgw} = P_g - P_w. \tag{1.23}$$

When oil and gas coexist in a reservoir, the oil phase typically behaves as the wetting phase relative to the gas phase. Therefore, the capillary pressure between the gas and oil phases is:

$$P_{cog} = P_g - P_o. \tag{1.24}$$

Capillary pressure and gradients arise due to the density differences between oil, gas, and water, as well as their immiscibility. These factors influence the distribution of fluids in the reservoir, as shown in Figure 1.27.

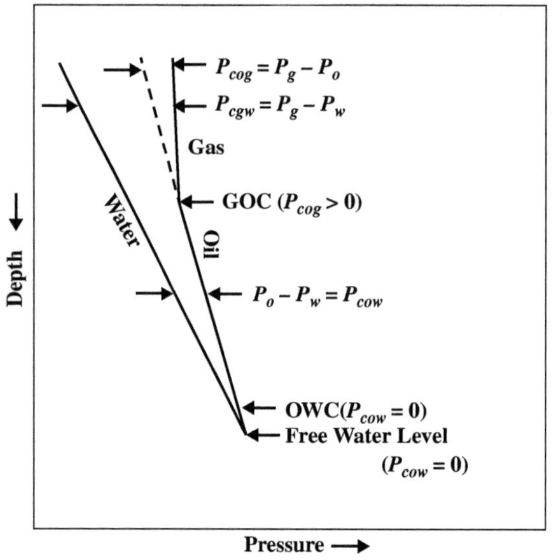

Figure 1.27 Capillary pressure in reservoirs.

1.10.2 Laboratory Capillary Pressure Measurements

While it would be ideal to measure capillary pressure using actual reservoir oil and water at reservoir temperature, practical difficulties, such as obtaining representative samples and handling them in the laboratory, make this challenging. Instead, substitute fluids are used, and results are adjusted for reservoir conditions (Table 1.1).

Table 1.1 Typical interfacial tension and contact angle constants.

System	Contact angle, θ (deg.)	Interfacial tension, σ (dynes/cm)
Laboratory		
Air–water	0	72
Oil–water	30	48
Air–mercury	140	480
Air–oil	0	24
Reservoir		
Water–oil	30	30
Water–gas	0	50[a]

a. Pressure and temperature dependent, reasonable value to depth of 5000 ft. The equation for correcting laboratory-measured capillary pressure to reservoir conditions is:

$$P_{c(res)} = P_{c(lab)} \frac{(\sigma \cos\theta)_{res}}{(\sigma \cos\theta)_{lab}}, \tag{1.25}$$

where $P_{c(res)}$ = capillary pressure corrected to reservoir conditions (psig), $P_{c(lab)}$ = capillary pressure measured in the laboratory (psig), σ_{res} = interfacial tension between reservoir fluids under reservoir conditions (dynes/cm), σ_{lab} = interfacial tension between laboratory fluids under laboratory conditions (dynes/cm), θ_{res} = contact angle for reservoir conditions and fluids (degrees), and θ_{lab} = contact angle for laboratory conditions and fluids (degrees).

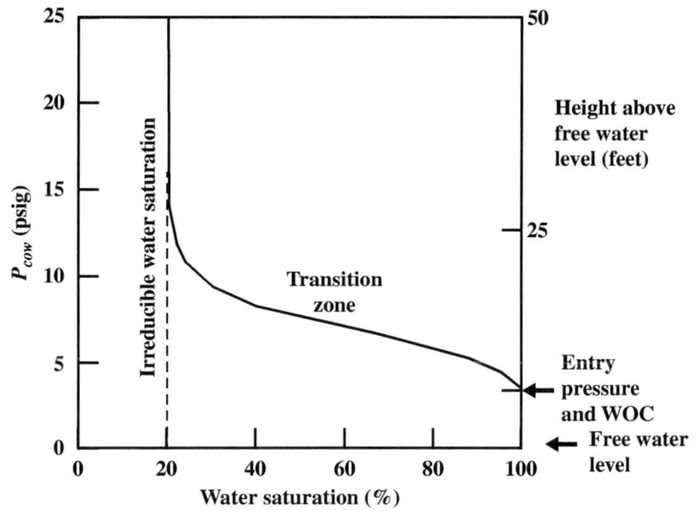

Figure 1.28 Conventional capillary pressure curve.

1.10.3 Entry Pressure

In water-wet reservoirs, water (the wetting phase) is more readily attracted to the rock surface, while the nonwetting phase (usually oil) requires additional pressure to enter the rock. This pressure is known as the **entry pressure**—the minimum capillary pressure required to force the first droplet of oil into the rock.

- Initially, oil enters the **largest pores** of the rock.
- As capillary pressure increases, oil enters smaller pores, reducing water saturation at increasing depths above the OWC.

Figure 1.28 illustrates a conventional capillary pressure curve for a water-wet oil reservoir. The FWL is the depth where the capillary pressure between oil and water is zero. However, the OWC, where oil saturation first appears, occurs at a shallower depth, where the capillary entry pressure is reached.

As capillary pressure rises, more oil enters smaller pores, displacing water and establishing fluid saturations in the reservoir.

1.10.4 Hysteresis—Imbibition Versus Drainage

The relationship between capillary pressure and saturation is dependent on the direction of saturation change, as demonstrated in Figure 1.29.

- *Drainage Curve*: Represents the process of oil migrating into a reservoir initially filled with water (100% water saturation). The nonwetting oil phase enters the reservoir at entry pressure. As capillary pressure increases, oil saturation also rises, causing a decrease in water saturation. This reduction in wetting-phase saturation is referred to as a drainage process, which continues until it reaches the irreducible water saturation level, S_{wi}.

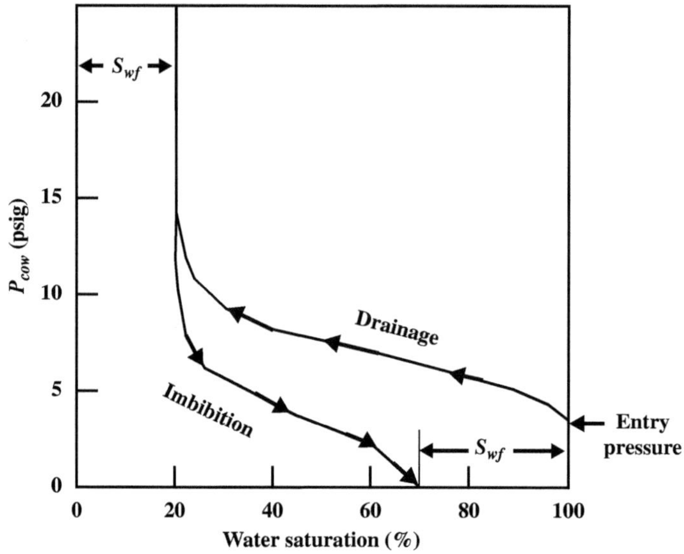

Figure 1.29 Imbibition and drainage capillary pressure curves illustrating hysteresis.

- *Imbibition Curve*: Illustrates the reverse process, where water is injected into the reservoir that was at irreducible water saturation (as in a waterflooding operation). The arrows on this curve indicate increasing water saturation, where water is being imbibed into the rock. This process is known as imbibition.
- *Hysteresis*: The difference between the drainage and imbibition curves is known as hysteresis, a key feature that shows how saturation behavior differs based on the direction of change.
- *Residual Oil Saturation* (S_{or}): Even after imbibition, some oil remains trapped in the reservoir and cannot be displaced by water. This trapped oil is referred to as the residual oil saturation, S_{or}.

1.10.5 Permeability Effects

Capillary pressure and irreducible water saturation are inversely related to pore size. As pore size increases, capillary pressure and irreducible water saturation decrease. Figure 1.30 demonstrates the relationship between permeability and capillary pressure, and Table 1.2 summarizes the key permeability effects. In larger pores, lower capillary pressure means that the reservoir rock holds less water at irreducible saturation levels.

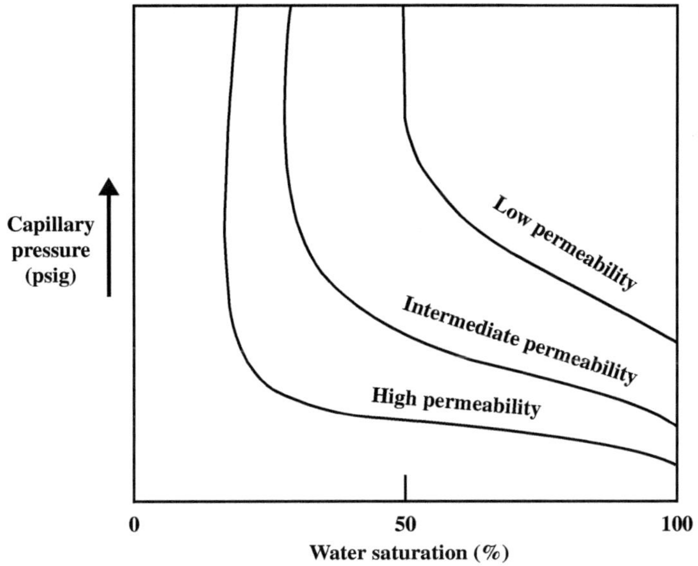

Figure 1.30 Effect of permeability on capillary pressure.

Table 1.2 Permeability effects.

Property	High permeability	Low permeability
Entry pressure	Lower	Higher
Irreducible water saturation	Lower	Higher
Slope of transition zone curve	Lower	Higher
Grain size distribution	Narrow	Wide
Grain and pore size	Large	Small

1.10.6 Relative Permeability–Capillary Pressure Relationship

Both relative permeability and capillary pressure are closely tied to pore size distribution and fluid saturation:

- *Relative permeability*: Refers to how easily a fluid moves through porous rock when more than one fluid is present.
- *Capillary pressure*: The pressure difference at the interface of immiscible fluids inside pores.

As capillary pressure rises, water saturation drops to irreducible levels (Figure 1.31). During this process, relative permeability to water decreases while relative permeability to oil increases. The pore geometry of the rock directly influences both relative permeability and capillary pressure.

- *Permeability influence*: Illustrated in Figure 1.32, an increase in permeability corresponds to a broader pore size distribution. Higher permeability results in lower irreducible water saturation and lower capillary pressure, which reflects improved fluid movement.
- *Production prediction*: The curves in Figure 1.33 reveal how capillary pressure and relative permeability are critical for identifying the productive interval in a reservoir (such as the OWC). These curves can predict changes in production over time as a result of multiphase fluid flow, with permeability variations affecting productivity.

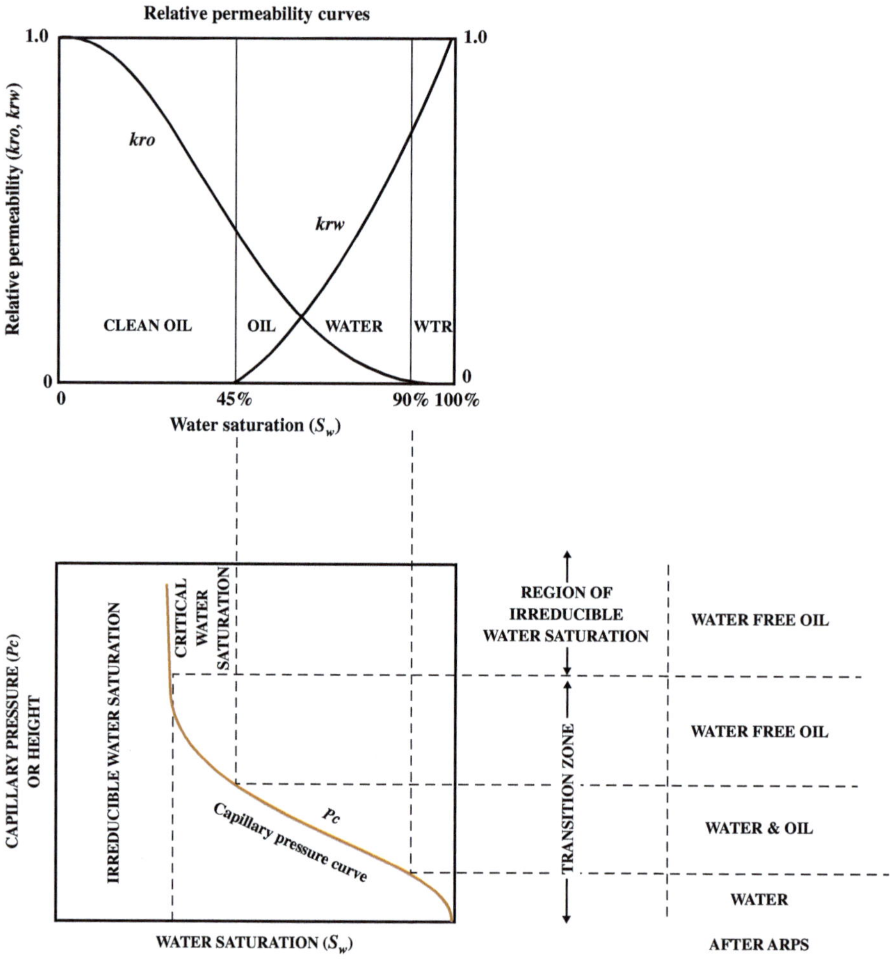

Figure 1.31 Relationship between capillary pressure and relative permeability.

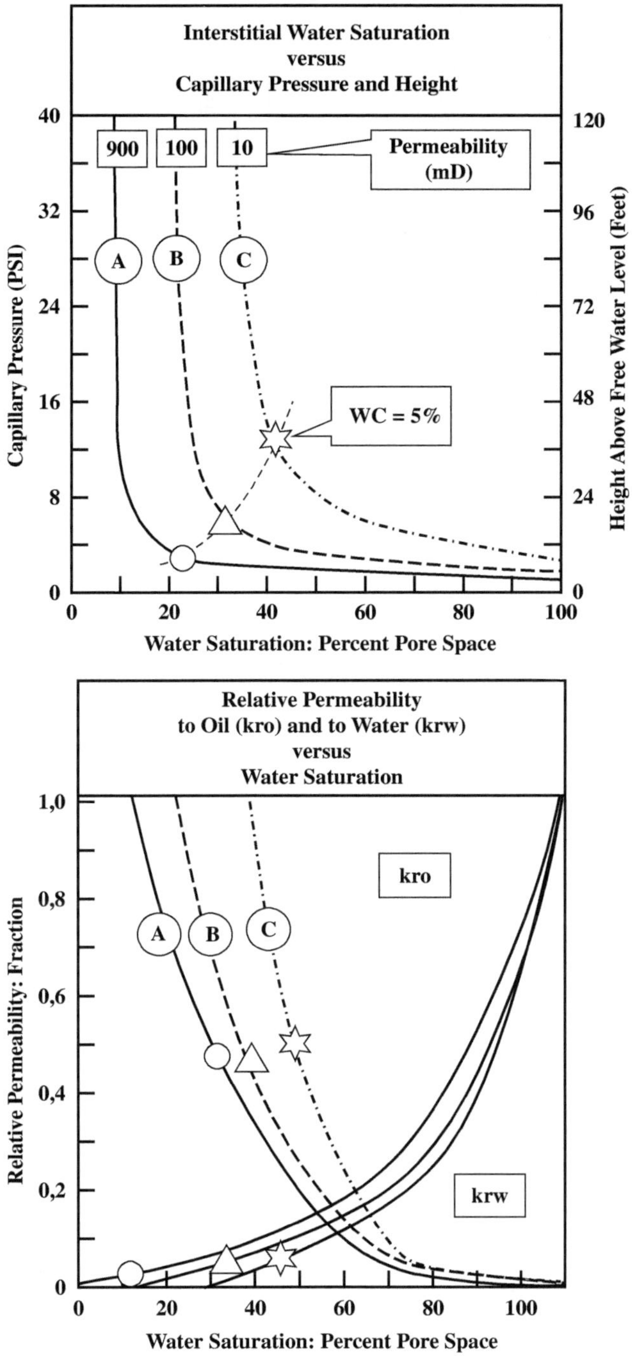

Figure 1.32 Relationship between capillary pressure and relative permeability.

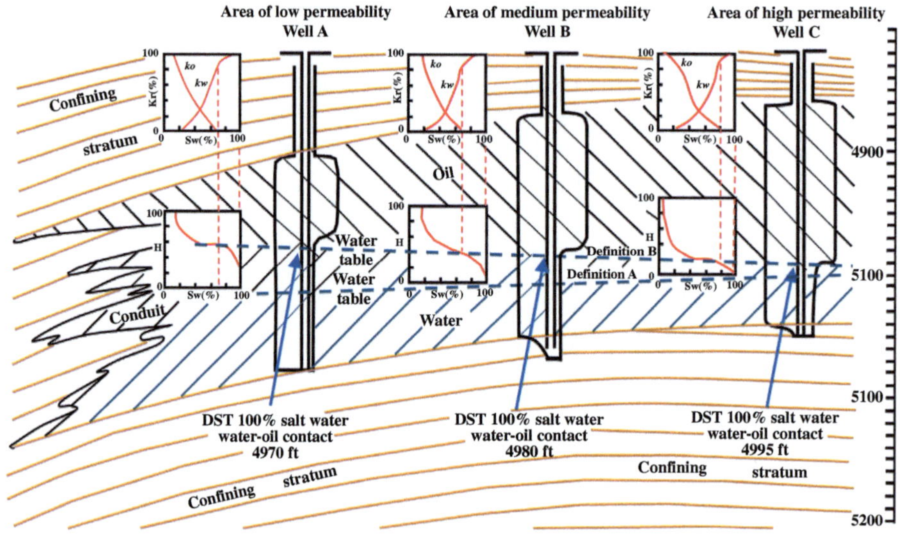

Figure 1.33 Spatial variation of P_c and k_r curves for a reservoir.

1.11 Types of Reservoir Fluids

Reservoir fluids are categorized based on their phase behavior and critical properties, which are essential for effective reservoir management. The five main types of reservoir fluids are:

1. Black Oil
2. Volatile Oil
3. Gas Condensate
4. Wet Gas
5. Dry Gas

1.11.1 Black Oil

- *Phase diagram characteristics*: The phase diagram shows pressure declining at constant temperature (isothermal; Figure 1.34).
- *Undersaturated state* (*Point 1 to Point 2*): The reservoir starts at a pressure much higher than the bubble point, where no gas has evolved. The oil can still absorb more gas.
- *Saturated state* (*Point 2 to Point 3*): Upon reaching the bubble point (Point 2), gas begins to evolve from the oil. Further pressure decline leads to more gas evolution.
- *Production behavior*: In a black oil reservoir, most of the oil phase is recoverable at the surface, with gas separating during production.

Black oil systems are characterized by relatively low gas-to-oil ratios (GOR) and exhibit significant liquid recovery.

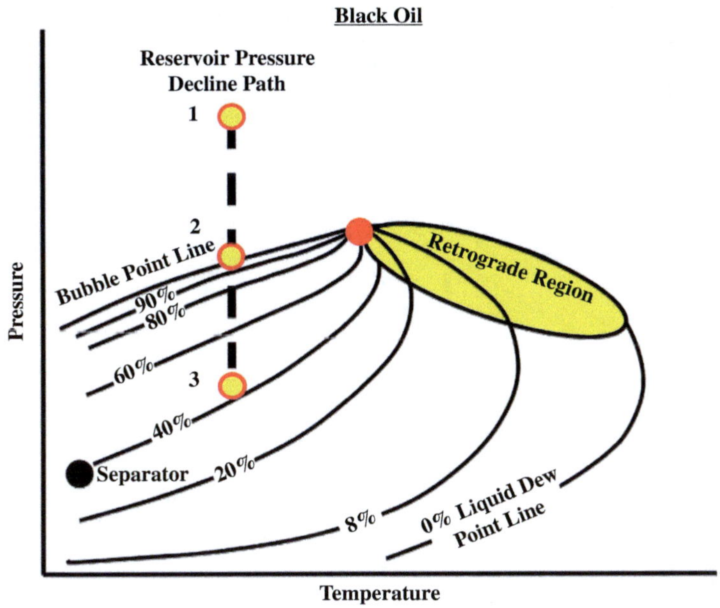

Figure 1.34 Phase diagram of black oil.

1.11.2 Volatile Oil

- *Phase diagram characteristics*: Similar to black oil but with the initial pressure closer to the bubble point. The reservoir contains more light hydrocarbons, making the phase diagram shift up and left (Figure 1.35).
- *Gas evolution*: A small pressure drop below the bubble point releases large amounts of gas. Volatile oil reservoirs produce more gas at the surface than black oils, resulting in less liquid recovery.

1.11.3 Gas Condensate

- *Temperature range*: Between the critical point and cricondentherm.
- *Liquid formation*: As the pressure drops below the dew point, heavier hydrocarbons condense in the reservoir, partially blocking gas flow near the wellbore. Liquid re-vaporizes once pressure falls below the retrograde region (Figure 1.36).

1.11.4 Wet Gas

- *Phase behavior*: The reservoir temperature is above the cricondentherm, meaning no liquid forms in the reservoir. Liquid only forms at surface conditions (Figure 1.37).
- *Production*: The gas stays as a single phase in the reservoir but forms condensate at the surface, hence "wet" gas.

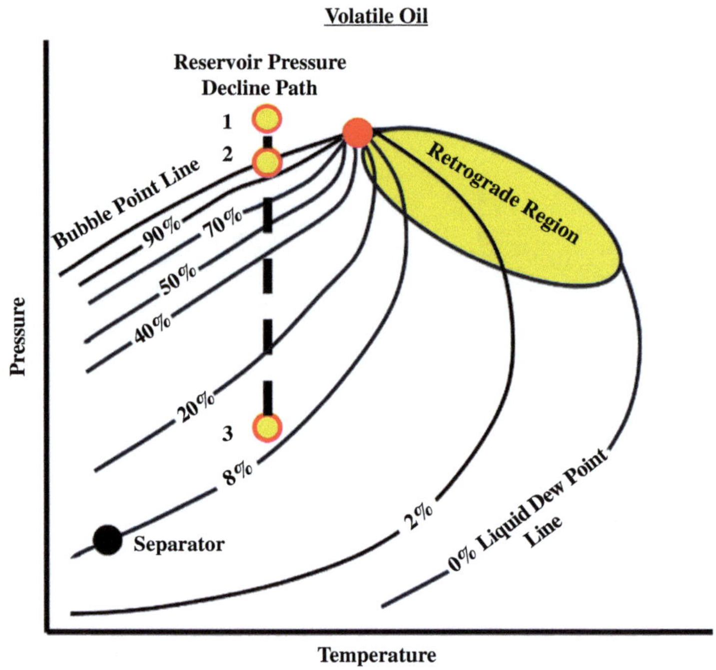

Figure 1.35 Phase diagram of volatile oil.

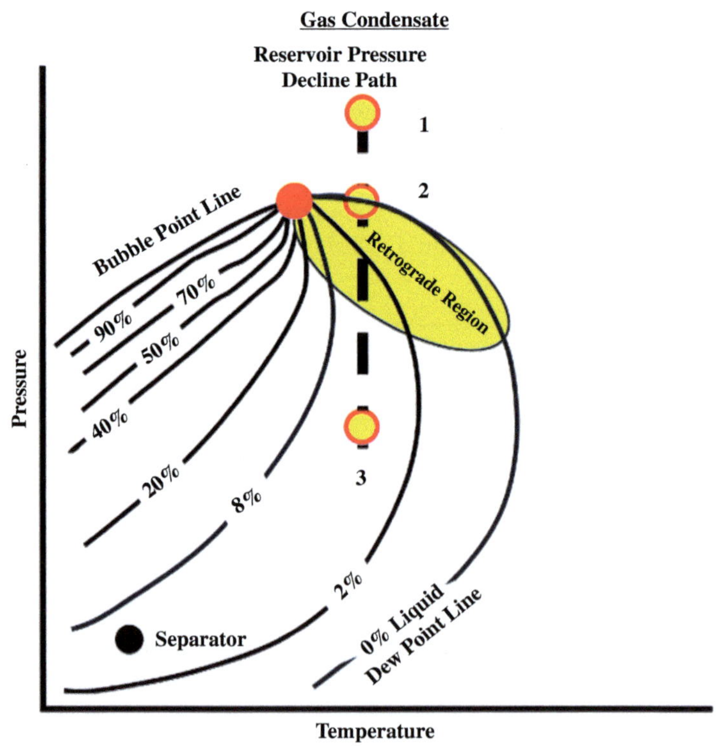

Figure 1.36 Phase diagram of gas condensate.

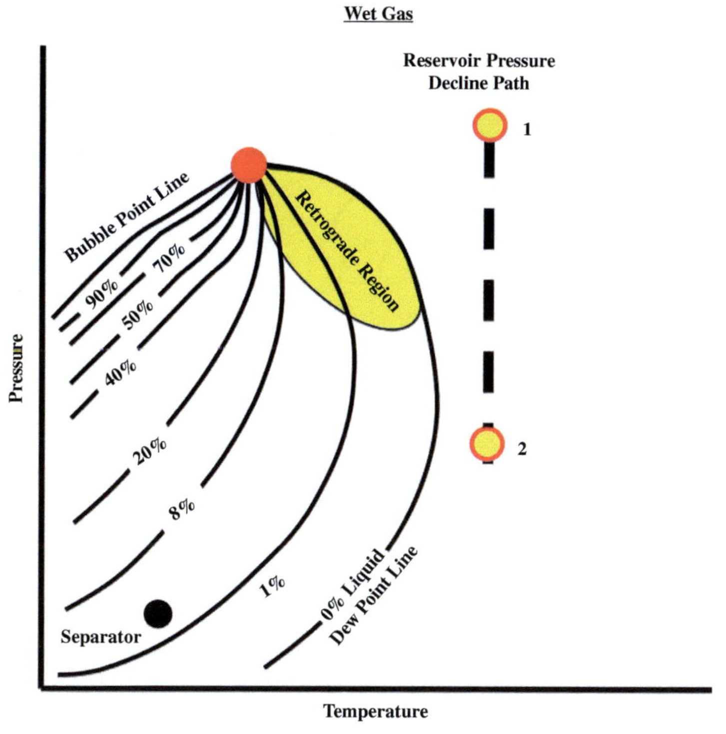

Figure 1.37 Phase diagram of wet gas.

1.11.5 Dry Gas

- *Composition*: Mostly methane, with a simple, condensed phase diagram.
- *Phase Behavior*: Both in the reservoir and at surface conditions, the fluid remains in the gas phase, never entering the two-phase region. Dry gas reservoirs are the easiest to model analytically (Figure 1.38).

For effective enhanced oil recovery (EOR), understanding the phase behavior of volatile oil, gas condensate, wet gas, and dry gas reservoirs is critical. Each fluid type presents unique challenges, from gas evolution in volatile oil reservoirs to condensation issues in gas condensate reservoirs. EOR techniques, such as gas injection, reinjection, and gas cycling, are employed to manage these challenges and maximize recovery.

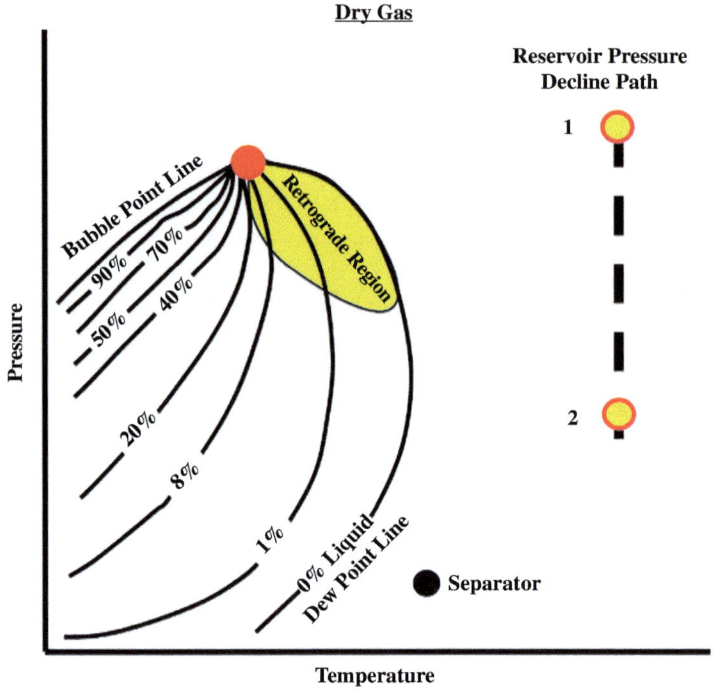

Figure 1.38 Phase diagram of dry gas.

Bibliography

Ahmed, T. (2019). *Reservoir Engineering Handbook* (5th ed.). Gulf Professional Publishing.

Amyx, J.W., Bass, D.M., & Whiting, R.L. (1960). *Petroleum Reservoir Engineering: Physical Properties*. McGraw-Hill.

Bazhenov, O.K., Burlin, Y.K., Sokolov, B.A., & Khain, V.E. (2012). *Geology and Geochemistry of Oil and Gas*. MSU Publishing House.

Craft, B.C., & Hawkins, M.F. (2014). *Applied Petroleum Reservoir Engineering* (3rd ed.). Pearson.

Eremenko, N.A. (1968). *Geology of Oil and Gas*. Nedra.

Levorsen, A. (1970). *Geology of Oil and Gas*. Mir.

Norman, H.J. (2008). *Geology, Exploration, Drilling and Oil Production*. Olimp-Business.

Zhdanov, M.A. (1970). *Oil and Gas Field Geology and Calculation of Oil and Gas Reserves*. Nedra.

2

Fluid Flow in Porous Media

2.1 Introduction

In the field of **enhanced oil recovery (EOR)**, understanding and predicting fluid flow behavior within reservoirs is essential for optimizing hydrocarbon production. This chapter delves into the fundamental principles governing fluid flow, with a particular focus on complex scenarios such as **boundary effects**, **ideal gas flow**, and **anisotropic media**, all of which play crucial roles in EOR processes. By examining these concepts through the lens of **diffusivity equations**, **streamline analysis**, and **image well theory**, this chapter offers valuable insights into how fluid dynamics can be manipulated to improve oil recovery rates.

The chapter begins with a detailed discussion on **simulating boundary effects**, where the concept of image wells is introduced to represent the influence of reservoir boundaries. This is critical in understanding how production wells respond to external constraints, allowing engineers to model and manage pressure distributions more effectively. The application of the **superposition principle** in this context offers a simplified yet powerful tool to mimic real-world scenarios where **wellbore** interactions and reservoir boundaries dictate the flow behavior.

Further, the discussion transitions into ideal gas flow, which is particularly important for reservoirs where gas is a dominant phase. Gas dynamics differ significantly from liquid flow due to factors such as **compressibility** and the Klinkenberg effect. Mastery of these concepts is essential for engineers involved in gas injection projects, a common EOR method where gas is injected into the reservoir to increase pressure and displace oil.

Finally, the chapter explores streamlines, isopotentials, and source/sink representations, providing a foundational understanding of how fluid moves through the porous media of a reservoir. These principles are indispensable in EOR applications, especially in designing flow nets and visualizing fluid movement during complex recovery processes like **steam flooding**, **water flooding**, and **miscible gas injection**.

Incorporating these advanced flow dynamics concepts into the broader framework of EOR strategies helps ensure that production is maximized while minimizing the risks and uncertainties associated with reservoir behavior. As the petroleum

Methods for Enhanced Oil Recovery: Fundamentals and Practice, First Edition.
Baghir A. Suleimanov and Elchin F. Veliyev.

industry continues to pursue more challenging and mature fields, a deep understanding of the interaction between fluid mechanics and reservoir boundaries will remain central to developing more efficient recovery techniques. This chapter provides the theoretical underpinning required for tackling these challenges, making it a critical component of the EOR knowledge base.

2.2 Applications of Darcy's Law

2.2.1 Radial Flow

In this section, we begin by discussing radial flow, which is a common flow pattern in reservoir conditions, especially when fluids are moving toward a wellbore. Unlike linear flow, radial flow occurs when fluid flows from a cylindrical drainage area surrounding the well, as depicted in Figure 2.1.

Two key pressures govern this system: the external boundary pressure, p_e, measured at the boundary radius r_e, and the internal bottom-hole flowing pressure, p_{wf}, measured at the wellbore radius r_w. These pressures define the driving force for fluid to flow toward the wellbore.

$$v = \frac{q}{A} = \frac{q}{2\pi rh} = \frac{k}{\mu}\frac{dp}{dr}. \tag{2.1}$$

When applying **Darcy's law** in this radial system, it's important to note that the radial direction is measured as positive opposite to the flow direction. As a result, the potential gradient in this case is positive. By integrating Darcy's law between the two boundary conditions, the following relationship can be established:

$$q = \frac{2\pi kh(p_e - p_{wf})}{\mu B_0 \ln \frac{r_e}{r_w}}. \tag{2.2}$$

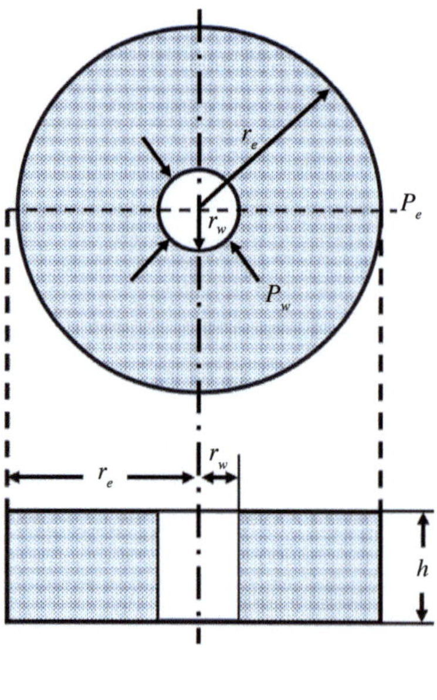

Figure 2.1 Radial fluid flow model toward the wellbore.

This equation, initially derived in Darcy units, is often converted to field units for practical use. The resulting expression is widely utilized to describe the radial flow of a single-phase fluid toward the wellbore in reservoir engineering.

$$q = \frac{kh(p_e - p_{wf})}{141.2\mu B_0 \ln \frac{r_e}{r_w}}. \tag{2.3}$$

2.2.2 Permeability of Combination Layers

In real-world reservoir conditions, most porous rocks exhibit spatial variations in permeability. These variations can be due to distinct layers, blocks, or concentric rings of constant permeability. To accurately assess the average permeability of such systems, we must analyze specific cases.

2.2.2.1 Case I: Interbedded Reservoir Rocks

In this scenario, the reservoir rocks are interbedded with impermeable layers, such as shales or silts, preventing crossflow between the sand beds (see Figure 2.2). Here, the total flow rate (q_t) is the sum of the individual flow rates, i.e., $q_t = q_1 + q_2 + q_3$, and the total pressure drop Δp_t remains constant across the layers.

By applying Darcy's law to the flow rates, we derive the following expression:

$$\frac{k_{avg} w h_t (p_1 - p_2)}{\mu L} = \frac{k_1 w h_1 (p_1 - p_2)}{\mu L} + \frac{k_2 w h_2 (p_1 - p_2)}{\mu L} + \frac{k_3 w h_3 (p_1 - p_2)}{\mu L}.$$

This simplifies to:

$$\frac{k_{avg} w h_t (p_1 - p_2)}{\mu L} = \frac{w(p_1 - p_2)}{\mu L}(k_1 h_1 + k_2 h_2 + k_3 h_3),$$

which can be further reduced to the general formula for average permeability in layered systems:

$$k_{avg} = \frac{\sum_{i=1}^{n} k_i h_i}{\sum_{i=1}^{n} h_i}. \tag{2.4}$$

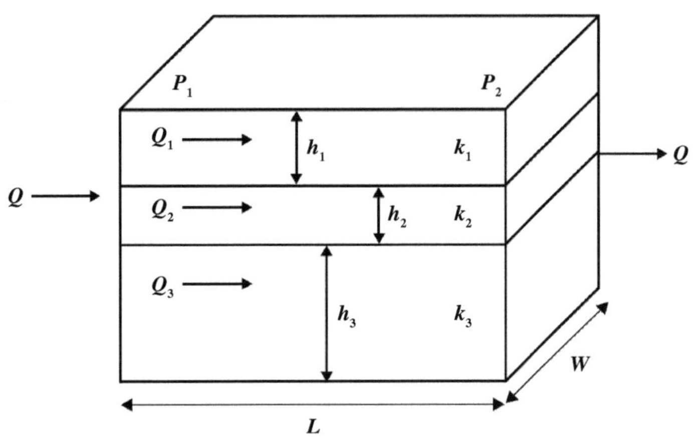

Figure 2.2 Linear flow model for parallel bed combinations.

This equation provides a method to calculate the average permeability when reservoir rocks are divided into distinct, impermeable layers.

2.2.2.2 Case II: Composite Reservoirs

A composite reservoir features variations in permeability away from the wellbore. These variations could result from operational activities like fluid invasion during drilling or water injection in water flooding operations, or they could occur naturally within the reservoir. In this scenario, the system is simplified to a set of blocks arranged in series (see Figure 2.3).

Here, the total flow rate remains constant, i.e., $q_t = q_1 = q_2 = q_3$, but the total pressure drop is the sum of individual pressure drops: $\Delta p_t = \Delta p_1 + \Delta p_2 + \Delta p_3$. Applying Darcy's law to the pressure drop gives:

$$\frac{q\mu}{wh}\frac{L}{k_{avg}} = \frac{q\mu}{wh}\left(\frac{L_1}{k_1} + \frac{L_2}{k_2} + \frac{L_3}{k_3}\right).$$

This can be generalized to:

$$k_{avg} = \frac{L}{\sum\limits_{i=1}^{n}\frac{L_i}{k_i}}. \tag{2.5}$$

This expression is useful for determining the average permeability of reservoirs that exhibit variations in permeability along the flow path.

2.2.2.3 Radial Flow in Multiple Beds

For radial flow through multiple layers (see Figure 2.4), we apply a similar approach. The general form of Darcy's law for this case is:

$$\frac{q\mu}{2\pi k_{avg}h}\ln\frac{r_e}{r_w} = \frac{q\mu}{2\pi k_1 h}\ln\frac{r_1}{r_w} + \frac{q\mu}{2\pi k_2 h}\ln\frac{r_2}{r_w}.$$

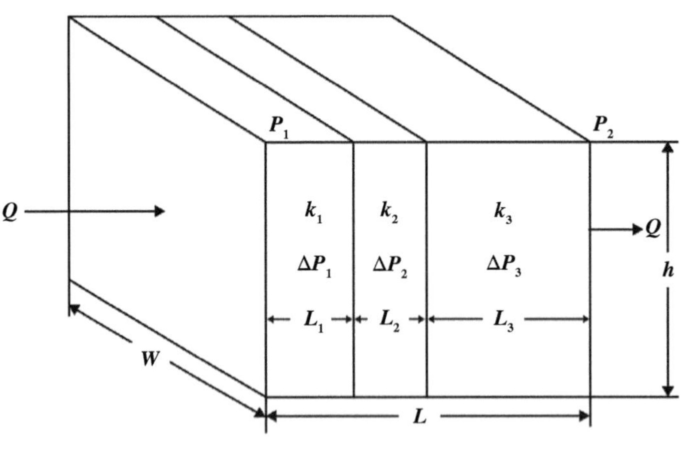

Figure 2.3 Linear flow model for beds in series combination.

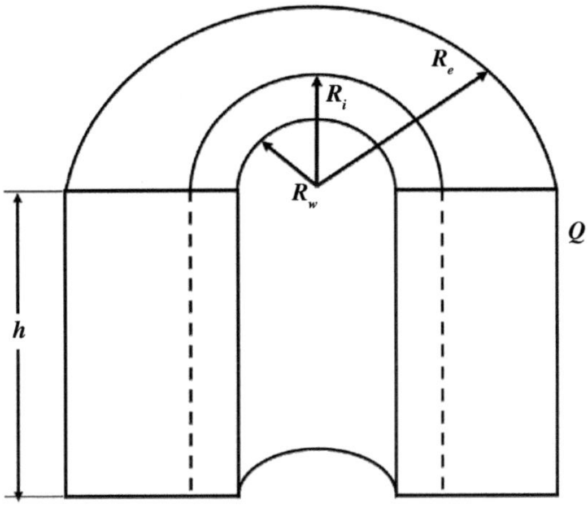

Figure 2.4 Radial flow model for beds in series combination.

This simplifies to:

$$k_{avg} = \frac{\ln \frac{r_e}{r_w}}{\sum_{i=1}^{n} \frac{\ln \frac{r_i}{r_{i-1}}}{k_i}}. \tag{2.6}$$

This equation allows us to calculate the effective permeability for radial flow through multiple beds, which is essential for accurately modeling fluid flow in reservoirs with varying layers of permeability.

2.2.3 High-velocity Flow

Darcy's experiments, and the equations derived from them, form the foundation of fluid flow analysis in petroleum engineering. These equations were developed under the assumption of strictly laminar flow, where the flow rate (q) is directly proportional to the pressure drop (Δp). However, as flow rates increase, this linear relationship no longer holds (see Figure 2.5). At high velocities, applying Darcy's law in its traditional form would result in an underestimation of permeability, which is lower than the true permeability of the rock.

One critical aspect to consider is the difference in gas and liquid velocities under the same pressure differential. Due to the lower viscosity of gases compared to liquids, gas flow velocities are higher. This high-velocity flow effect is particularly important in gas well testing, such as isochronal and deliverability tests. Accurate lab measurements must account for both the Klinkenberg effect at low pressures and non-Darcy flow at high pressures.

To address the issue of non-Darcy flow, Forchheimer, in 1901, proposed a quadratic flow equation that accounts for both laminar and turbulent flow regimes:

$$\frac{dP}{dL} = \frac{\mu v}{k} + \beta \rho v^2, \tag{2.7}$$

where P is pressure in atmospheres (atm), L is length in centimeters (cm), k is permeability in Darcy, ρ is density in grams per milliliter (g/mL), μ is viscosity in

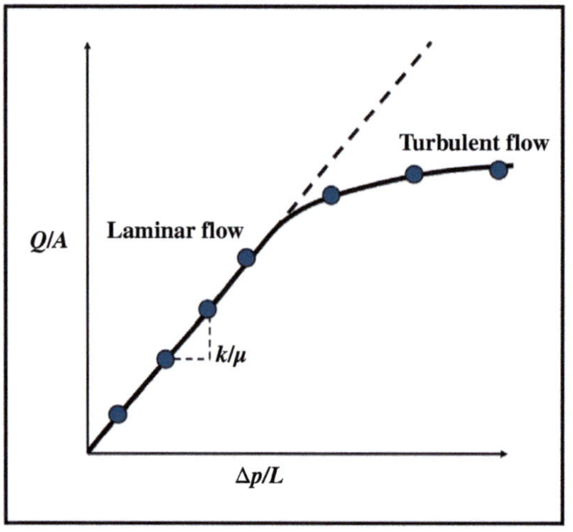

Figure 2.5 Impact of high-velocity flow on the Darcy equation.

centipoise (cP), β is the non-Darcy flow coefficient in atm-sec^2/g, v is velocity in centimeters per second (cm/sec).

The non-Darcy flow coefficient, β, plays a crucial role in correcting for high-velocity flow effects. Dimensional analysis reveals that β has dimensions of $\frac{1}{\text{length}}$, and its conversion factor is:

$$\beta\left\{\frac{1}{ft}\right\} = \beta\left\{\frac{\text{atm-sec}^2}{\text{g}}\right\} \times 3.01889 \times 10^7.$$

2.2.3.1 Estimating the Non-Darcy Flow Coefficient (β)

A key challenge in applying the Forchheimer equation is estimating the value of β. Several methods for calculating β have been developed, which rely on experimental measurements, empirical correlations, and direct use of the Forchheimer equation. To determine β experimentally, the absolute permeability of core samples is first measured. A series of increasing pressure differentials is then applied across the core samples by flowing air through the core plugs at progressively higher rates.

The results are typically presented as a curve, with β plotted against the absolute permeability for the range of core samples tested. From this data, a relationship is often derived in the following form:

$$\beta = \frac{\text{constant}}{k^a}. \tag{2.8}$$

In this equation, the exponent a is a constant determined from the experimental data.

2.2.4 Fracture Flow

In many reservoirs, the majority of fluid storage occurs in the rock matrix. However, in terms of fluid flow and flow capacity, fractures dominate, and the matrix typically contributes only a minor portion to the overall flow. To illustrate this, consider a matrix block with negligible porosity that is intersected by a fracture with width (w_f)

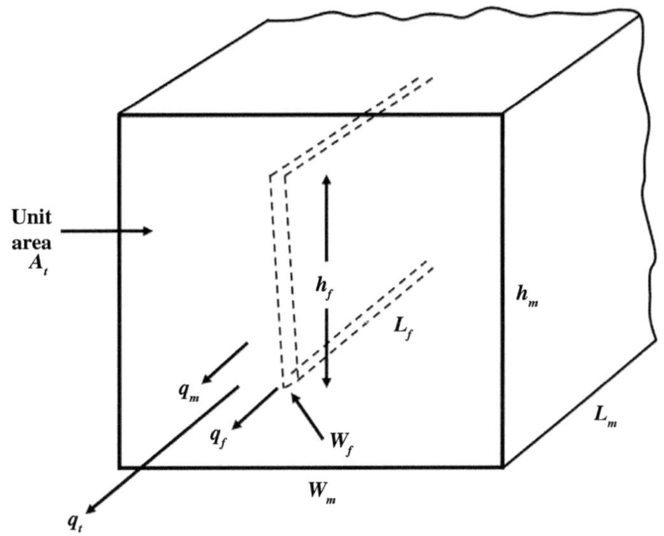

Figure 2.6 Representative model for fracture permeability.

and height (h_f), as shown in Figure 2.6. The flow occurs parallel to the fracture length, which aligns with the principal flow direction. The volumetric flow through such a fracture can be described by **Buckingham's equation** for flow through fine slots:

$$q_f = \frac{n_f A h_f w_f^3}{12}\left(\frac{\Delta p}{\mu L_f}\right), \tag{2.9}$$

where n_f is the number of fractures per unit area, A is the area, Δp is the pressure differential, μ is the fluid viscosity, and L_f is the length of the fracture.

Tortuosity can be included in this flow equation by multiplying and dividing the right-hand side by the unit length, L. Thus, Eq. (2.9) becomes:

$$q_f = \frac{n_f A h_f w_f^3}{12\sqrt{\tau}}\left(\frac{\Delta p}{\mu L}\right), \tag{2.10}$$

where τ is the tortuosity factor. Furthermore, by substituting an expression for porosity, we can modify the equation to include the porosity of the fracture:

$$q_f = \frac{A\varphi_f w_f^2}{12\tau}\left(\frac{\Delta p}{\mu L}\right). \tag{2.11}$$

This modified equation describes flow through multiple fractures with tortuous paths. By comparing it with Darcy's law:

$$q = -\frac{kA}{\mu}\left(\frac{\Delta p}{L}\right). \tag{2.12}$$

we can derive an expression for permeability:

$$k = \frac{\varphi_f w_f^2}{12\tau}. \tag{2.13}$$

In Darcy units, permeability can be expressed as:

$$k_f\,\{\text{Darcy}\} = 84.4 \times 10^5 w_f^2\{cm^2\} \times \frac{\varphi_f}{\tau_f}$$
$$k_f\,\{\text{Darcy}\} = 54.5 \times 10^6 w_f^2\{in^2\} \times \frac{\varphi_f}{\tau_f}. \quad (2.14)$$

The effect of multiple fractures on permeability is accounted for in the porosity term. When this relationship is substituted into the expression for permeability (Eq. 2.13), we derive the **Carman-Kozeny** type equation for fractures:

$$k = \frac{\varphi_f}{3\tau S_{pv}^2}. \quad (2.15)$$

This equation can be further generalized by introducing a **fracture shape factor** k_{sf}, which allows the effective permeability to account for different fracture shapes. Consequently, the **effective zoning factor** k_{Tf} is defined as:

$$k_{Tf} = k_{sf} \times \tau. \quad (2.16)$$

Substituting this into the previous equation, we get the generalized Carman-Kozeny equation for fractures:

$$k = \frac{\varphi_f}{k_{Tf} S_{pv}^2}. \quad (2.17)$$

This generalized form allows for flexibility in modeling fracture flow with different geometries and provides a more accurate representation of the permeability-porosity relationship in fractured reservoirs.

2.2.4.1 Effect of Fracture Shape

The constant in Eq. (2.15) is valid only for fractures with a rectangular parallelepiped shape. To explore the behavior of different fracture geometries, consider an ellipsoidal fracture, as shown in Figure 2.7. Assuming that flow occurs only in the

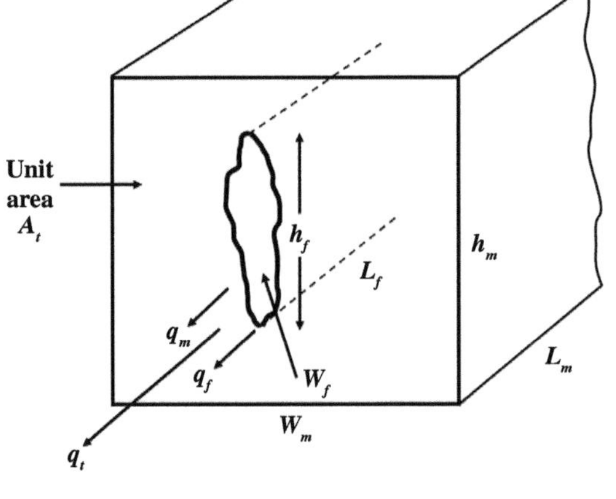

Figure 2.7 Influence of geometric shape on the permeability-porosity relationship.

fracture network and that matrix porosity is negligible, the specific surface area per unit pore volume for an ellipsoidal cylinder is given by:

$$V_p = \frac{\pi}{4} w_f h_f L_f. \tag{2.18}$$

$$A_S = \pi L_f \times \left[\frac{3}{4}(w_f + h_f) - \frac{1}{2}\sqrt{w_f h_f}\right]. \tag{2.19}$$

By combining these equations and simplifying, the specific surface area becomes:

$$S_{pv} = 3\left(\frac{1}{h_f} + \frac{1}{w_f}\right) - \frac{2}{\sqrt{w_f h_f}}. \tag{2.20}$$

If the fracture width is much smaller than the fracture height ($\frac{1}{w_f} \gg \frac{1}{h_f}$) and assuming $\sqrt{w_f h_f} \gg w_f$ (which introduces less than 5% error), the specific surface area simplifies to:

$$S_{pv} = \frac{3}{w_f}. \tag{2.21}$$

This result shows how fracture shape influences hydraulic characterization and emphasizes the dependence of fracture geometry in determining hydraulic units.

2.2.4.2 Hydraulic Radius of a Fracture

An effective hydraulic radius for the fracture can be derived by equating the **Hagen–Poiseuille equations** for a capillary tube and a fracture. The resulting relationship is:

$$\frac{\pi r_{hf}^4}{8} = \frac{h_f w_f^3}{12}. \tag{2.22}$$

Solving for the hydraulic radius r_{hf}:

$$r_{hf} = \sqrt[4]{\frac{2}{3\pi} h_f w_f^3}. \tag{2.23}$$

We can relate this hydraulic radius to permeability through the Kozeny equation for capillary tubes:

$$k = \frac{\varphi r^2}{8} = \frac{\varphi}{8}\sqrt{\frac{2}{3\pi} h_f w_f^3}. \tag{2.24}$$

If we treat the fracture as a slit, we can substitute $S_{pv} = \frac{2}{w_f}$ into Eq. (2.24):

$$k = \frac{\varphi}{2}\sqrt{\frac{1}{3\pi}\frac{h_f}{S_{pv}^3}}. \tag{2.25}$$

The significance of Eqs. (2.24) and (2.25) is in how they relate fracture height to the permeability-porosity relationship. For example, knowing the fracture height allows us to estimate permeability if the other parameters have been determined through core, log, or other analyses.

2.3 Differential Equations for Fluid Flow

A mathematical description of fluid flow in porous media can be obtained by combining several key physical principles:

1. The law of conservation of mass:

$$\begin{pmatrix}\text{mass into}\\ \text{the element}\end{pmatrix} - \begin{pmatrix}\text{mass out of}\\ \text{the element}\end{pmatrix} + \begin{pmatrix}\text{net amount of}\\ \text{mass introduced}\end{pmatrix} = \begin{pmatrix}\text{rate of}\\ \text{accumulation}\end{pmatrix}$$

This principle can be applied to an arbitrary volume element within the flow region. For instance, consider a three-dimensional element, as illustrated in Figure 2.8. The volumetric flow components in the x, y, and z directions are denoted by v_x, v_y, and v_z, respectively, which represent volumetric flow rates per unit cross-sectional area.

The mass flow rate into the element in the x-direction is:

$$\rho u_x \Delta y \Delta z.$$

The mass flow rate out of the element in the x-direction is:

$$(\rho u_x + \Delta(\rho u_x))\Delta y \Delta z,$$

where $\Delta(\rho u_x)$ represents the change in mass flux within the element. The net mass flow rate in the x-direction (amount in minus amount out) is:

$$-\Delta(\rho u_x)\Delta y \Delta z.$$

Similar expressions apply for the y- and z-directions. Assuming no mass is generated or lost, the net change in mass in the element over a time interval Δt is:

$$-\Delta t\left[\Delta(\rho u_x)\,\Delta_y \Delta_z + \Delta(\rho u_y)\,\Delta_x \Delta_z + \Delta(\rho u_z)\,\Delta_x \Delta_y\right] = [\varphi \rho|_{t+\Delta t} - \varphi \rho|_t]. \quad (2.26)$$

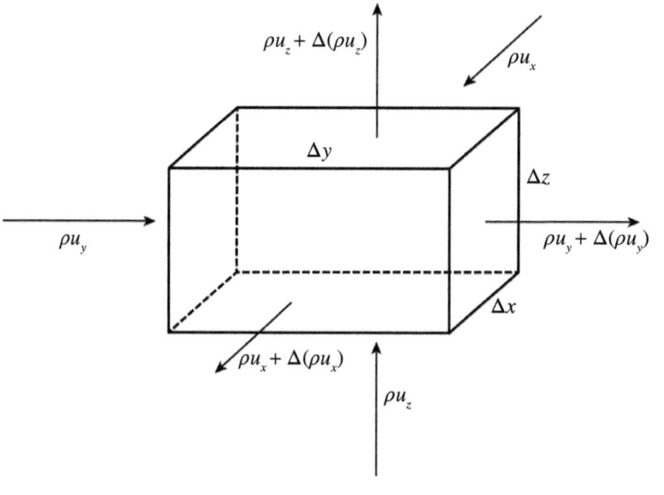

Figure 2.8 Cartesian coordinate system volume element.

Dividing by the volume (V) and Δt, and taking the limit as Δx, Δy, Δz, and Δt approach zero, we obtain the **continuity equation** for fluid flow in porous media:

$$-\nabla\left(\rho\vec{v}\right) + G = \frac{\partial}{\partial t}(\rho\varphi), \tag{2.27}$$

where G represents the mass introduced (source) or removed (sink) from the system.

2. Momentum Equation: Darcy's Law
Darcy's law states that the volumetric flow rate per unit cross-sectional area ($\bar{v}$) at any point in a uniform porous medium is proportional to the gradient of potential ($\nabla\psi$) in the direction of flow. For laminar flow, Darcy's law is expressed as:

$$\bar{v} = -\bar{\bar{k}}\frac{\rho}{\mu}\nabla\psi, \tag{2.28}$$

where k is the permeability tensor, and μ is the fluid viscosity. Flow occurs in the direction of decreasing potential.

For isotropic media, combining Darcy's law with the continuity equation results in:

$$\nabla\left(\rho\frac{k_i}{\mu}\nabla \mathrm{p}\right) \pm G = \frac{\partial}{\partial t}(\rho\varphi). \tag{2.29}$$

3. Equation of State
In the context of fluid flow through porous media, various **equations of state** are essential for accurately describing the behavior of different fluids. These equations define how fluid density (ρ) depends on pressure (p) and temperature (T). The equation of state that should be used depends on the type of fluid being considered.

A simple scenario to start with is the case of an incompressible fluid, where the density remains constant regardless of any changes in pressure. Mathematically, this can be expressed as:

$$\frac{d\rho}{dP} = 0. \tag{2.30}$$

In this case, density can be treated as a constant and can be eliminated from the governing flow equations, such as Eq. (2.29).

A more complex situation arises with **isothermal flows** of fluids that exhibit small, yet constant compressibility, like ideal liquids. The compressibility (c) of a fluid is defined as:

$$c = \frac{1}{\rho}\frac{\partial\rho}{\partial P}. \tag{2.31}$$

This leads to the governing differential equation for fluid flow, which becomes:

$$\nabla^2 P = \frac{\varphi\mu c}{k}\frac{\partial P}{\partial t}. \tag{2.32}$$

To arrive at Eq. (2.32), several key assumptions are made, including:

- the compressibility of the fluid is small and remains constant,
- permeability is isotropic (the same in all directions) and constant,
- porosity does not vary and is constant,
- the squared terms of the gradient are assumed to be negligible, and
- other potential assumptions specific to the fluid or system behavior.

Relaxing these assumptions or altering the constraints on the system can lead to alternative formulations and solutions to the flow equations. Some examples of such variations include:

- pressure-dependent porosity and permeability,
- incompressible fluid flow,
- anisotropy, where permeability differs along principal directions,
- compressible fluid flow, and
- multiphase flow, where multiple fluids coexist and interact.

2.3.1 Real Gas Flow in Porous Media

Another important set of equations deals with the flow of real gases in porous media. For a real, compressible gas, the equation of state is given by:

$$\rho = \frac{PM}{z_{(p,T)}RT}. \tag{2.33}$$

Here, P is the pressure, M is the molar mass of the gas, $z_{(p,T)}$ is the gas compressibility factor, R is the gas constant, and T is the temperature.

The corresponding governing equation for real gas flow in porous media becomes:

$$\nabla\left(\frac{p}{\mu z}\nabla p\right) = \frac{\varphi\mu c}{k}\frac{\partial}{\partial t}\left(\frac{p}{\mu z}\right). \tag{2.34}$$

Since the governing equation for real gas flow is non-linear, there are three common approaches to handling pressure-dependent fluid properties:

1. **Low-pressure approximation**, where the behavior of the gas can be linearized for low pressures.
2. **High-pressure approximation**, for cases where the pressure is significantly high.
3. **Real gas potential or pseudo-pressure function**, which adjusts the equations to more accurately describe the behavior of real gases over a wider range of pressures.

2.3.2 Conservation Principle in Fluid Flow

In any fluid flow system, the conservation principle applies, meaning that physical quantities, such as mass, are conserved—they cannot be created or destroyed within the system. For fluid flow in porous media, the most important conserved quantity is mass.

To solve the differential equations governing fluid flow through porous media, simply having the equations is not enough. Boundary and initial conditions specific to the flow problem must also be provided. These conditions define the physical limits and behavior of the system, making it possible to find unique solutions for different scenarios. Some commonly used boundary conditions in fluid flow problems are described below.

2.3.2.1 Initial and Boundary Conditions

When solving flow equations in porous media, it is crucial to define both the initial condition and the boundary conditions (Figure 2.9). The initial condition is typically expressed as $P(r,0) = P$, where the pressure P at time $t=0$ is defined across the spatial domain.

- **Infinite-acting reservoir:** In this scenario, the well is assumed to be located in a reservoir of infinite extent, with the potential at infinity remaining constant. This is represented as:

$$\psi(\infty, t) = \psi_i. \tag{2.35}$$

- **Constant potential outer boundary:** In this case, the well is located in the center of an area where the pressure along the outer boundary is held constant, represented as:

$$\psi(r_e, t) = \psi_e. \tag{2.36}$$

- **Bounded reservoir:** Here, the well is situated in the center of a reservoir with no flow across the external boundary, implying a zero flux condition. Mathematically, this is expressed as:

$$\frac{\partial \psi}{\partial l_n} = 0. \tag{2.37}$$

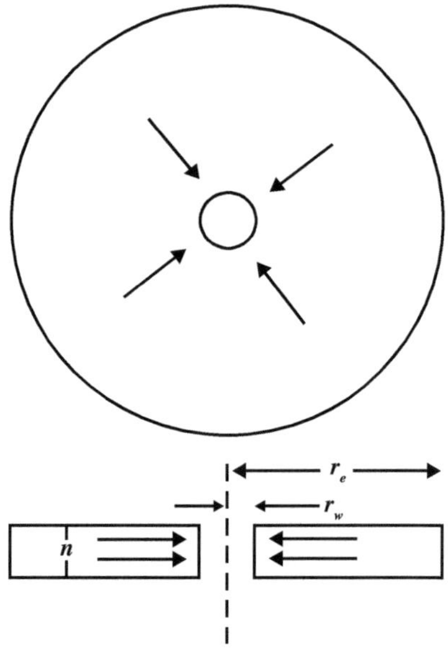

Figure 2.9 Diagram illustrating boundary conditions for radial flow with constant flow rate (Matthews and Russel 1967).

- **Constant flux boundary:** This boundary condition specifies that the fluid flux across the boundary is held constant. It can be described by:

$$v_n = -\frac{k\rho}{\mu}\frac{\partial \psi}{\partial l_n}. \tag{2.38}$$

where v_n is the normal velocity across the boundary and l_n is the distance measured normal to the boundary.

2.3.3 Discontinuities in Porous Media

Flow problems often involve **discontinuities in permeability** across the flow domain. The proper boundary conditions at such interfaces are derived from physical principles. Suppose the flow domain consists of two regions with different permeabilities, k_1 and k_2. Since pressure must be single-valued at any point, the first requirement is:

$$p_1 = p_2.$$

Additionally, the normal velocities on both sides of the boundary must be equal to satisfy mass conservation. This condition can be written as:

$$\rho\frac{k_1}{\mu}\frac{\partial \psi_1}{\partial l_n} = \rho\frac{k_2}{\mu}\frac{\partial \psi_2}{\partial l_n},$$

where l_n is the distance measured normal to the boundary.

At the wellbore, the boundary conditions depend on whether the production rate or potential is constant. If the production rate is constant, Darcy's law provides the expression:

$$q = -\frac{kA}{\mu}\nabla\psi. \tag{2.39}$$

Alternatively, if the potential is held constant at the wellbore, it is expressed as:

$$\psi(r_w, t) = \psi_w. \tag{2.40}$$

2.4 Steady-state Flow

In a steady-state flow regime, pressure and fluid velocity distributions do not vary with time. Therefore, the continuity equation simplifies to:

$$\nabla \cdot \left(\rho\vec{v}\right) = 0. \tag{2.41}$$

For example, for linear, horizontal flow through an isotropic porous medium, the diffusivity equation becomes:

One practical application of this equation is the measurement of permeability in laboratory samples. For **linear, horizontal flow** of an incompressible fluid through isotropic porous media, the diffusivity equation simplifies to:

$$\nabla\left(\frac{k}{\mu}\rho\nabla\psi\right) = 0. \tag{2.42}$$

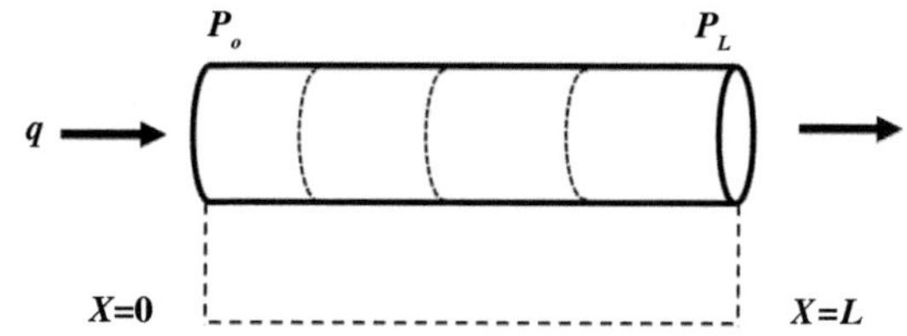

Figure 2.10 Schematic of flow through a horizontal core.

For example, consider a long cylindrical core with length L and cross-sectional area A, where a pressure drop $\Delta p = p(0) - p(L)$ is applied (Figure 2.10). The flow rate q remains constant, following Darcy's law. In this case, permeability may vary with position or pressure, leading to the flow equation:

$$q = -\frac{\vec{k}A}{\mu}\frac{\Delta p}{L}, \tag{2.43}$$

where the harmonic mean permeability $\bar{k}$ of the sample is given by:

$$\frac{L}{\bar{\bar{k}}} = \int_0^L \frac{dx}{k(x)}. \tag{2.44}$$

2.5 Basic Solutions of the Constant Terminal Rate Case for Radial Models

In this flow regime, one of the key conditions for solving the diffusivity equation is that the flow rate remains constant. This condition applies to situations such as oil flowing toward a fully perforated well or water flowing into a producing reservoir from an aquifer. The flow can be approximated as radial, where a single phase flows from the outer radius, "b," of a right hollow cylinder toward its inner radius, "a." The cylinder is assumed to be a homogeneous porous medium.

In the case of well drainage, the inner radius "a" represents the well's radius, r_w, and the outer radius "b" corresponds to the external boundary radius, r_e. The flow rate q at radius $r = r_w$ represents the well's production rate. For natural water influx into a reservoir, the inner radius "a" is the mean reservoir radius, while the outer radius "b" is the mean aquifer radius, with q denoting the volumetric flow rate of water across the initial oil-water contact.

The radial constant terminal rate (CTR) case is governed by the following system of equations:

$$\frac{1}{r}\frac{\delta}{\delta r}\left(r\frac{\delta P}{\delta r}\right) = \frac{\varphi\mu c}{k}\frac{\delta P}{\delta t}; \; a \le r \le b. \tag{2.45}$$

$$q = \left(\frac{2\pi rkh}{\mu}\right)\left(\frac{\delta P}{\delta r}\right); \; r = a. \tag{2.46}$$

2.5.1 Initial Condition

Initially, the pressure at all points is assumed to be constant throughout the radial system:

$$a \leq r \leq b,\ t = 0;\ P = P_i = \text{constant}. \tag{2.47}$$

2.5.2 Boundary Conditions

At the wellbore, the flow rate is assumed to remain constant after production begins:

$$r = a,\ t \geq 0:\ q = \text{constant}. \tag{2.48}$$

At the **outer boundary**, two cases are considered:

- **Pressure maintenance case**: The pressure remains constant and equal to the initial pressure at the outer boundary:

$$r = b,\ t \geq 0:\ P = P_i = \text{constant}. \tag{2.49}$$

- **Sealing boundary case**: There is no flow across the boundary, and the pressure gradient at the outer boundary is zero:

$$r = b,\ t \geq 0;\ \frac{\delta P}{\delta r} = 0. \tag{2.50}$$

2.5.3 Solution and Flow Regimes

The solutions to Eqs. (2.45–2.48) combined with either boundary condition (Eqs. 2.49 or 2.50) are well-documented in Matthews and Russell (1967). However, these solutions are complex and often impractical for direct use. Instead, asymptotic solutions, which serve as reasonable approximations of the full solutions, are typically applied in practical scenarios. These simplified solutions correspond to specific flow regimes, and the main challenge is accurately identifying the prevailing flow regime to select the appropriate asymptotic solution.

The **steady-state solution** is the simplest case and corresponds to Darcy's law. In contrast, the **non-steady-state solutions** incorporate a time element and are usually expressed in dimensionless form to facilitate application in various scenarios.

2.5.4 The Steady-state Solution

When a well is produced at a constant flow rate (q), and the pressure at the external boundary radius (r_e) is maintained constant, the system eventually reaches steady-state conditions. In this state, the flow rate remains constant $q = \text{constant}$ and the pressure gradient with respect to time, $\frac{\partial P}{\partial t}$, becomes zero for all values of radius (r) and time (t).

As a result, the pressure gradient with respect to radius simplifies to $\frac{dP}{dr}$, and the flow equation becomes:

$$q\frac{dr}{r} = \left(\frac{2\pi kh}{\mu}\right)dP. \tag{2.51}$$

By integrating this expression between the limits r_w (the wellbore radius) and r_e (the external boundary radius), we obtain:

$$P - P_w = \left(\frac{q\mu}{2\pi kh}\right)\ln\frac{r}{r_w}. \tag{2.52}$$

This is identical to the relationship for radial flow as described by Darcy's law. In this scenario, measuring the pressure at the external radius (r_e) of the reservoir is essential, but it would require drilling a well at the boundary, which is economically unfeasible. Therefore, a **mean reservoir pressure** $\overline{P}$ is used instead.

The mean reservoir pressure $\overline{P}$ is determined from routine **bottom-hole pressure measurements** and well tests. It reflects the effect of the area drained by each well. Conceptually, the bottom-hole pressure measurements from different wells are weighted according to the volume each well drains, and these weighted pressures are averaged to obtain $\overline{P}$.

Figure 2.11 illustrates a well in a reservoir, highlighting the well's area of influence. The volume drained by the well is used to calculate the mean reservoir pressure $\overline{P}$ in relation to the pressure (P) of individual elements of volume (dV) being drained. The total volume is (V), and the mean pressure is given by:

$$\overline{P} = \frac{1}{V}\int_{r_w}^{r_e} P dV, \tag{2.53}$$

where

$$dV = 2\pi rh\varphi dr. \tag{2.54}$$

The total volume of the well's drainage zone is:

$$V = \pi\left(r_e^2 - r_w^2\right)h\varphi.$$

Assuming $r_w \ll r_e$, this simplifies to:

$$V \cong \pi r_e^2 h\varphi.$$

Thus, the mean reservoir pressure is:

$$\overline{P} = \frac{2}{r_e^2}\int_{r_w}^{r_e} Pr dr. \tag{2.55}$$

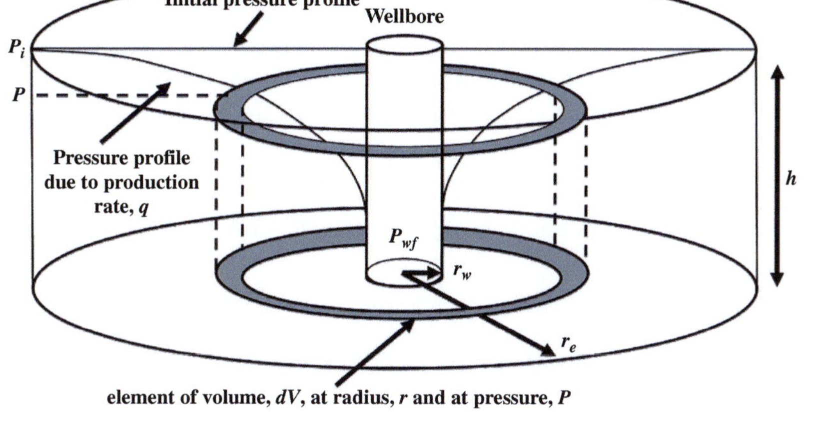

Figure 2.11 Pressure profile around a well.

2.5.4.1 Solution Using Darcy's Law

Substituting Eq. (2.52) for P in the integral, we get:

$$P = P_w + \left(\frac{q\mu}{2\pi kh}\right)\ln\frac{r}{r_w}.$$

$$\overline{P} = \frac{2}{r_e^2}\int_{r_w}^{r_e}\left[P_w + \left(\frac{q\mu}{2\pi kh}\right)\ln\left(\frac{r}{r_w}\right)\right]rdr.$$

Simplifying the terms, we find:

$$\overline{P} - P_w = \frac{2}{r_e^2}\left(\frac{q\mu}{2\pi kh}\right)\int_{r_w}^{r_e}\ln\left(\frac{r}{r_w}\right)rdr.$$

The integral results in:

$$\overline{P} - P_w = \frac{2}{r_e^2}\left(\frac{q\mu}{2\pi kh}\right)\left[\frac{1}{2}r^2\ln\frac{r}{r_w}\right]_{r_w}^{r_e} - \int_{r_w}^{r_e}\frac{1}{r}\left(\frac{r^2}{2}\right)dr.$$

After further simplifications and neglecting small terms, we obtain:

$$\overline{P} - P_w = \frac{2}{r_e^2}\left(\frac{q\mu}{2\pi kh}\right)\left[\left[\frac{r_e^2}{2}\ln\frac{r_e}{r_w} - \frac{r_w^2}{2}\ln\frac{r_w}{r_w}\right] - \left[\frac{r_e^2}{4} - \frac{r_w^2}{4}\right]\right].$$

Assuming $\frac{r_w^2}{4}$ is negligible

$$\begin{aligned}\overline{P} - P_w &= \frac{2}{r_e^2}\left(\frac{q\mu}{2\pi kh}\right)\left[\frac{r_e^2}{2}\ln\frac{r_e}{r_w} - \frac{r_e^2}{4}\right]\\ \overline{P} - P_w &= \left(\frac{q\mu}{2\pi kh}\right)\left[\ln\frac{r_e}{r_w} - \frac{1}{2}\right]\end{aligned} \tag{2.56}$$

This final expression relates the mean reservoir pressure $\overline{P}$ to the well's bottom-hole pressure (P_w), considering the steady-state flow regime.

2.5.5 Non-steady-state Flow Regimes and Dimensionless Variables

Using **dimensionless forms** of the diffusivity equation has become essential in modeling flow through porous media. By transforming the equations into dimensionless variables, they can be applied to various reservoirs, allowing for general solutions that can later be tailored to specific cases. For instance, universal plots showing the pressure difference between the reservoir and the wellbore over time can be generated, which can then be used to compute the actual pressure difference for a particular reservoir.

However, it is crucial to remember that these solutions are only valid when the dimensionless variables and the diffusivity equation are formulated specifically for **radial flow** reservoirs.

The dimensionless variables are defined as follows:

- **Dimensionless radius** (r_D), which relates the distance from the wellbore to the well's radius:

$$r_D = \frac{r}{r_w}, \tag{2.57}$$

where (r) is the radius of interest, and (r_w) is the radius of the wellbore.

- **Dimensionless time** (t_D), which normalizes time with respect to reservoir properties:

$$t_D = \frac{kt}{\varphi \mu c r_w^2}, \tag{2.58}$$

where t is the time of interest, k is permeability, φ is porosity, μ is the fluid's viscosity, c is the compressibility, and r_w is the radius of the wellbore.

- **Dimensionless pressure** (P_D), which compares pressure differences between the initial reservoir pressure and the current pressure at a specific time and radius:

$$P_D(r_D, t_D) = \left(\frac{2\pi kh}{q\mu}\right)(P_i - P_{r,t}), \tag{2.59}$$

where h is the thickness of the reservoir, P_i is the initial pressure in the reservoir, $P_{r,t}$ is the pressure at the radius r and time t, q is the flow rate, and μ is viscosity.

With these dimensionless variables, the **radial diffusivity equation** can now be rewritten as:

$$\frac{1}{r_D}\frac{\delta}{\delta r_D}\left(r_D \frac{\delta P_D}{\delta r_D}\right) = \frac{\delta P_D}{\delta t_D}. \tag{2.60}$$

In addition, other dimensionless parameters, like a **dimensionless external radius**, may be introduced depending on the specific case being studied. These dimensionless forms provide a framework for deriving solutions that are applicable to a wide variety of reservoir systems.

2.5.6 Unsteady State Solution

The CTR solution can be derived in multiple forms, depending on the assumptions made and the method of mathematical analysis used. Although these various solutions may overlap, each has specific applications and limitations.

2.5.6.1 General Considerations

Figure 2.12a,b illustrates the response of a reservoir at the wellbore when a flow rate (q) is suddenly imposed. The pressure of the fluid in the wellbore (P_{wf}), drops from the initial static equilibrium pressure (P_i) over time. The CTR solution of the diffusivity equation captures this pressure drop as a function of time. Thus, the CTR solution provides the relationship between (P_{wf}) and (t) for a constant production rate, regardless of how long the well has been flowing.

The pressure decline, as shown in Figure 2.12b, typically divides into three phases depending on the flowing time and the geometry of the reservoir or the region of the reservoir being drained. This figure reflects the pressure change at the wellbore over time, akin to the pressure change (or water level change) in the cylinder closest to the outlet in the model shown in Figure 2.13.

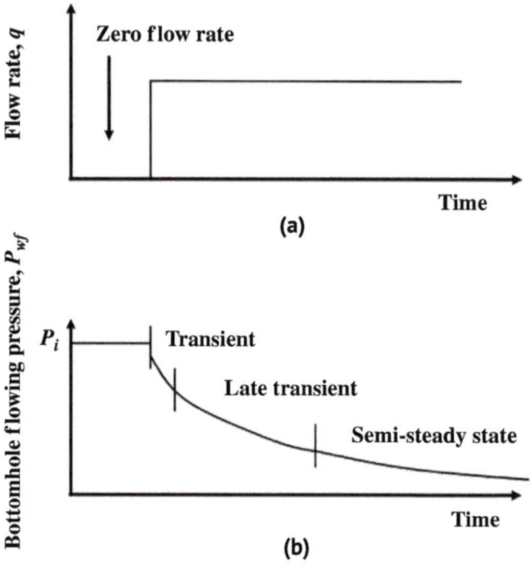

Figure 2.12 Wellbore pressure response to a change in flow rate.

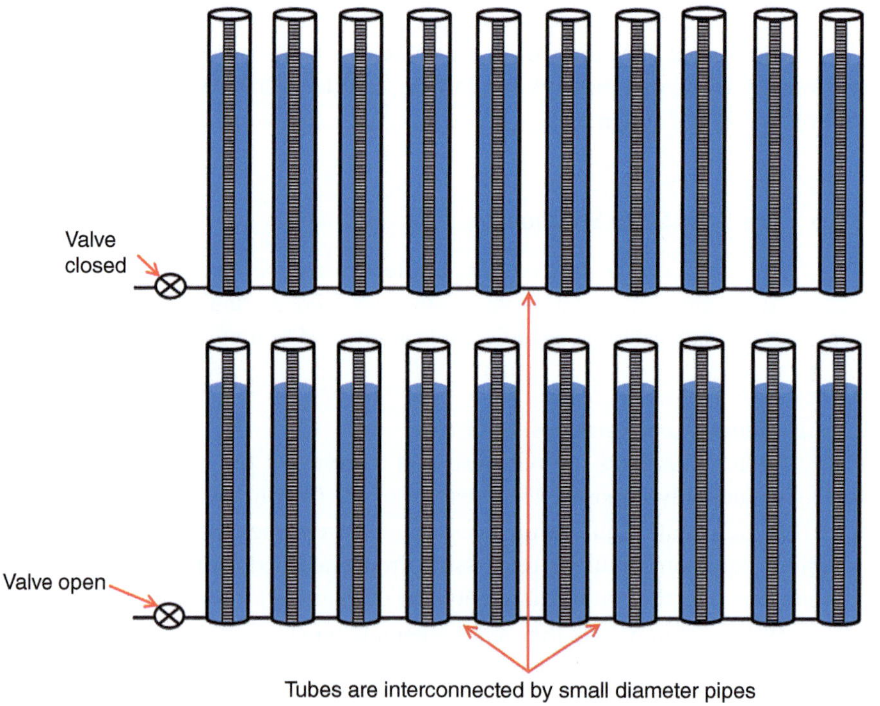

Figure 2.13 Model illustrating the pressure response of a reservoir to a flowing well.

Initially, the pressure response can be described by a **transient solution**, which assumes that during this early period, the wellbore's pressure response is unaffected by the reservoir boundaries, and vice versa. This scenario is often referred to as the **infinite reservoir case** since, during this transient flow period, the reservoir behaves as though it has infinite extent, with an unlimited amount of fluid available to expand and drive the system.

After the transient period, the pressure response enters the **late-transient phase**, during which the boundaries begin to influence the pressure behavior. This is analogous to the pressure disturbance moving along the series of tubes shown in Figure 2.13. The geometry of the reservoir boundaries affects the type of solution applied, as wells may drain irregularly shaped areas where boundaries are neither symmetrical nor equidistant from the well.

The final phase of pressure decline is known as the semi-steady state or pseudo-steady state. In this phase, the pressure profile within the reservoir stabilizes, meaning it no longer changes over time, but the wellbore pressure continues to decline at a constant rate. This situation mirrors the model in Figure 2.13, where the water levels in all the tubes drop consistently without any additional water being introduced into tube 10 to maintain the pressure.

If the pressure profile around the well had remained constant, true steady-state conditions would have been reached, and the steady-state solutions discussed in the previous section would apply.

2.5.6.2 Hurst and Van Everdingen Solution

The CTR solution for all flowing times was developed by Hurst and van Everdingen. They solved the radial diffusivity equation using the **Laplace transform** for both the CTR and constant terminal pressure cases. Their solution includes an infinite summation of **Bessel functions**, which are essential for accurately describing the complex pressure response at the wellbore during the late transient period. However, simpler solutions can be derived for the transient and semi-steady state flow regimes.

This solution provides the pressure drop as a function of time and radius, given fixed values of the external boundary radius (r_e), wellbore radius (r_w), and the properties of the rock and fluid. It is expressed in terms of **dimensionless variables** and parameters, represented as:

$$P_D = f(t_D, r_D, r_{eD}), \tag{2.61}$$

where t_D is the dimensionless time, r_D is the dimensionless radius, and $r_{eD} = \frac{r_e}{r_w}$ is the dimensionless external radius.

When the reservoir has a fixed size, i.e., r_{eD} has a specific value, the dimensionless pressure drop (P_D) becomes a function of only the dimensionless time (t_D) and dimensionless radius (r_D). Using this framework, the pressure in any particular reservoir can be calculated for any time and/or radius. Of particular importance is the pressure at the **wellbore**, since this is routinely measured during production operations and can be directly compared to theoretical solutions. While determining the pressure at a location far from the well might be of technical interest, it cannot be measured unless a well is drilled at that specific location.

At the **wellbore radius** ($r = r_w$; or equivalently $r_D = 1.0$), the dimensionless pressure is expressed as:

$$P_D = f(t_D, r_{eD}), \tag{2.62}$$

which is explicitly written as:

$$P_D(t_D) = \frac{2t_D}{r_{eD}^2} + \ln r_{eD} - \frac{3}{4} + 2\sum_{m=1}^{\infty} \frac{e^{-\alpha_m^2 t_D} J_1^2(\alpha_m r_{eD})}{\alpha_m^2\left(J_1^2(\alpha_m r_{eD}) - J_1^2(\alpha_m)\right)} \tag{2.63}$$

where α_m are the roots of:

$$J_1(\alpha_m r_{eD})\, Y_1(\alpha_m) - J_1(\alpha_m)\, Y_1(\alpha_m r_{eD}) = 0.$$

J_1 and Y_1 are **Bessel functions** of the first and second kind, respectively.

This series has been computed for several values of dimensionless external radius (r_{eD}), covering a wide range of dimensionless times (t_D). The results of these calculations are presented in tables, such as those found in Chatas (1953) and Lee (1982). A summary of how to use these tables for CTR problems is shown in Table 2.1. This table provides the dimensionless pressure for various time configurations in the reservoir and serves as a solution to Eq. (2.63).

Table 2.1 Dimensionless pressure values at various dimensionless times for different reservoir configurations under CTR conditions.

Table		Presents	Valid for
2	i.	P_D as a function of $t_D < 1000$ (from table)	Infinite-acting reservoirs
	ii.	$P_D = 2\sqrt{\frac{t_D}{\pi}}$ for $t_D < 0.01$ (an extension of the table)	Infinite-acting reservoirs
	iii.	$P_D = 0.5(\ln t_D + 0.80907)$ for $100 < t_D < 0.25\, r_{eD}^2$ (an extension of the table)	Infinite-acting reservoirs
	iv.	Valid for finite reservoirs with $t_D < 0.25\, r_{eD}^2$	Finite reservoirs
3	i.	P_D as a function of t_D for $1.5 < r_{eD} < 10$ (from table)	Finite reservoirs, but if the value of t_D is smaller than that listed for a given value of r_{eD}, the reservoir is infinite acting, and therefore Table 2 is used.
	ii.	$P_D = \frac{2(t_D + 0.25)}{r_{eD}^2 - 1} - \frac{(3r_{eD}^4 - 4r_{eD}^4 \ln r_{eD} - 2r_{eD}^2 - 1)}{4(r_{eD}^2 - 1)^2}$ $25 < t_D$ and $0.25\, r_{eD}^2 < t_D$	Finite reservoirs
	iii.	$P_D \cong \frac{2t_D}{r_{eD}^2} + \ln r_{eD} - \frac{3}{4}$ for $r_{eD}^2 \gg 1$	Finite reservoirs

Table 2.2 Dimensionless pressure (P_D) vs. dimensionless time (t_D) – infinite radial system, constant rate at inner boundary.

t_D	P_D	t_D	P_D	t_D	P_D	t_D	P_D
0	0	0.06	0.2500	3.0	1.1665	150.0	3.0636
0.0005	0.0250	0.07	0.2680	4.0	1.2750	200.0	3.1726
0.001	0.0352	0.08	0.2845	5.0	1.3625	250.0	3.2630
0.002	0.0495	0.09	0.2999	6.0	1.4362	300.0	3.3394
0.003	0.0603	0.1	0.3144	7.0	1.4997	350.0	3.4057
0.004	0.0694	0.15	0.3750	8.0	1.5557	400.0	3.4641
0.005	0.0774	0.2	0.4241	9.0	1.6057	450.0	3.5164
0.006	0.0845	0.3	0.5024	10.0	1.6509	500.0	3.5643
0.007	0.0911	0.4	0.5645	15.0	1.8294	550.0	3.6067
0.008	0.0971	0.5	0.6167	20.0	1.9601	600.0	3.6476
0.009	0.1028	0.6	0.6622	30.0	2.1470	650.0	3.6842
0.01	0.1081	0.7	0.7024	40.0	2.2824	700.0	3.7170
0.015	0.1312	0.8	0.7387	50.0	2.3884	750.0	3.7461
0.02	0.1503	0.9	0.7716	60.0	2.4758	800.0	3.7725
0.025	0.1669	1.0	0.8019	70.0	2.5501	850.0	3.7961
0.03	0.1818	1.2	0.8672	80.0	2.6147	900.0	3.8355
0.04	0.2077	1.4	0.9160	90.0	2.7233	950.0	3.8355
0.05	0.2301	2.0	1.0195	100.0	2.9212	1000.0	3.8584

The results in Table 2.1 are applicable for wells flowing at a constant rate or for systems where the reservoir and aquifer maintain a constant flow rate across the oil-water contact. Most well flow problems align with relationships 2(iii) and 3(iii) in the table. The procedure for using Table 2.1 is straightforward: start by determining whether the reservoir behaves as **infinite acting** by checking the dimensionless time (t_D). If (t_D) is less than $0.25\, r_{eD}$, then the reservoir is infinite acting, and you should use Table 2.2. Otherwise, refer to Table 2.3, as the reservoir is **finite acting**.

In some situations, finite reservoirs can be approximated by the infinite acting table (Table 2.2), and equations can extend the tables for use if (t_D) falls within certain limits. These solutions allow for pressure calculations anywhere within the reservoir, provided the flow rate is known. However, if pressure is required at a location where the flow rate is unknown, an alternative solution, such as the **line source solution**, must be employed.

2.5.6.3 The Line Source Solution

The line source solution assumes that the wellbore radius is vanishingly small in comparison to the mean radius of the reservoir. This solution allows for the

Table 2.3 Dimensionless pressure (P_D) vs. dimensionless time (t_D) – finite radial system with a closed exterior boundary and constant rate at the inner boundary.

r_{eD} = 1.5		r_{eD} = 2.0		r_{eD} = 2.5		r_{eD} = 3.0		r_{eD} = 3.5		r_{eD} = 4.0	
t_D	P_D	t_D	P_D	t_D	P_D	t_D	P_D	t_D	P_D	t_D	P_D
0.06	0.251	0.22	0.443	0.40	0.565	0.52	0.627	1.0	0.802	1.5	0.927
0.08	0.288	0.24	0.459	0.42	0.576	0.54	0.636	1.1	0.830	1.6	0.948
0.10	0.322	0.26	0.476	0.44	0.587	0.56	0.645	1.2	0.857	1.7	0.968
0.12	0.355	0.28	0.492	0.46	0.598	0.60	0.662	1.3	0.882	1.8	0.988
0.14	0.387	0.30	0.507	0.48	0.608	0.65	0.683	1.4	0.906	1.9	1.007
0.16	0.420	0.32	0.522	0.50	0.618	0.70	0.703	1.5	0.929	2.0	1.025
0.18	0.452	0.34	0.536	0.52	0.628	0.75	0.721	1.6	0.951	2.2	1.059
0.20	0.484	0.36	0.551	0.54	0.638	0.80	0.740	1.7	0.973	2.4	1.092
0.22	0.516	0.38	0.565	0.56	0.647	0.85	0.758	1.8	0.994	2.6	1.123
0.24	0.548	0.40	0.579	0.58	0.657	0.90	0.776	1.9	1.014	2.8	1.154
0.26	0.580	0.42	0.593	0.60	0.666	0.95	0.791	2.0	1.034	3.0	1.184
0.28	0.612	0.44	0.607	0.65	0.688	1.0	0.806	2.25	1.083	3.5	1.255
0.30	0.644	0.46	0.621	0.70	0.710	1.2	0.865	2.50	1.130	4.0	1.324
0.35	0.724	0.48	0.634	0.75	0.731	1.4	0.920	2.75	1.176	4.5	1.392
0.40	0.804	0.50	0.648	0.80	0.752	1.6	0.973	3.0	1.221	5.0	1.460
0.45	0.884	0.60	0.715	0.85	0.772	2.0	1.076	4.0	1.401	5.5	1.527
0.50	0.964	0.70	0.782	0.90	0.792	3.0	1.328	5.0	1.579	6.0	1.594
0.55	1.044	0.80	0.849	0.95	0.812	4.0	1.578	6.0	1.757	6.5	1.660
0.60	1.124	0.90	0.915	1.00	0.832	5.0	1.828			7.0	1.727
0.65	1.204	1.00	0.982	2.00	1.215					8.0	1.861
0.70	1.284	2.00	1.649	3.00	1.506					9.0	1.994
0.75	1.364	3.00	2.316	4.00	1.977					10.0	2.127
0.80	1.444	5.00	3.649	5.00	2.398						

calculation of pressure at any point in an **unbounded** (infinite-acting) reservoir, using only the flow rate at the well. The advantage of this method is that no additional flow rates, other than those measured at the well, are required to calculate pressures throughout the reservoir. However, this approach has limitations; it only works for infinite-acting reservoirs, and if the reservoir encounters boundaries, the solution no longer accurately represents the flow regime.

In cases where multiple wells exist in an infinite-acting reservoir, the **superposition principle** can be used to account for their combined effects. This principle can also be applied to simulate the presence of a boundary. In such cases, the boundary acts like a second, imaginary well producing at the same rate with the same production history as the real well, both in an infinite-acting reservoir. The Line Source solution is widely used in **well test analysis**.

In CTR problems, the flow rate at the well is given by:

$$q = \left(\frac{2\pi rhk}{\mu}\right)\left(\frac{\delta P}{\delta r}\right)_{r=r_w}. \tag{2.64}$$

For the line source solution, the boundary conditions must satisfy:

$$\lim_{y\to 0} 2y\frac{\delta p}{\delta y} = \frac{q\mu}{2\pi kh} \text{ for } t > 0.$$

By applying the **Boltzmann transformation** $y = \frac{\varphi\mu c r^2}{4kt}$ and substituting it into the diffusivity equation:

$$\frac{1}{r}\frac{\delta}{\delta r}\left(r\frac{\delta P}{\delta r}\right) = \frac{\varphi\mu c}{k}\frac{\delta P}{\delta t}. \tag{2.65}$$

We get the transformed equation:

$$y\frac{d^2p}{dy^2} + \frac{dp}{dy}(1+y) = 0 \tag{2.66}$$

The boundary conditions for this transformed equation are:

- $p \to p_i$ as $y \to \infty$,
- $\lim_{y\to 0} 2y\frac{\delta p}{\delta y} = \frac{q\mu}{2\pi kh}$.

Let $p' = \frac{dp}{dy}$. The transformed equation becomes:

$$y\frac{dp'}{dy} + (1+y)p' = 0.$$

Solving this by separation of variables and integrating yields:

$$\ln p' = -\ln y - y + C$$

or,

$$p' = \frac{dp}{dy} = \frac{C_1}{y}e^{-y} \tag{2.67}$$

Here, C and C_1 are integration constants. Since:

$$\lim_{y\to 0} 2y\frac{\delta p}{\delta y} = \frac{q\mu}{4\pi kh} = \lim_{y\to 0} 2C_1 e^{-y}$$

we get:

$$C_1 = \frac{q\mu}{4\pi kh}$$

and Eq. (2.67) becomes:

$$\frac{dp}{dy} = \frac{q\mu}{4\pi kh}\frac{e^{-y}}{y}$$

Integrating this gives:

$$p = \frac{q\mu}{4\pi kh}\int_{\infty}^{y}\frac{e^{-y}}{y}dy + C_2$$

or equivalently:

$$p = -\frac{q\mu}{4\pi kh}\int_{y}^{\infty}\frac{e^{-y}}{y}dy + C_2$$

which simplifies to:

$$p = \frac{q\mu}{4\pi kh}Ei(-y) + C_2. \tag{2.68}$$

Applying the boundary condition $p \to p_i$ as $y \to \infty$ gives $C_2 = p_i$, resulting in the Line Source solution:

$$p_i - p_{(r,t)} = -\frac{q\mu}{4\pi kh}\left[Ei\left(-\frac{\varphi\mu cr^2}{4kt}\right)\right]. \tag{2.69}$$

Here, $Ei(-y)$ is the **exponential integral** of y, expressed as:

$$Ei(-y) = -\int_{y}^{\infty}\frac{e^{-y}}{y}dy. \tag{2.70}$$

The exponential integral can be approximated as:

$$Ei(-y) = \gamma + \ln y - \left[\frac{y^n}{n!n}\right], \tag{2.71}$$

where $\gamma = 0.5772157$ is **Euler's Constant**. For small values of y (i.e., $y < 0.01$), the Ei function can be simplified to:

$$Ei(-y) = \ln 1.781y = \ln\bar{\gamma}y, \tag{2.72}$$

where $(\bar{\gamma} = 1.781 = e^{\gamma} = e^{0.5772157})$.

Thus, for practical purposes, the Ei can be approximated by a logarithmic function. The condition $y < 0.01$ corresponds to the time $t > \frac{25\varphi\mu cr^2}{k}$. Using this approximation, the line source solution becomes:

$$P = P_i + \frac{q\mu}{4\pi kh}\left(\ln\frac{\bar{\gamma}\varphi\mu cr^2}{4kt}\right)$$

or equivalently:

$$P = P_i - \frac{q\mu}{4\pi kh}\left(\ln\frac{4kt}{\bar{\gamma}\varphi\mu cr^2}\right). \tag{2.73}$$

At the wellbore radius ($r = r_w$), the pressure is given by:

$$P_{wf} = P_i - \frac{q\mu}{4\pi kh}\left(\ln\frac{4kt}{\bar{\gamma}\varphi\mu cr_w^2}\right). \tag{2.74}$$

2.5.6.3.1 Range of Application and Limitations

The Ei function has limitations. It cannot represent the initial flow into the wellbore because the assumption that the wellbore is a line is unrealistic. Some time must pass for the wellbore's relative size to become negligible, allowing the fluid to expand throughout most of the reservoir. Additionally, the solution is only valid for infinite-acting reservoirs.

Empirical analysis of real reservoirs suggests that the *Ei* function is valid when:

1. **Flowing time:** $t > \dfrac{100\,\varphi\mu\, c r_w^2}{k}$, (2.75)

where r_w is the wellbore radius. The constant 100 is derived from real reservoir data and can vary depending on the well and reservoir characteristics.

2. **Time limit:** $t < \dfrac{\varphi\mu\, c r_e^2}{4k}$, (2.76)

where r_e is the external reservoir boundary. After this time, the reservoir boundary begins to affect the pressure distribution, and the infinite-acting assumption no longer holds, making the Line Source solution invalid.

2.5.6.4 The Skin Factor

The fluid flow analysis we have discussed so far assumes a constant permeability throughout the reservoir, from the wellbore to the external boundary. However, in reality, the permeability around the wellbore may differ from that of the rest of the reservoir. This change in permeability can be caused by formation damage, which may occur during drilling or completion. For example, wellbore fluids may alter the wettability of the near-wellbore formation, allowing fluids to leak off or allowing solids suspended in the drilling fluids to deposit in pore spaces, reducing permeability by physically blocking fluid flow. Damage may also occur during production, where sand or precipitates from hydrocarbons or formation brines clog the pore spaces or change the wettability.

In contrast, some wellbores intersecting natural fractures may experience increased permeability, as these fractures provide larger, more conductive paths for fluids to flow, thereby enhancing permeability. **Reservoir management techniques**, such as hydraulic fracturing or acidizing workovers, may also deliberately alter permeability, allowing fluid to bypass areas of reduced permeability that developed during production.

In cases where the near-wellbore permeability differs from that of the larger reservoir, the *Ei* equation cannot accurately model the pressure drop at the wellbore. This is because the *Ei* equation assumes uniform permeability throughout the entire drainage area of the well. Figure 2.14 illustrates the impact of reduced permeability around the wellbore, showing that the **skin zone** affects only the pressure drop near the wellbore and not the pressures in the formation farther from the well.

If the skin zone is modeled as a region with altered permeability (k_s) and an outer radius (r_s), the additional pressure drop (ΔP_s) across this zone can be described by the **steady-state radial flow equation**. It is assumed that after the pressure disturbance caused by the start of production moves into the reservoir, the skin zone reaches a steady-state flow regime. The pressure drop due to the skin is then the difference in **bottom-hole flowing pressures** with and without the skin:

$$\Delta P_s = \frac{q\mu}{2\pi k_s h}\ln\left(\frac{r_s}{r_w}\right) = \frac{q\mu}{2\pi k h}\left(\frac{k}{k_s} - 1\right)\ln\left(\frac{r_s}{r_w}\right). \tag{2.77}$$

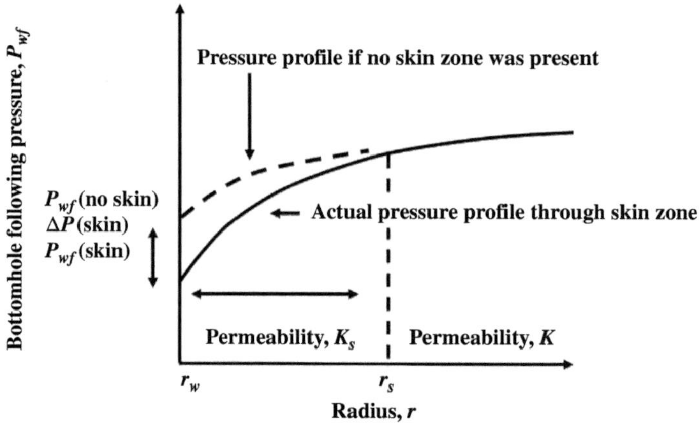

Figure 2.14 Variation of the permeability around the wellbore changes the local pressure profile.

This equation shows that the pressure drop in the skin zone is inversely proportional to the permeability (k_s) in that region, rather than to the permeability (k) of the rest of the reservoir. A correction for the pressure drop in the skin zone is necessary.

When the skin factor is included in the line source solution, the total pressure drop at the wellbore becomes:

$$\Delta P_s = \frac{q\mu}{2\pi k_s h}\ln\left(\frac{r_s}{r_w}\right) - \frac{q\mu}{2\pi kh}\ln\left(\frac{r_s}{r_w}\right) = \frac{q\mu}{2\pi kh}\left(\frac{k}{k_s} - 1\right)\ln\left(\frac{r_s}{r_w}\right). \tag{2.78}$$

If we apply the logarithmic approximation for the *Ei* function near the wellbore, the equation becomes:

$$P_i - P_{wf} = -\frac{q\mu}{4\pi kh}\left[\ln\frac{\bar{\gamma}\varphi\mu c r_w^2}{4kt} - 2\left(\frac{k}{k_s} - 1\right)\ln\frac{r_s}{r_w}\right]. \tag{2.79}$$

A skin factor (s) can be defined as:

$$s = \left(\frac{\mathrm{k}}{k_s} - 1\right)\ln\frac{r_s}{r_w}. \tag{2.80}$$

Therefore, the **drawdown** equation becomes:

$$P_i - P_{wf} = -\frac{q\mu}{4\pi kh}\left[\ln\frac{\bar{\gamma}\varphi\mu c r_w^2}{4kt} - 2s\right]. \tag{2.81}$$

Equation 2.81 indicates that a **positive skin factor** reflects a reduction in near-wellbore permeability, likely due to some form of formation damage. The magnitude of the skin factor indicates the degree of permeability contrast between the skin zone and the rest of the reservoir, as well as the extent of the damage. Understanding the skin factor is essential in well testing, as it provides insight into the formation damage caused during drilling and completion.

Conversely, a **negative skin factor** suggests that the skin zone permeability is higher than that of the unaltered reservoir, potentially due to the creation of highly conductive fractures or channels in the rock.

It is important to note that the extent of the damage zone may not be uniform, and there could be variations in permeability at different depths. This simple model, therefore, may not fully capture the near-wellbore permeability profile.

The altered zone near a well affects only the pressure in the immediate vicinity of the well. The pressure in the unaltered reservoir farther from the well remains unaffected by the presence of the skin zone.

2.5.7 Semi-steady State Solution

After a well is brought into production, the initial pressure disturbance propagates through the reservoir. Once it reaches the boundaries, the fluid flow transitions from infinite-acting to finite-acting behavior. This flow regime is referred to as pseudo steady state or semi-steady state because the pressure drop with respect to time is the same at every point around the flowing well, expressed as:

$$\frac{\delta P}{\delta t} = \frac{dP}{dt} = \text{constant}.$$

At the outer boundary of the drainage area, $r = r_e$, there is no flow across the boundary, which implies:

$$\frac{\delta P}{\delta r} = 0 \text{ at } r = r_e.$$

Similar to the **steady-state flow regime**, the pressure difference between the wellbore and a location such as the external radius, the initial pressure, or the **average reservoir pressure** can be calculated depending on the available physical measurements. Typically, the average reservoir pressure is known, and it is used to determine the pressure drop. Figure 2.15 depicts the pressure profile in the reservoir and the relevant pressure values.

In semi-steady state conditions, the pressure profile can be averaged across the entire reservoir volume. This gives the average reservoir pressure at any given time during the reservoir's depletion. For reservoirs with multiple wells, each well drains a specific portion of the reservoir. Once conditions stabilize, the volume drained by each well remains constant, allowing the entire reservoir to be divided into separate

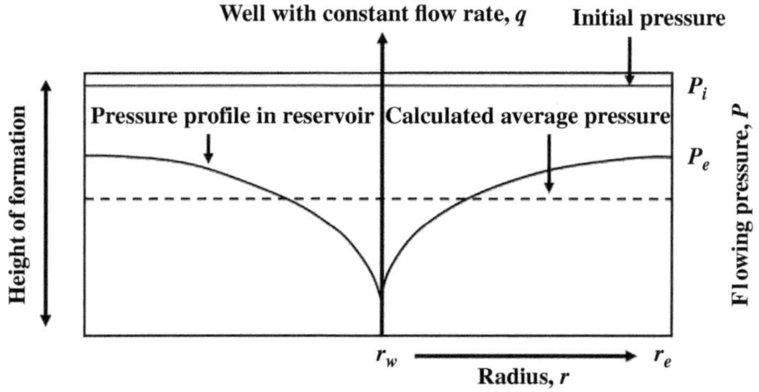

Figure 2.15 Pressure profile in a reservoir under semi-steady state flow conditions.

sections or **cells**. The average pressure in each cell can be calculated based on the stabilized pressure profile. This average pressure is determined by the **material balance equation**, which accounts for the initial pressure, the fluid volume, and its **isothermal compressibility**. The expansion of fluid in each cell is observed as a flow rate at the well, expressed as:

$$cV(P_i - \overline{P}) = qt \tag{2.82}$$

where V = pore volume of the radial cell, q = constant production rate, t = total flowing time, and c = isothermal compressibility.

Given that:

$$q = \frac{dV}{dt} \text{ and } c = -\frac{1}{V}\left(\frac{dV}{dP}\right)_T$$

the flow rate can be expressed as:

$$q = -cV\frac{dP}{dt}$$

or equivalently:

$$\frac{dP}{dt} = -\frac{q}{cV} \tag{2.83}$$

For radial drainage of a cell, this becomes:

$$\frac{dP}{dt} = -\frac{q}{c\pi r_e^2 h\varphi}. \tag{2.84}$$

Substituting Eq. (2.84) into the **radial diffusivity equation**:

$$\frac{1}{r}\frac{\delta}{\delta r}\left(r\frac{\delta P}{\delta r}\right) = \frac{\varphi\mu c}{k}\frac{\delta P}{\delta t}$$

yields:

$$\frac{1}{r}\frac{\delta}{\delta r}\left(r\frac{\delta P}{\delta r}\right) = -\frac{q\mu}{\pi r_e^2 hk}$$

Integration results in:

$$r\frac{dP}{dr} = \frac{q\mu r^2}{2\pi r_e^2 hk} + C. \tag{2.85}$$

At the outer boundary, where the pressure gradient is zero $\left(r\frac{dP}{dr} = 0\right)$, we find that $C = \frac{q\mu}{2\pi kh}$. Substituting this into Eq. (2.85) gives:

$$\frac{dP}{dr} = \frac{q\mu}{2\pi kh}\left(\frac{1}{r} - \frac{r}{r_e^2}\right). \tag{2.86}$$

Integrating this yields:

$$P_r - P_{wf} = \frac{q\mu}{2\pi kh}\left(\ln\frac{r}{r_w} - \frac{r^2}{2r_e^2}\right). \tag{2.87}$$

For practical purposes, the term $\frac{r_w^2}{2r_e^2}$ is negligible. If we consider the pressure at the external radius r_e, including the skin factor (s) around the well, the equation becomes:

$$P_e - P_{wf} = \frac{q\mu}{2\pi kh}\left(\ln\frac{r_e}{r_w} - \frac{1}{2} + s\right). \tag{2.88}$$

When using the average pressure, the **volume-weighted average pressure** of the drainage cell can be calculated in the same way as in the steady-state flow regime:

$$\overline{P} = \frac{2}{r_e^2}\int_{r_w}^{r_e} Prdr. \tag{2.55}$$

In this case:

$$\overline{P} - P_{wf} = \frac{2}{r_e^2}\frac{q\mu}{2\pi kh}\int_{r_w}^{r_e}\left(\ln\frac{r}{r_w} - \frac{r^2}{2r_e^2}\right)dr.$$

After integrating, we get:

$$\overline{P} - P_{wf} = \frac{q\mu}{2\pi kh}\left(\ln\frac{r_e}{r_w} - \frac{3}{4} + s\right). \tag{2.89}$$

In summary, the pressure differences $(P_r - P_{wf})$, $(P_e - P_{wf})$, $\left(\overline{P} - P_{wf}\right)$ remain constant over time, while the individual pressures P_r, P_e, P_w, and $\overline{P}$ continue to change.

2.5.7.1 Pressure Drop from Initial Reservoir Pressure

When calculating the pressure drop from the initial reservoir pressure, P_i, we can modify Equation 2.82 to describe the pressure at any time (t) after flow has begun. This is expressed as:

$$P = P_0 + \left(\frac{q}{cV}\right)(t_0 - t), \tag{2.90}$$

and for the average reservoir pressure:

$$\overline{P} = P_i - \left(\frac{qt}{cV}\right). \tag{2.91}$$

In these equations:

- q is the volume flow rate,
- c represents the isothermal compressibility,
- V is the original reservoir volume,
- t_0 is the reference time after which flow begins,
- t is the time after flow starts,
- P_0 is the pressure at the reference time (t_0),
- P is the pressure at time (t),
- $\overline{P}$ is the average reservoir pressure at time (t).

By subtracting Eq. (2.91) from Eq. (2.89), we derive the pressure drop between the initial reservoir pressure and the wellbore pressure:

$$P_i - P_{wf} = \frac{q\mu}{2\pi kh}\left(\ln\frac{r_e}{r_w} - \frac{3}{4} + \frac{2kt}{\varphi\mu c r_{e^2}}\right). \tag{2.92}$$

2.5.7.2 Generalized Reservoir Geometry: Flowing Equation under Semi-steady State Conditions

In semi-steady state conditions, the flow regime is influenced by the fact that the reservoir has finite boundaries, limiting the amount of fluid available. While the equations developed so far apply to radial geometries, the semi-steady state flow in non-radial reservoirs can still be examined using the radial flow equation if the reservoir geometry can be represented by a shape factor. This factor accounts for the relative position of the producing well in the reservoir. A commonly used shape factor for non-symmetrical geometries is the **Dietz shape factor** (denoted as C_A).

For a reservoir with no skin factor, the pressure drop, using the average reservoir pressure, is given by:

$$\overline{P} - P_{wf} = \frac{q\mu}{2\pi kh}\left(\ln\frac{r_e}{r_w} - \frac{3}{4}\right). \tag{2.89}$$

The term $\left(\ln\left(\frac{r_e}{r_w}\right) - \frac{3}{4}\right)$ can be expressed as:

$$\frac{1}{2}\left(2\ln\left(\frac{r_e}{r_w}\right) - \frac{3}{2}\right) = \frac{1}{2}\ln\left[\frac{r_e^2}{r_w^2 e^{\frac{3}{2}}}\right].$$

Since the area drained by the well for a radial geometry is πr_e^2, the logarithmic term becomes:

$$\frac{\pi r_e^2}{\pi r_w^2 e^{\frac{3}{2}}} = \left[\frac{4A}{1.781 \times 31.6 \times r_w^2}\right],$$

where (A) is the drained area, $\gamma = 1.781$, and the Dietz shape factor (C_A) for a radial drainage area is 31.6.

Thus, the generalized semi-steady state inflow equation, for the average reservoir pressure, can be written as:

$$\overline{P} - P_{wf} = \frac{q\mu}{2\pi kh}\left(\frac{1}{2}\ln\frac{4A}{\gamma C_A r_w^2} + s\right). \tag{2.93}$$

For the pressure drop between the initial reservoir pressure (P_i) and the bottom-hole flowing pressure during semi-steady state flow, Eq. (2.92) can be modified as:

$$P_i - P_{wf} = \frac{q\mu}{2\pi kh}\left(\frac{1}{2}\ln\frac{4A}{\gamma C_A r_w^2} + \frac{2\pi kt}{\varphi\mu cA}\right),$$

or

$$P_{wf} = P_i - \frac{q\mu}{2\pi kh}\left(\frac{1}{2}\ln\frac{4A}{\gamma C_A r_w^2} + \frac{2\pi kt}{\varphi\mu cA}\right). \tag{2.94}$$

In **dimensionless form**, this becomes:

$$\frac{2\pi kh}{q\mu}(P - P_{wf}) = \frac{1}{2}\ln\frac{4A}{\gamma C_A r_w^2} + 2\pi\left(\frac{kt}{\varphi\mu c r_w^2}\right)\frac{r_w^2}{A},$$

or equivalently

$$P_D(t_d) = \frac{1}{2}\ln\frac{4A}{\bar{\gamma} C_A r_w^2} + 2\pi t_D \frac{r_w^2}{A}. \tag{2.95}$$

To handle the term involving the wellbore radius, a **modified dimensionless time** can be introduced:

$$t_{DA} = t_D \frac{r_w^2}{A},$$

and thus

$$P_D(t_d) = \frac{1}{2}\ln\frac{4A}{\bar{\gamma} C_A r_w^2} + 2\pi t_{DA}. \tag{2.96}$$

The calculation of Dietz shape factors and their limitations are detailed in Lee's work (1982) and reproduced in Table 2.4. These factors apply to several common reservoir geometries, including cases where wells are located near barriers. The values of (t_{DA}) indicate when the shape factors should be used.

- For the **infinite system solution** with less than 1% error, $t_{DA} < X$, where X represents the maximum elapsed time during which the reservoir behaves as infinite-acting, and the *Ei* function can be applied. The time (t) is calculated as:

$$t < t_{DA} \frac{\varphi \mu c A}{k} \tag{2.97}$$

- For $t_{DA} > X$, the semi-steady state solution can be used with an error of less than 1% over the elapsed time:

$$t > t_{DA} \frac{\varphi \mu c A}{k} \tag{2.98}$$

- For $t_{DA} > X$, the semi-steady state solution can also be used as an **exact solution** for this elapsed time:

$$t > t_{DA} \frac{\varphi \mu c A}{k} \tag{2.99}$$

In a real reservoir under semi-steady state conditions, the volume of the reservoir drained by a well can be estimated from the well's flow rate. This volume can then be correlated with the reservoir's structural map to determine the shape of the drainage area. Using the calculated shape factors, the position of the well relative to the reservoir boundaries can be determined. While this is not an exact process due to the reservoir's heterogeneity, it provides an analytical tool for characterizing the reservoir.

2.5.8 The Application of the CTR Solution in Well Testing

In the study of fluid flow, we have primarily focused on predicting the pressure drop resulting from a well's production in a reservoir. If key parameters—such as porosity, permeability, and fluid viscosity—are known, the pressure drop at a certain distance from the well and at a specific time after production begins can be calculated for a particular flow regime, like an unsteady state.

However, in practice, **flow rates** and **well pressures** are the only directly measurable quantities. One of the most critical unknowns in the **diffusivity equation**

Table 2.4 Drainage area shape factors for different single-well configurations.

In bounded reservoirs	C_A	Exact for $t_{DA}>$	Less than 1% error for $t_{DA}>$	Use infinite system solution with less than 1% error for $t_{DA}<$
	31.62	0.1	0.06	0.10
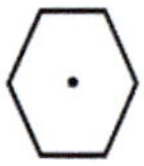	31.6	0.1	0.06	0.1
	27.6	0.2	0.07	0.09
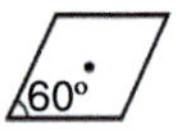	27.1	0.2	0.07	0.09
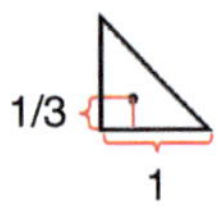	21.9	0.4	0.12	0.08
	0.098	0.9	0.6	0.015
	30.8828	0.1	0.05	0.09
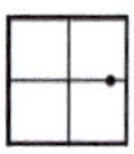	12.9851	0.7	0.25	0.03
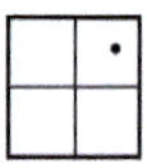	4.5132	0.6	0.3	0.025
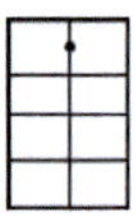	3.3351	0.7	0.25	0.01
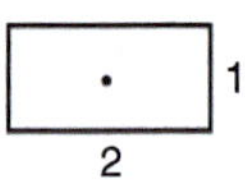	21.8369	0.3	0.15	0.025
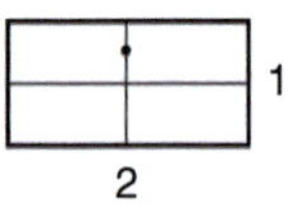	10.8374	0.4	0.15	0.025
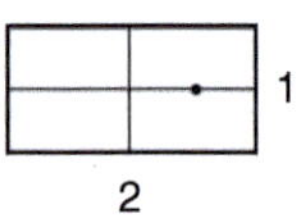	4.5141	1.5	0.5	0.06

In bounded reservoirs	C_A	Exact for $t_{DA}>$	Less than 1% error for $t_{DA}>$	Use infinite system solution with less than 1% error for $t_{DA}<$
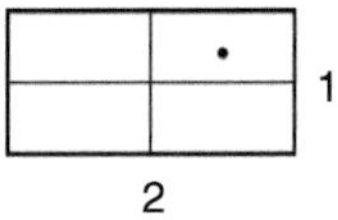	2.0769	1.7	0.5	0.02
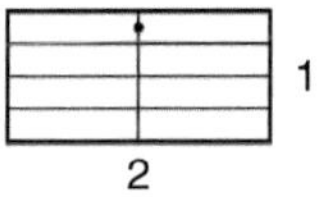	3.1573	0.4	0.15	0.005
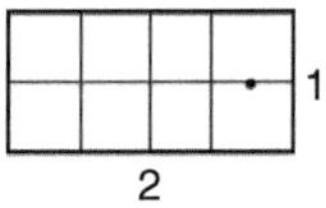	0.5813	2.0	0.6	0.02
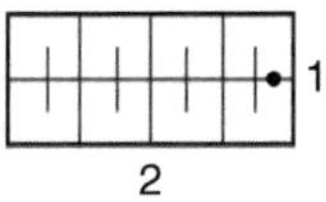	0.1109	3.0	0.6	0.005
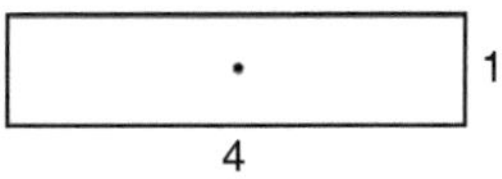	5.3790	0.8	0.3	0.01
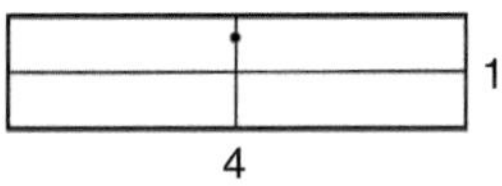	2.6896	0.8	0.3	0.01
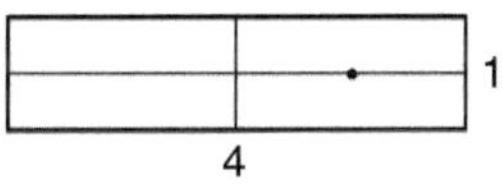	0.2318	4.0	2.0	0.03
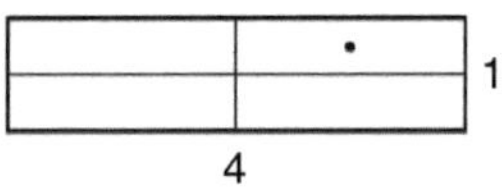	0.1155	4.0	2.0	0.01
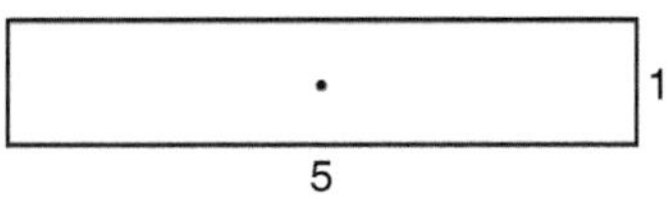	2.3606	1.0	0.4	0.025

is permeability. Therefore, instead of calculating pressure drops for known conditions, well testing involves measuring the pressure drop continuously to determine the permeability.

This is one of the primary goals of well testing. As an illustration, the following example calculates both permeability and the skin factor for a well in a reservoir. It is crucial to note that these examples assume the well is being brought into production for the first time in an **undisturbed reservoir**, meaning the pressure is at its initial value. In real-world well test analysis, a well's previous production history must be considered. The concept of **superposition** will be introduced in later sections to account for a well's multi-rate production history.

2.6 The Constant Terminal Pressure Solution

In the CTR solution of the diffusivity equation, the flow rate is known to remain constant in a certain part of the reservoir, and pressures are calculated throughout the reservoir. In contrast, in the constant terminal pressure solution, the pressure at a specific point in the reservoir is known to be constant, and the cumulative flow at any given radius can be calculated.

The constant terminal pressure solution is generally less complex than the CTR solution, mainly because fewer variations exist. In fact, only one constant terminal pressure solution is widely recognized, eliminating the need to choose between different solutions, as is the case with CTR solutions.

The constant terminal pressure solution was developed by **Hurst and Van Everdingen** for two primary cases:

1. An **infinite radial system** with constant pressure at the inner boundary.
2. Constant pressure at the inner boundary with **no flow across the outer boundary**.

These cases can be used to model scenarios such as a wellbore where the bottom-hole flowing pressure remains constant while fluid flows in the reservoir, or a reservoir surrounded by an **aquifer**.

The same geometric and property assumptions apply as with the CTR solutions, including:

- Radial geometry of constant thickness,
- A well located at the center,
- Fixed rock and fluid properties throughout the reservoir.

In this solution, however, there is a **pressure drop** from an initial pressure to a constant value. For cases involving **aquifer encroachment**, the "well" radius corresponds to the radius of the initial **oil-water contact**.

The constant terminal pressure solution is particularly useful for calculating **water encroachment** (or **natural water influx**) into the original **oil and gas zone** due to **water drive** in the reservoir.

2.7 Superposition

In the previous analyses, we assumed that the well's flow rate changed instantaneously from zero to a constant value. However, in reality, well flow rates often fluctuate during normal production operations, and wells may be periodically **shut in** for testing or operational reasons. Additionally, most reservoirs have more than a single well draining them, and this must be accounted for when analyzing flow behavior.

In essence, there may be a combination of **multiple wells** in a reservoir, each producing at different flow rates. Despite these complexities, reservoir pressures can still be calculated using the same analytical techniques previously discussed if the effects of each flow rate change are **superposed**. In other words, the total pressure drop at a wellbore can be determined by summing the effects of multiple flow rate changes within the well, or by summing the pressure effects caused by production from nearby wells.

Furthermore, superposition can also be used to simulate the impact of barriers within the reservoir. By introducing **imaginary wells** (sometimes called image wells), the pressure response near a barrier can be mimicked using infinite-acting solutions, offering a practical way to model such scenarios.

Mathematically, superposition is based on the linearity of the differential equations that describe fluid flow in reservoirs, and it relies on two key properties:

1. If P is a solution, then any multiple $C \times P$ (where C is a constant) is also a solution.
2. If both P_1 and P_2 are solutions, then their sum, $P_1 + P_2$, is also a solution.

These properties form the basis for generating both the CTR and constant terminal pressure solutions. By adding the solutions together, we can determine the cumulative effect on pressure in the reservoir, for example, from multiple wells or varying production rates. This approach is especially useful when dealing with multiple wells in a reservoir.

2.7.1 Effects of Multiple Wells

In a reservoir where more than one well is produced, the pressure perturbation caused by each well is evaluated independently, as if the other wells and their flow rate/pressure history do not exist. The total pressure drop at a particular well at a specific time is then calculated by **superimposing** the effects of all the individual wells.

For instance, consider three wells—X, Y, and Z—which start producing simultaneously in an infinite-acting reservoir (Figure 2.16). The total pressure drop at well Y due to the presence of all three wells is calculated as follows:

$$\left(P_i - P_{wf}\right)_{\text{Total at Well Y}} = (P_i - P)_{\text{Due to Well X}} + (P_i - P)_{\text{Due to Well Y}} + (P_i - P)_{\text{Due to Well Z}}.$$

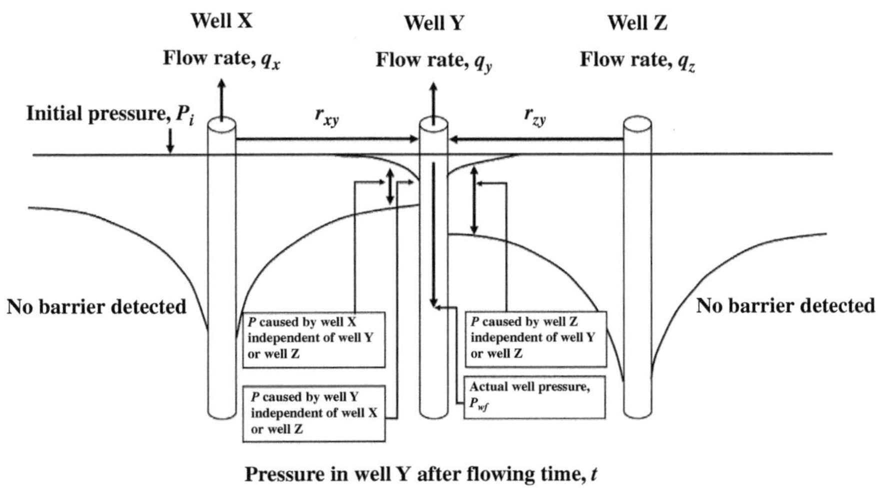

Figure 2.16 Superposition of pressure variations from multiple wells.

Assuming unsteady-state flow conditions, the line source solution is used to determine the pressure at well Y. For well Y itself, the logarithmic function is applied, and the presence of a skin factor around well Y is taken into account. However, for wells X and Z, the *Ei* function is used to describe their effects on the pressure at well Y. Any skin factors associated with wells X and Z only affect the pressure drop around those wells and do not influence the calculation for well Y.

Thus, the total pressure drop at well Y is expressed as:

$$\begin{aligned}(P_i - P_{wf})_{\text{Total at Well Y}} &= \frac{-q_Y\mu}{4\pi kh}\left(\left(\ln\frac{\bar{\gamma}\varphi\mu c r_{\text{wY}}^2}{4kt}\right) - 2S_{\text{Y}}\right)\\ &+ \frac{-q_X\mu}{4\pi kh}\left(Ei\left(\frac{\varphi\mu c r_{\text{XY}}^2}{4kt}\right)\right)\\ &+ \frac{-q_Z\mu}{4\pi kh}\left(Ei\left(\frac{\varphi\mu c r_{\text{ZY}}^2}{4kt}\right)\right)\end{aligned} \tag{2.100}$$

where:

- q_Y is the flow rate from well Y,
- q_X is the flow rate from well X,
- q_Z is the flow rate from well Z,
- r_{wY} is the radius of well Y,
- r_{XY} is the distance between wells X and Y,
- r_{ZY} is the distance between wells Z and Y.
- The remaining symbols retain their usual meanings.

This superposition technique can be applied to analyze the effects of any number of wells in an infinite-acting reservoir. It can be used to predict possible flowing well pressures within a group of wells or to deliberately evaluate the interaction between wells to assess **reservoir continuity**. Such tests, known as **interference tests**, are designed to characterize the reservoir on a large scale, rather than merely determining the permeability and skin factor around individual wells.

2.7.2 Principle of Superposition and Approximation of Variable-rate Pressure Histories

The previous section demonstrated the impact of production from multiple wells on the bottom-hole flowing pressure of a particular well. Equally important, however, is understanding how multiple **flow rate changes** within a single well affect its bottom-hole pressure. This is a more realistic scenario than the constant flow rate cases previously discussed, where a well is brought into production at a fixed rate over a specific time period.

In practice, a newly completed well may experience several flow rate changes, starting with initial **cleanup after completion**, followed by **production testing**, and finally, adjustments made during **regular production** to meet reservoir management objectives (e.g. limiting the gas-oil ratio). A simple plot of **pressure** and **flow rate versus time** would resemble Figure 2.17.

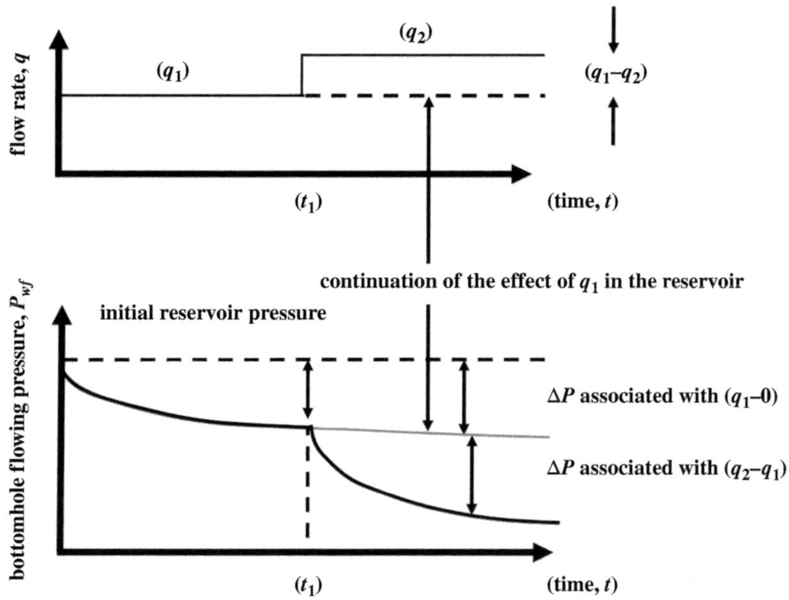

Figure 2.17 Effect of flow rate variations on wellbore pressure.

In this example, the well starts production at an initial flow rate q_1. Over time, the bottom-hole flowing pressure drops according to the boundary conditions and flow regime. At time t_1, the flow rate increases to q_2, causing a change in bottom-hole pressure. The total bottom-hole pressure at any given time is calculated by superposing the pressure drop caused by the initial flow rate (q_1) and the pressure drop resulting from the change in flow rate from q_1 to q_2 after t_1.

During the first period (with flow rate q_1), the pressure drop at time (t) is given by:

$$\Delta P(t) = P_i - P_{wf} = \Delta P_D(t)\frac{q_1\mu}{2\pi kh}, \tag{2.101}$$

where $\Delta P_D(t)$ is the **dimensionless pressure drop** at the well, determined by the applicable boundary condition.

For times greater than t_1, the pressure drop is:

$$\Delta P(t) = \frac{q_1\mu}{2\pi kh}\Delta P_D(t) + \frac{(q_2 - q_1)\mu}{2\pi kh}\Delta P_D(t - t_1) \tag{2.102}$$

In this case, the total pressure drop is the sum of the pressure caused by flow rate q_1 over time t, and the additional pressure drop resulting from the rate change ($q_2 - q_1$) over time ($t - t_1$). The initial pressure perturbation due to q_1 continues to affect the reservoir, while the second perturbation caused by $q_2 - q_1$ is superposed on top of it to give the total pressure drop at the wellbore.

In mathematical terms:

$$0 \leq t \leq t_1; \Delta P(t) = \Delta P_D(t)\frac{q_1\mu}{2\pi kh}, \tag{2.103}$$

and for $t > t_1$:

$$\Delta P(t) = \frac{q_1\mu}{2\pi kh}\Delta P_D(t) + \frac{(q_2 - q_1)\mu}{2\pi kh}\Delta P_D(t - t_1). \tag{2.104}$$

In Eq. (2.104), the first term represents the pressure drop due to flow rate q_1, while the second term represents the additional pressure drop from the rate change $q_2 - q_1$. This formulation holds true whether q_2 is larger or smaller than q_1, so even if the well is shut in, the effects of the previous flow rate history are still valid.

The dimensionless pressure drop function ΔP_D depends on the flow regime and boundary conditions. If unsteady-state flow is assumed, and the line source solution is applied, then:

$$\Delta P_D = \frac{P_i - P_{wf}}{\frac{q\mu}{2\pi kh}} = -\frac{1}{2}Ei\left(\frac{-\varphi\mu c r_w^2}{4kt}\right). \tag{2.105}$$

For times $t \leq t_1$, the pressure drop would be:

$$\Delta P(t) = -\frac{q_1\mu}{4\pi kh}Ei\left(\frac{-\varphi\mu c r_w^2}{4kt}\right). \tag{2.106}$$

For times $t > t_1$, the additional pressure drop is superposed, giving:

$$\Delta P(t) = -\frac{q_1\mu}{4\pi kh}Ei\left(\frac{-\varphi\mu c r_w^2}{4kt}\right) - \frac{(q_2 - q_1)\mu}{4\pi kh}Ei\left(\frac{-\varphi\mu c r_w^2}{4k(t - t_1)}\right). \tag{2.107}$$

This method can be extended to multiple flow rate changes, as shown in Figure 2.18. The general equation becomes:

$$\begin{aligned}\Delta P(t) &= \frac{q_1\mu}{2\pi kh}\Delta P_D(t) + \frac{(q_2 - q_1)\mu}{2\pi kh}\Delta P_D(t - t_1) + \frac{(q_3 - q_2)\mu}{2\pi kh}\Delta P_D(t - t_2) \\ &+ \cdots + \frac{(q_n - q_{n-1})\mu}{2\pi kh}\Delta P_D(t - t_{n-1}).\end{aligned} \tag{2.108}$$

Or, in a more compact form:

$$\Delta P(t) = \frac{q_1\mu}{2\pi kh}\left[\Delta P_D(t) + \sum_{i=2}^{n}\frac{q_i - q_{i-1}}{q_1}\Delta P_D(t - t_{i-1})\right]. \tag{2.109}$$

This is the general form of the **principle of superposition** for wells with multiple rate histories.

In the special case where the well is **shut in** and pressure builds up, an additional term is included to account for the pressure increase. Assuming the well is shut in during the ***n*-th flow rate period**, the pressure builds back toward the initial reservoir pressure according to:

$$\begin{aligned}P_i - P_{ws} = &\frac{q_1\mu}{2\pi kh}\left[\Delta P_D(t) + \sum_{i=2}^{n}\frac{q_i - q_{i-1}}{q_1}\Delta P_D(t_n - t_{i-1} + \Delta t)\right] \\ &- \frac{q_n\mu}{2\pi kh}\Delta P_D(\Delta t),\end{aligned} \tag{2.110}$$

where P_{ws} is the shut-in bottom-hole pressure, t_{n-1} is the total production time before shut-in, and Δt is the shut-in time from the moment the well is shut in.

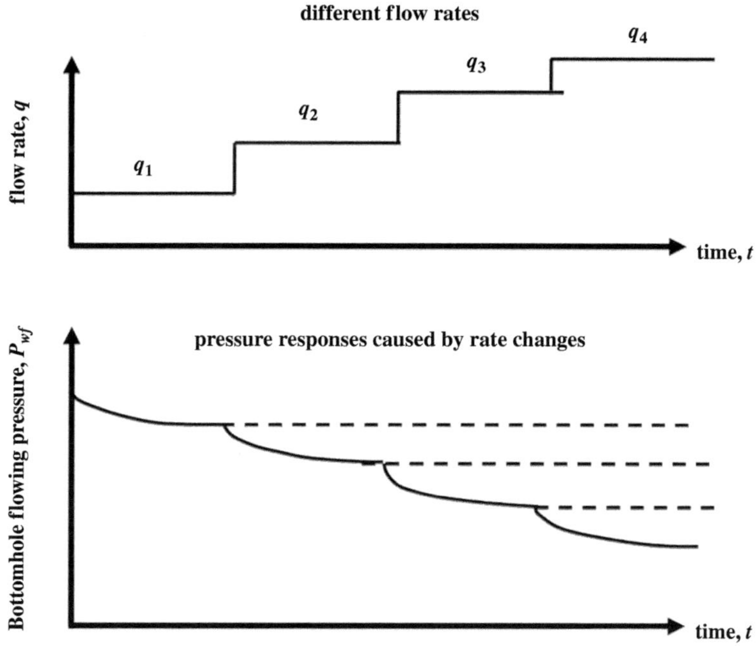

Figure 2.18 Wellbore pressure response to multiple flow rates.

2.7.3 Effects of Rate Changes

The application of superposition to a well with several rate changes can be illustrated as follows. Consider a well with the flow rate history presented in Figure 2.19. Initially, the well is brought into production at a flow rate q_1, which remains constant until time t_1, at which point the flow rate is decreased to q_2. This second flow rate continues until time t_2, when the flow rate is increased to q_3.

Assuming the reservoir is in an unsteady-state flow regime, the line source solution is used to describe the pressure drop resulting from the flow rate changes. When the well is first brought into production, the change from zero to q_1 creates the first **pressure perturbation** that moves into the reservoir.

The focus here is on the bottom-hole flowing pressure, P_{wf}, which can be calculated using the line source solution. If a skin zone exists around the well, it must also be considered in the calculation. If no further flow rate changes occur, the unsteady state would eventually transition to either a semi-steady or steady-state condition, causing the bottom-hole flowing pressure to either decline steadily or (in the case of steady-state) stabilize at a constant level.

However, in this case, unsteady-state conditions persist when the flow rate changes from q_1 to q_2 at time t_1. This change creates a second pressure perturbation, which moves into the reservoir, following the first one. Since the reservoir remains in an unsteady state, the first pressure perturbation has not encountered any barriers, and the reservoir fluid behaves as though it is part of an infinite volume. This behavior

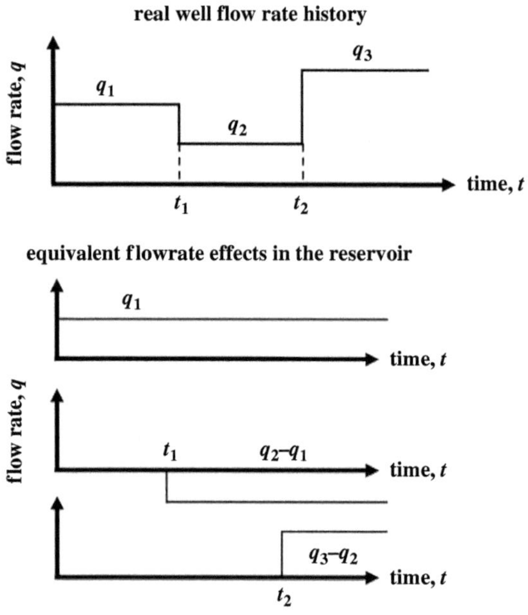

Figure 2.19 Equivalence of flow rate variations in a reservoir.

continues to cause a pressure decline at the wellbore, even as the second pressure perturbation is created and moves through the reservoir.

The pressure drop caused by this flow rate change can be calculated using the line source solution and added to the initial pressure drop from the well being brought into production. At time t_2, the flow rate changes again to q_3, and a third pressure perturbation is created. This third perturbation follows the first two into the reservoir, and as long as the reservoir fluid still behaves as though it is infinite in volume, the pressure drop resulting from this flow rate change can be added to the others to give the total pressure drop.

The pressure drop from the initial flow rate q_1 when the well is brought into production can be calculated using the **logarithmic approximation** of the *Ei* function (assuming the *Ei* function and its approximation have been validated for this application):

$$\Delta P_1 = (P_i - P_{wf})_1 = \frac{-q_1 \mu}{4\pi kh}\left[\ln\left(\frac{\bar{\gamma}\varphi\mu c r_w^2}{4kt}\right) - 2s\right]. \tag{2.111}$$

Next, the pressure drop caused by the flow rate change $q_2 - q_1$ at time t_1 is calculated. Since we are still determining the bottom-hole flowing pressure, the skin zone still needs to be accounted for:

$$\Delta P_2 = (P_i - P_{\mathrm{wf}})_2 = \frac{-(q_2 - q_1)\mu}{4\pi kh}\left[\ln\left(\frac{\bar{\gamma}\varphi\mu c r_w^2}{4k(t - t_1)}\right) - 2s\right]. \tag{2.112}$$

Finally, the pressure drop caused by the flow rate change $q_3 - q_2$ at time t_2 is calculated as:

$$\Delta P_3 = (P_i - P_{wf})_3 = \frac{-(q_3 - q_2)\mu}{4\pi kh}\left[\ln\left(\frac{\bar{\gamma}\varphi\mu c r_w^2}{4k(t - t_2)}\right) - 2s\right]. \tag{2.113}$$

The total pressure drop at the wellbore, considering all flow rate changes, is the sum of the individual pressure drops:

$$P_i - P_{wf} = \Delta P_1 + \Delta P_2 + \Delta P_3. \tag{2.114}$$

2.7.4 Simulating Boundary Effects (Image Wells)

One of the intriguing applications of the principle of superposition in reservoir flow is simulating boundary effects. Initially, when a well begins production, the flow regime is unsteady, with the reservoir fluid reacting to the pressure perturbation as if the reservoir were infinite in volume (i.e. an infinite-acting reservoir). However, once the reservoir's boundaries are detected, there is a finite limit to the available fluid, and the pressure response transitions to a different regime, such as semi-steady-state or steady-state flow.

For a well located at the center of a **circular reservoir**, the pressure perturbation reaches the external boundary uniformly around the circumference (assuming homogeneity). However, for wells that are not centrally located, some parts of the boundary are detected earlier than others, meaning parts of the reservoir remain in an unsteady-state while others transition to a different flow regime. This seemingly complicates the use of the line source solution, but we can use superposition to simulate boundaries through the concept of image wells.

If an imaginary well is placed at a distance from the real well equal to twice the distance to the boundary, with an identical flow rate history, the principle of superposition can couple the effects of the imaginary well to the real well. This allows for the calculation of the real well's bottom-hole flowing pressure. Figure 2.20 illustrates this concept, and Figure 2.21 simplifies the model.

In Figure 2.20, we have a **plane-fault boundary** in an otherwise infinite-acting reservoir. The line source solution can be used to model the pressure response until the perturbation reaches the fault. Afterward, there are no solutions for this complex geometry. However, by placing an image well at twice the distance to the fault, the infinite-acting reservoir can be simulated. No flow occurs across the plane midway between the real and image well, and the flow configuration is the same on both sides. The real well reacts to the pressure drop created by both itself and the imaginary well across the fault.

The pressure drop is given by:

$$P_i - P_{wf} = -\frac{q\mu}{4\pi kh}\left[\ln\left(\frac{\bar{\gamma}\varphi\mu c r_w^2}{4kt}\right) - 2s\right] - \frac{q\mu}{4\pi kh} Ei\left(\frac{-\varphi\mu c (2L)^2}{4kt}\right), \tag{2.115}$$

where L is the distance from the real well to the fault, and s is the skin factor for the real well (but not for the image well).

There are other instances where image wells are used to simulate boundary effects in reservoir flow. Larger networks of image wells may require computer modeling for complex geometries. For example, if a well is located near the **corner of a rectangular boundary**, more image wells are required to balance the flow from the real well. Figure 2.22 shows such a boundary and the required image wells.

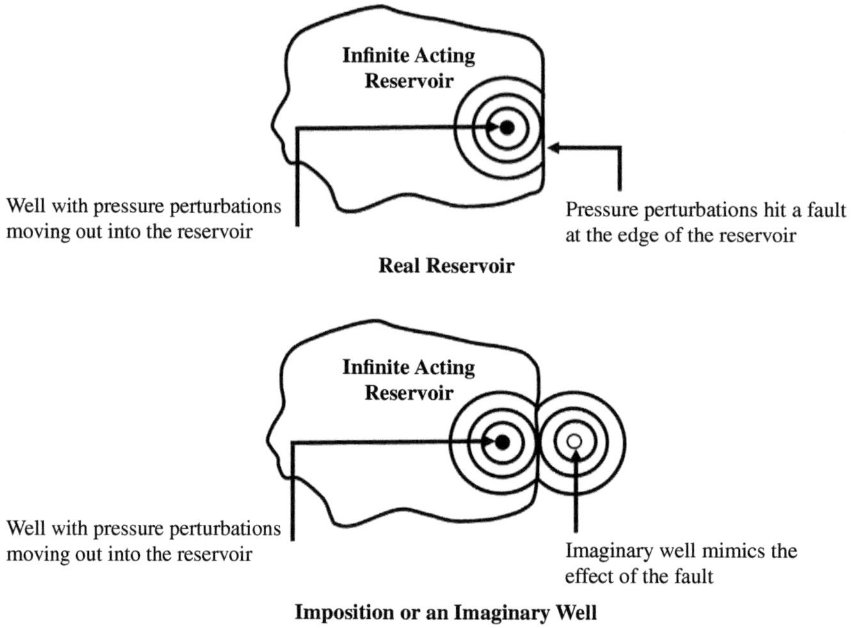

Figure 2.20 The pressure impact of a barrier in a real reservoir represented by an imaginary well.

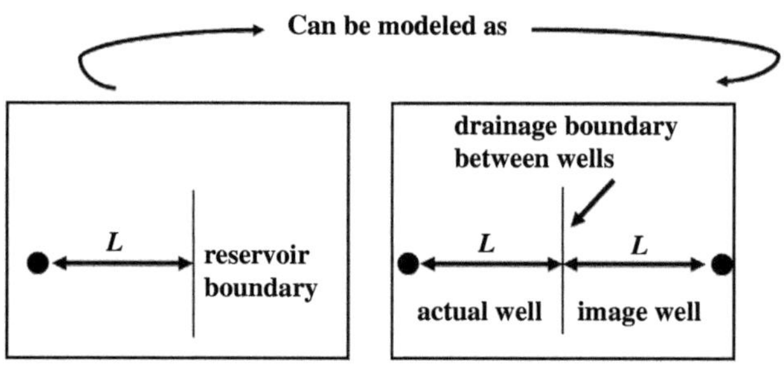

Figure 2.21 Representation of a boundary using a real well and an image well.

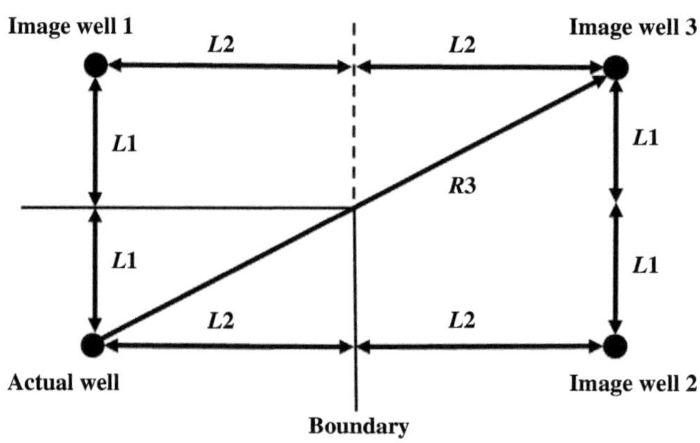

Figure 2.22 Representation of a well at the intersection of two boundaries.

In this case, four pressure drop terms are needed to calculate the pressure at the real well:

$$P_i - P_{wf} = \Delta P_{rw} + \Delta P_{2L1} + \Delta P_{2L2} + \Delta P_{r3}, \tag{2.116}$$

where

$$\begin{aligned}(P_i - P_{wf})_{\text{Total at the actual well}} &= (P_i - P)_{\text{at the actual wellbore radius, } r_w} \\ &+ (P_i - P)_{\text{Due to image well 1 at distance } 2L1} \\ &+ (P_i - P)_{\text{Due to image well 2 at distance } 2L2} \\ &+ (P_i - P)_{\text{Due to image well 3 at distance } R3}.\end{aligned}$$

The number and position of image wells can become quite complex.

In a seemingly simpler case, such as a well located between two equidistant barriers (as shown in Figure 2.23), the flow can still be balanced by defining an image well, i_1, on one side. However, on the opposite side, additional image wells, i_2 and i_3, are needed to account for the presence of the real and image wells across the other barrier. This process continues indefinitely as new imbalances are created, requiring more image wells (i_4, i_5, i_6, etc.). Fortunately, since the line source solution has minimal influence beyond a certain distance, a limited number of image wells can provide an accurate approximation.

Even more complex geometrical configurations can be devised. For example, Matthews et al. (1954) studied the pressure behavior of wells surrounded by boundaries in **rectangular-shaped reservoirs**. Figure 2.24 illustrates the network of image wells used to simulate the effect of these boundaries.

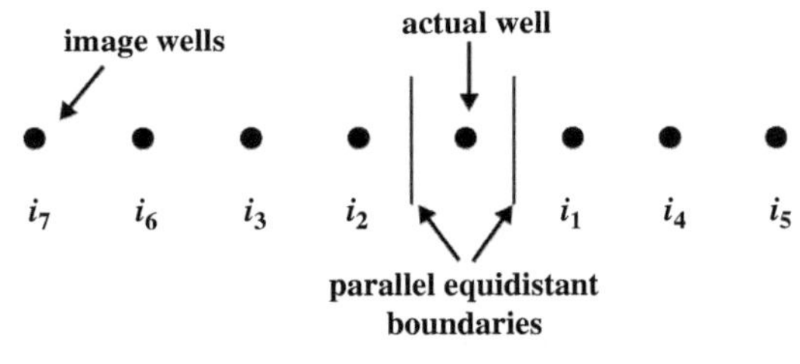

Figure 2.23 Representation of an actual well between two barriers.

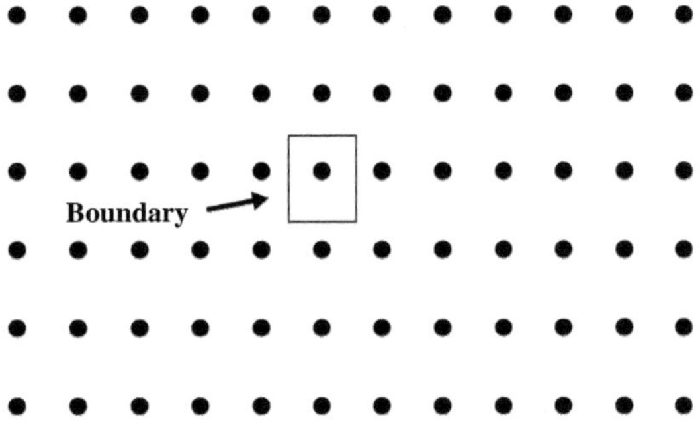

Figure 2.24 Representation of a well surrounded by boundaries.

2.8 Ideal Gas Flow

Consider gas as the flowing fluid, which introduces the need to account for both **compressibility** and **gas-slip effects** (Collins, 1990). The governing equation for gas flow, incorporating the Klinkenberg effect, is represented by the diffusivity equation:

$$\frac{d}{dx}\left[k_{\infty}p\left(1+\frac{b}{p}\right)\frac{dp}{dx}\right]=0. \tag{2.117}$$

Here, the Klinkenberg effect is given by the relationship:

$$k=k_{\infty(x)}\left(1+\frac{b}{p}\right). \tag{2.118}$$

The **equivalent liquid permeability** depends on the distance (x). By integrating the equation twice with respect to x, gives:

$$\bar{k}_{\infty}\bar{p}\left(1+\frac{b}{\bar{p}}\right)\frac{\Delta p}{L}=c, \tag{2.119}$$

where k_{∞} represents the equivalent liquid permeability, which varies as a function of distance (x). After integrating this equation twice, we obtain an expression where c is a constant, and $\bar{p}$ represents the average pressure, which is the mean of the pressure at the boundaries $p(0)$ and $p(L)$. The average permeability is given by:

$$\frac{L}{\bar{k}_{\infty}}=\int_0^L\frac{dx}{k(x)}.$$

By comparing this result with Darcy's law, we obtain the constant:

$$c=\frac{\bar{q}\mu\bar{p}}{A},$$

where the flow rate is evaluated at the mean pressure.

2.8.1 Streamlines, Isopotentials, and Source/Sink Representation

Streamlines indicate the direction of flow for a fluid particle within the system at any given time. For steady-state flow, the direction remains constant over time, but for **transient conditions**, the streamline is valid only for a specific instant. Streamlines are always **orthogonal to equipotential lines** throughout the flow region, except in the case of anisotropic media, as will be discussed later. **Isopotential** or **equipotential lines** represent areas of **constant potential**. A **streamtube** is the area between two adjacent streamlines, and a **flow net** consists of both equipotential lines and streamlines. An example of a flow net is illustrated in Figure 2.25.

In this figure, the **hydraulic head** is held constant at 100 m on the left boundary and 40 m on the right boundary, while **no-flow boundaries** exist at the top and bottom. Flowlines show the direction of fluid movement from higher potential to lower, toward the right. The equipotential lines (dashed lines) are perpendicular to the flowlines, consistent with the following rules for homogeneous, isotropic media:

- Flowlines and equipotential lines must be **normal** to each other.
- Equipotential lines must intersect **impermeable boundaries** at **right angles**.
- Equipotential lines must be **parallel** to **constant potential boundaries**.

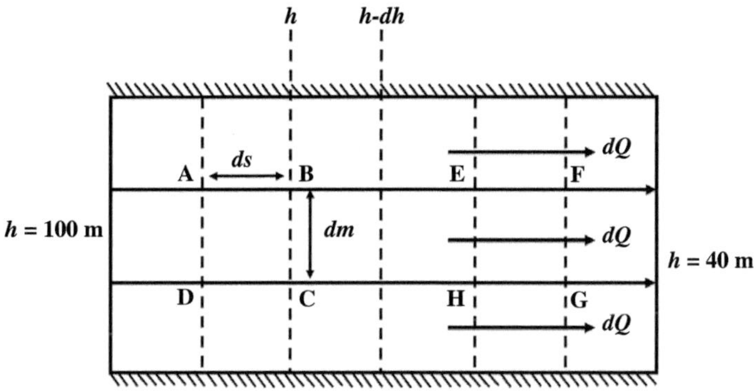

Figure 2.25 Flow net for a simple flow system. *Source:* Freeze and Cherry, 1979 / Prentice-Hall.

The boundary conditions are special forms of either flowlines or equipotential lines. For **closed boundaries**, the velocity of the fluid normal to the boundary is zero, meaning there is no flow across the boundary (Figure 2.26). The adjacent flowlines must be parallel to the boundary, and the equipotential lines meet the boundary at right angles (see Figure 2.25 for an example).

Mathematically, this can be expressed as:

$$\vec{v}_n = 0$$
$$\frac{\partial \psi}{\partial l_n} = 0, \tag{2.120}$$

where n is the unit vector normal to the boundary, and l_n is the distance measured parallel to n. For **open boundaries**, the flux is continuous, and the boundary becomes an **isopotential line** (ψ = constant) with **no flux tangential to the boundary**. Figure 2.27 illustrates this condition.

Figure 2.26 Schematic representation of a closed boundary.

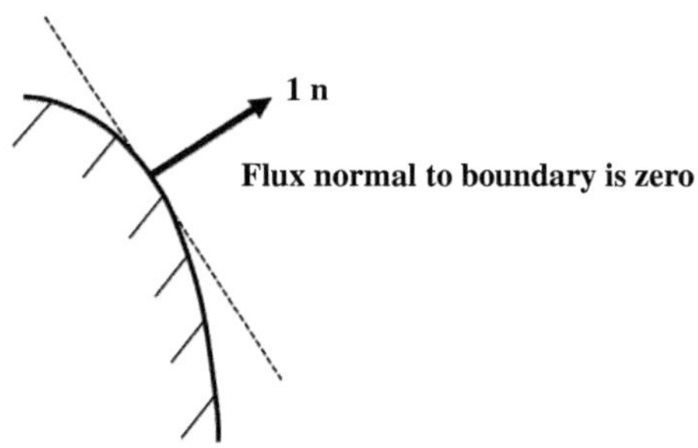

Figure 2.27 Schematic representation of an open boundary.

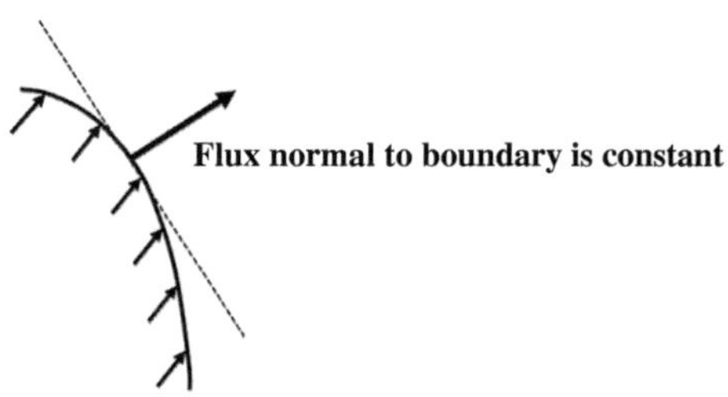

Bibliography

Chatas, A.T. (1953). A practical treatment of nonsteady-state flow problems in reservoir systems. *Journal of Petroleum Engineering, 25*(5), B44–B56.

Collins, R.E. (1990). *Flow of Fluids Through Porous Materials*. Research & Engineering Consultants, Inc.

Feder, J., Flekkøy, E.G., & Hansen, A. (2022). *Physics of Flow in Porous Media*. Cambridge University Press.

Freeze, R.A., & Cherry, J.A. (1979). *Groundwater*. Prentice-Hall.

Lee, J. (1982). *Well Testing* (SPE Textbook Series, Vol. 1, p. 159). Society of Petroleum Engineers.

Matthews, C.S., Brons, F., & Hazebroek, P. (1954). A method for determination of average pressure in a bounded reservoir. *Transactions of the AIME, 201*(01), 182–191.

Matthews, C.S., & Russell, D.G. (1967). *Pressure Buildup and Flow Tests in Wells* (Vol. 1, p. 27). Henry L. Doherty Memorial Fund of AIME.

Ph, F. (1901). Wasserbewegung durch boden. *Zeitschrift des Vereines Deutscher Ingenieure, 45*(50), 1781–1788.

Van Everdingen, A.F., & Hurst, W. (1949). The application of the Laplace transformation to flow problems in reservoirs. *Journal of Petroleum Technology, 1*(12), 305–324.

Zimmerman, R.W. (2018). *Fluid Flow in Porous Media* (Vol. 5). World Scientific.

3

Classification of Hydrocarbons and Oil Reserves

3.1 Common Classification of Hydrocarbons

There are a large number of hydrocarbon compounds, the molecules of which consist of the chemical elements hydrogen and carbon in various proportions.

Hydrocarbons, depending on the thermobaric conditions of the formation, are in different states of aggregation (gaseous, solid, or liquid).

Hydrocarbon fluids are typically produced by wells and are divided into liquid hydrocarbons and natural hydrocarbon gases. The classification and nomenclature of hydrocarbon fluids are shown in Figure 3.1.

Liquid hydrocarbons are divided into the following categories:

1. Crude oil: A mixture of hydrocarbons in a liquid aggregate state, either in the reservoir or in atmospheric conditions, after passing through surface separation equipment.
   - Liquefied natural gas: The portion of natural gas that is liquefied in onshore separators, production facilities, and gas processing plants. Natural gas liquids include (but are not limited to) ethane, propane, butane, pentane, natural gas, and condensate.
2. Natural gas is divided into the following categories:
   - Free gas: Natural gas accumulated at the top of an oil reservoir (gas cap) or found in reservoirs with little oil content.
3. Associated gas: Natural gas dissolved in oil and released during its production and treatment.
   - Dissolved gas: Natural gas dissolved in oil at reservoir conditions.
   - Injected gas: Hydrocarbon gases injected into a reservoir for reservoir pressure maintenance or storage.

Methods for Enhanced Oil Recovery: Fundamentals and Practice, First Edition.
Baghir A. Suleimanov and Elchin F. Veliyev.

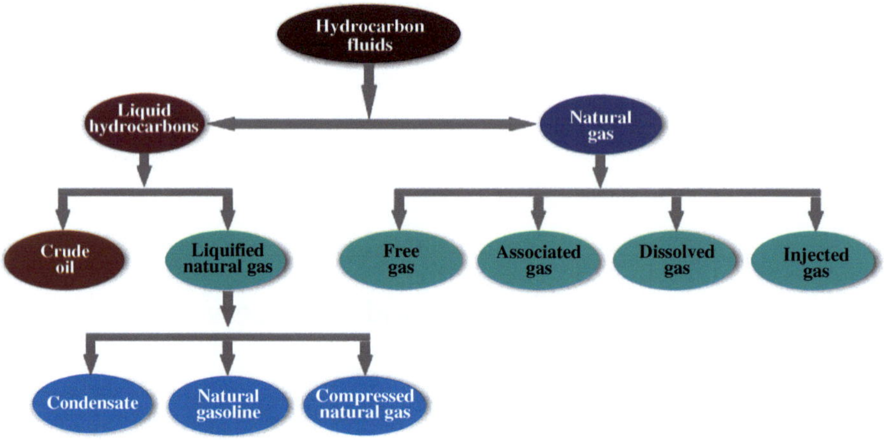

Figure 3.1 Classification and nomenclature of hydrocarbon fluids.

3.2 Classification of Oil Reserves

The classification of oil reserves takes into account the following criteria:

- possible oil recovery factor (ORF),
- degree of proof of reserves,
- current state of production and field development, and
- energy resource.

3.2.1 Possible ORF

The potential oil recovery rate is the amount of oil recovered in relation to the original geological reserves, taking into account the limitations imposed by the existing oil recovery mechanism, technological efficiency, and economic conditions.

3.2.2 Degree of Proof of Reserves

Accurate data on geological oil reserves are very difficult to determine at the exploration stage due to the lack of initial information. For this purpose, data on geological reserves are systematically updated using the volumetric method or the material balance method directly during the operation of the field. However, the volume of cumulative oil production measured in surface (atmospheric) conditions is determined quite accurately.

If V is the volume of geological oil reserves, and ΔV is the volume of accumulated oil production for the current period, the ratio $K = (\Delta V/V)\cdot 100\%$ is the ORF in the current period, or the actual ORF.

It is important for a petroleum engineer to predict the ultimate ORF. At an early stage of field development, in conditions of limited information, the value of the final ORF must be estimated with the greatest possible accuracy.

Arps identifies three stages in the life cycle of an oil field. At the first stage, before drilling begins, geological reserves are calculated based on previous experience in the development of similar deposits. At the second stage, the first wells are drilled and put into operation. The volume of geological oil reserves V is calculated either by the volumetric method or using the material balance method. The final ORF is predicted based on the prevailing development regime. At the third stage, the oil production decline curve is analyzed by extrapolation or the construction of mathematical models, which makes it possible to clarify the geological reserves and ORF of the field.

3.2.3 Current State of Production and Field Development

It is also necessary to take into account the current state of development of the field in order to distinguish between reserves that can be recovered using existing wells and reserves located in undeveloped areas of the field. The second category of reserves is so related to the first that it can be predicted with the same confidence as in the presence of drilled wells.

Oil reserves determined during the production period are the most accurate. The production process is largely determined by the type of reservoir energy inherent in a particular field initially or created artificially during its development.

Figure 3.2 shows the classification of oil reserves, taking into account the possible ORF, the degree of proof of reserves, and the current state of field development.

Proven oil reserves are reserves that can be recovered using various methods (innovative and traditional methods of enhancing oil recovery [EOR]). The fundamental factor when choosing one or another EOR is the predominant type of reservoir energy in the field. Thus, it is important that oil reserves are also classified by reservoir energy type.

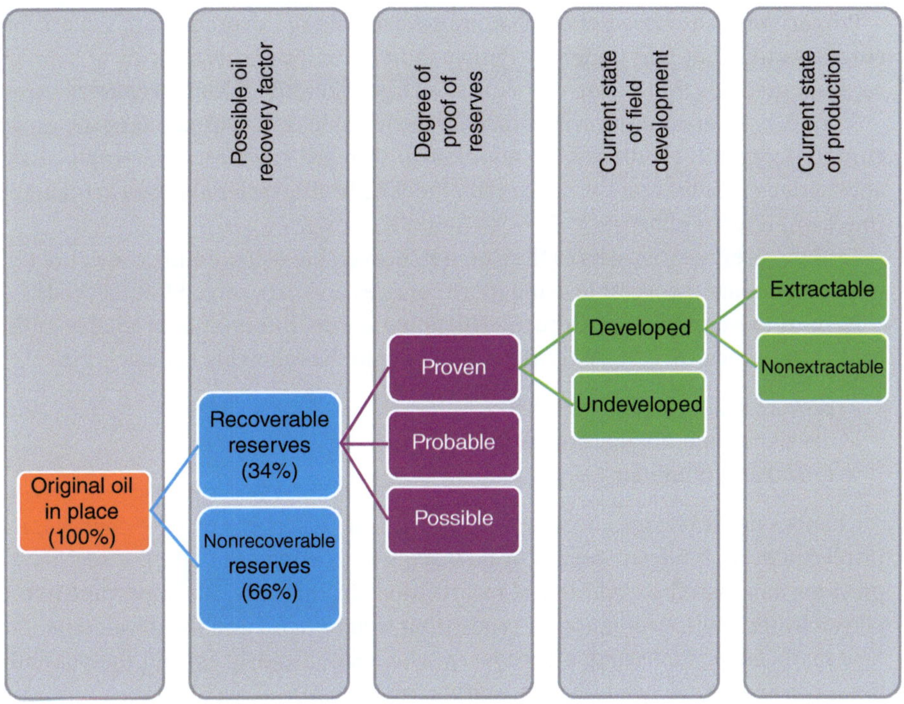

Figure 3.2 Classification of oil reserves.

3.2.4 Energy Resource

In a reservoir, the main source of energy for oil extraction is reservoir pressure. Depending on the predominant energy source during drainage in the reservoir, the following operating modes of the oil field are distinguished.

1. Dissolved gas mode: Oil drainage occurs due to the release and expansion of dissolved gas.
2. Gas pressure (gas) regime: OIL drainage occurs under the influence of the expansion of the gas cap.
3. Water-pressure regime: Oil drainage occurs under the influence of advancing marginal water.

The force of gravity, as well as various combinations of the above modes, can act as the driving energy. As a rule, in practice, various driving forces appear simultaneously, forming the so-called mixed regime.

Methods of maintaining (increasing) reservoir pressure by injecting gas or water into the reservoir are called secondary or traditional methods of oil extraction. Despite the effectiveness of these methods, a significant amount of oil remains unrecovered. Retention of residual oil by the rock occurs mainly due to capillary forces, interfacial tension, and also due to incomplete displacement of oil during injection of displacing agents. Residual reserves are the goal of using more complex and expensive technologies, the so-called tertiary methods of EOR.

Thus, taking into account the energy resources of the reservoir, the following conclusions can be drawn for the classification of oil reserves.

Primary reserves are reserves that are recoverable on a commercial scale using current equipment and under current economic conditions as a result of primary recovery methods that utilize the natural energy inherent in the reservoir.

Secondary reserves are reserves that are recoverable on a commercial scale under current economic conditions, in addition to primary reserves, as a result of the application of traditional (secondary) methods (injecting water and gas) to maintain (increase) reservoir energy.

Tertiary reserves are reserves that are not recoverable using traditional extraction methods, but are recoverable using more complex and expensive EOR methods.

Each of these categories, in turn, depending on the information available at the time the reserves are assessed, can be divided into the following subcategories:

- proven, probable, and possible;
- developed and not developed; and
- mined and unmined.

Changes in oil reserves occur annually with the discovery of new deposits, the involvement in the development of new areas in existing ones and the revision of previous data based on additional information. In this case, an important role is played by the improvement of oil production technologies and the development of new methods for increasing oil recovery, which significantly expand the possibilities for increasing recoverable hydrocarbon reserves.

3.3 Oil Recovery Factor

The amount of oil recovered in relation to the original geological reserves is called the ORF or oil recovery factor.

The amount of additional oil produced as a result of EOR techniques can be expressed as a percentage of the original oil reserves in place. To assess this indicator, it is necessary to determine the potential of the field in question, which consists of the characteristics of the formation and its inherent development regime.

For this purpose, oil fields are classified depending on the average final ORF achieved under a particular development mode:

- 5–10%—poorly permeable deposits, slightly fractured or reservoirs with heavy oil;
- 10–25%—in dissolved gas mode;
- 25–40%—with partial water pressure and gravity modes, as well as a gas injection;
- 40–55% are oil deposits developed with waterflooding.

Possible assessments of the prospects of a particular EOR at a specific field can be made using Taber tables.

The choice of any EOR method should be based on a careful study of relevant field data (especially residual oil saturation), laboratory, and pilot tests. Studies have shown that there is a potential for an increase in recoverable reserves by 40% of current proven reserves using existing EOR methods.

3.4 SPE/WPC/AAPG Classification of Reserves

In February 2000, the World Petroleum Council (WPC), the Society of Petroleum Engineers (SPE), and the American Association of Petroleum Geologists (AAPG) approved the new petroleum resources classification system and definitions. These resource definitions supplement the existing "SPE/WPC Petroleum Reserves Definitions" (1997) without modifying them, and both sets should be viewed as companion documents.

The principles of resource subdivision were established by McKelvey in 1972 and continue to form the basis of the new system (Figure 3.3).

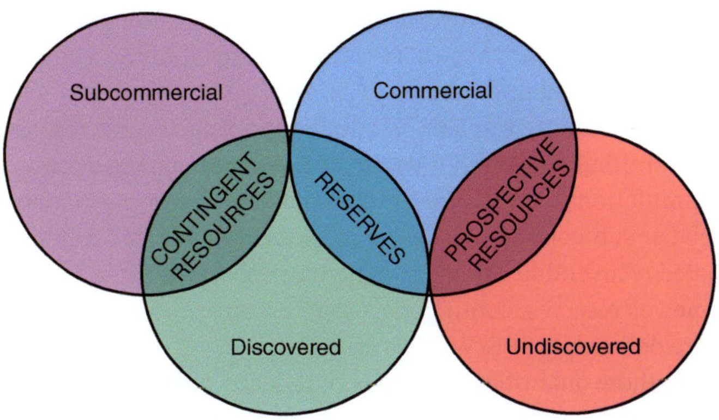

Figure 3.3 Adapted McKelvey box demonstrating terminology for recoverable resources.

In 1987, the WPC published the report "Classification and Nomenclature Systems for Petroleum and Petroleum Reserves," which included definitions for all categories of resources. This report, along with definitions from other industry organizations and recognition of current industry practices, forms the basis for the SPE/WPC/AAPG system.

The resource definitions primarily address the classification of estimated recoverable quantities from known but currently subcommercial accumulations (contingent resources) and from undiscovered accumulations (prospective resources). These can be summarized as follows:

1. *Reserves*: Quantities of petroleum expected to be commercially recovered from known accumulations from a specific date forward.
2. *Contingent resources*: Quantities of petroleum potentially recoverable from known accumulations as of a specific date, but not currently considered commercially recoverable.
3. *Prospective resources*: Quantities of petroleum potentially recoverable from undiscovered accumulations as of a specific date.

The fundamental subdivisions of the system—what constitutes a known accumulation and the criteria for commerciality—are discussed in the subsequent section.

Resource volume estimates are derived for an accumulation, and reserves are only quoted for known accumulations. Thus, it's essential to define these terms, recognizing that commerciality decisions, and thus development likelihood, are based on specific projects. First, defining "reservoir" and "field" is beneficial.

Since reserves are a subset of resources, it is necessary to define resources first to place reserves in context. The entire resource base, defined as total petroleum-initially-in-place, includes all estimated quantities of petroleum in the subsurface and quantities already produced. Previously defined by the WPC as "Petroleum-in-place," this is sometimes referred to as the "Resource Base" by others. To avoid confusion, the approved classification system includes specific definitions for quantities initially in place and those estimated to be recoverable.

All petroleum-initially-in-place quantities may constitute potentially recoverable resources, as the proportion that may be recoverable is subject to significant uncertainty and can change over time. Unrecoverable quantities may become recoverable resources in the future as commercial circumstances change and technological developments occur. These quantities include volumes left in the ground when a field is abandoned.

Total petroleum-initially-in-place is subdivided into discovered and undiscovered petroleum-initially-in-place, reflecting whether the quantities are contained in "known accumulations." Discovered petroleum-initially-in-place may be classified as commercial or sub-commercial based on the criteria for commerciality discussed later. Estimated recoverable volumes are further subsets of these in-place categories and are defined as reserves, contingent resources, and prospective resources.

Reserves are defined by SPE/WPC as follows:

Reserves are those quantities of petroleum that are anticipated to be commercially recovered from known accumulations from a given date forward.

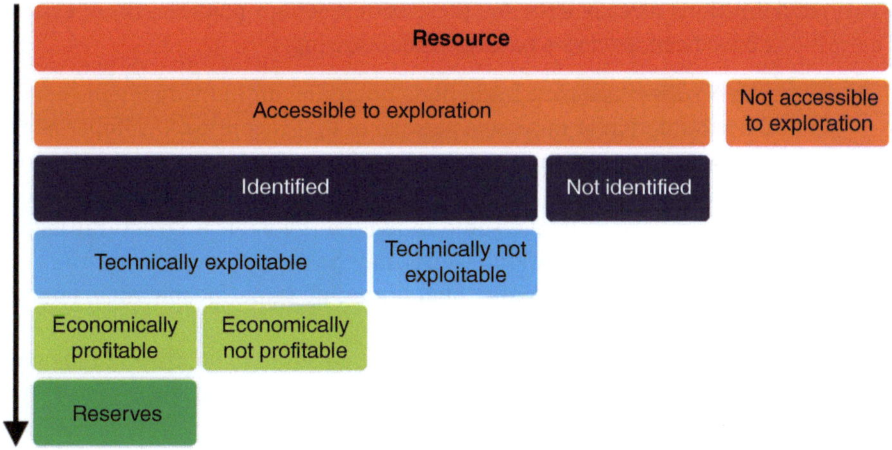

Figure 3.4 Graphical representation of resources and reserves.

Thus, reserves must satisfy four criteria: they must be discovered, recoverable, commercial, and remaining (Figure 3.4).

Despite this clarity, the term reserves is often misused, and it is strongly urged that the following expressions are not used:

1. Geologic reserves—sometimes used to denote petroleum-initially-in-place.
2. Technical reserves—sometimes used to classify sub-commercial discovered volumes, defined here as contingent resources.
3. Prospective or speculative reserves—sometimes used for undiscovered volumes, defined here as prospective resources.
4. Initial or ultimate reserves—sometimes used instead of estimated ultimate recovery (EUR), defined here as estimated remaining recoverable quantities plus cumulative production.

Contingent Resources are those discovered and potentially recoverable quantities that currently do not meet the criteria for commerciality and are defined as follows:

- Contingent resources are those quantities of petroleum that are estimated, on a given date, to be potentially recoverable from known accumulations but are not currently considered to be commercially recoverable.

Prospective Resources are those potentially recoverable quantities in accumulations yet to be discovered and are defined as follows:

- Prospective resources are those quantities of petroleum that are estimated, on a given date, to be potentially recoverable from undiscovered accumulations.

Resource volume estimates are derived for an accumulation, and reserves may only be quoted for a known accumulation; thus, it is important to define these terms. Decisions on commerciality and the likelihood of development are based on specific projects. First, however, it is beneficial to define reservoir and field.

Various definitions already exist for these terms; where possible, these existing definitions have been used as a basis for the following:

1. *Reservoir*: A subsurface rock formation containing an individual and separate natural accumulation of moveable petroleum that is confined by impermeable rock or by water barriers and is characterized by a single-pressure system.
2. *Field*: An area consisting of a single reservoir or multiple reservoirs all grouped on or related to the same individual geological structural feature and/or stratigraphic condition. There may be two or more reservoirs in a field that are separated vertically by intervening impermeable rock, laterally by local geologic barriers, or both.
3. *Known accumulation*: The term accumulation is used to identify an individual body of moveable petroleum in a reservoir. However, to be considered known and hence contain reserves or contingent resources, each accumulation/reservoir must have been penetrated by a well. Generally, the well must have clearly demonstrated the existence of moveable petroleum in that reservoir by flow to the surface or at least some recovery of a petroleum sample from the well. However, where log and/or core data exist, this may suffice, provided there is a good analogy to a nearby and geologically comparable known accumulation.

Two types of accumulations have been defined that may require different evaluation approaches:

1. *Conventional deposit*: A discrete accumulation related to a localized geological structural feature and/or stratigraphic condition, typically with each accumulation bounded by a down-dip contact with an aquifer, and significantly affected by hydrodynamic influences, such as the buoyancy of petroleum in water.
2. *Continuous-type deposit*: A petroleum accumulation that is pervasive throughout a large area and is not significantly affected by hydrodynamic influences. Examples include "basin-center" gas and gas hydrate accumulations.

   The resources and reserves definitions, together with the classification system, are intended to be appropriate for all types of petroleum accumulations. However, it is recognized that continuous-type deposits pose additional challenges in their evaluation, and generally accepted methods have yet to be established.

   - *Project*: Represents the link between petroleum accumulation and the decision-making process, including budget allocation. A project may involve the development of a single reservoir or field, an incremental development for a producing field, or the integrated development of several fields. Generally, an individual project represents the level at which a decision is made whether or not to proceed (i.e., spend money), and there should be an associated range of estimated recoverable volumes for that project.

Note that there are no commercial connotations associated with the terms reservoir, known accumulation, and field.

The distinction between commercial and sub-commercial known accumulations—and thus between reserves and contingent resources—is crucial for maintaining consistency in reserves reporting. According to the classification system outlined, an accumulation must be assessed as commercial before any reserves can be assigned.

Although the "SPE/WPC Petroleum Reserves Definitions" acknowledge some uncertainty in commercial criteria through reserve categories (proved, probable, and possible), it is explicitly stated that all reserves must be commercial. Thus, contingent resources may include quantities estimated to be recoverable from accumulations that currently lack a viable market or where commercial recovery hinges on the development of new technology. Additionally, new discoveries should be classified as contingent resources rather than reserves if they are at an early evaluation stage and commerciality has not yet been confirmed.

Once an accumulation is deemed commercial, reserves can be assigned. However, reserves must still be categorized according to the specific criteria outlined in the SPE/WPC definitions. Proved reserves are restricted to those quantities deemed commercial under current economic conditions, while probable and possible reserves can be based on future economic scenarios.

The responsibility for assessing the commerciality of an accumulation generally falls to the country or company considering its development, and this assessment will vary based on local conditions and circumstances. However, some guidelines can help ensure consistency in practice. It is recommended that if the commitment level does not indicate that the accumulation will be developed and brought into production within a reasonable timeframe, the estimated recoverable volumes should be classified as contingent resources. A reasonable timeframe for initiating development is typically around five years. A longer timeframe may be acceptable in cases where, for example, a group of gas fields is committed to a sales contract (making them clearly commercial), but some fields will not be developed until required to meet contractual obligations.

To assign reserves of any category, a project must be defined with a commercially viable development plan, and there should be clear evidence of intent to proceed with that plan. Reserve quantities then reflect the estimated recovery resulting from implementing that plan.

Within individual countries and companies, more specific guidelines are often established and applied. These guidelines are primarily based on economic evaluations but should also consider political, environmental, and market risks that could delay or prevent development. The economic evaluation generally includes the country's or company's best estimate of future costs and prices, along with a best-estimate production profile. For offshore environments, this profile often equates to a "proved plus probable" scenario for the field. Onshore, the decision may depend on the viability of completing the discovery well.

Even if the "proved plus probable" scenario appears commercially viable, significant uncertainty in the estimate might still prevent classification as reserves. There is typically a check to ensure limited downside risk, often by confirming that a "low" case scenario at least breaks even. This aligns with the requirement that proved reserves be viable "under current economic conditions" (see Chapter 3). In practice, if no proved reserves can be assigned to any reservoirs in a field, all estimated recoverable quantities would be classified as contingent resources rather than probable or possible reserves.

If an accumulation is deemed commercial, reserves of all categories can be assigned provided that the quantities in each category meet the relevant criteria of the "SPE/WPC Petroleum Reserves Definitions." Some countries and companies choose not to assign proved reserves until the development plan has received all necessary formal approvals. However, under the SPE/WPC definitions, this is not strictly necessary, as proved reserves may be booked if there is a "reasonable expectation" that the required facilities to process and transport those reserves will be installed.

3.4.1 Resource Uncertainty Categories

Estimating resource quantities for an accumulation or a group of accumulations (a project) inherently involves uncertainty and should generally be expressed as a range. The primary categories of reserves—proved, probable, and possible—within the SPE/WPC definitions, are designed to illustrate the degree of uncertainty in estimating the potentially recoverable volume of petroleum from a known accumulation. These estimates can be derived deterministically or probabilistically (see Chapter 4) and then aggregated for the accumulation or project as a whole (see Chapter 5). By applying similar methodologies to all volumetric estimates, including contingent and prospective resources, the estimate of uncertainty for each accumulation can be tracked throughout the exploration, discovery, development, and production phases. This method effectively evaluates the validity of the estimation approach for potentially recoverable volumes.

The range of uncertainty represents a reasonable spectrum of estimated recoverable volumes for an individual accumulation or project. For reserves, this uncertainty is reflected in estimates for Proved Reserves (1P), Proved plus Probable Reserves (2P), and Proved plus Probable plus Possible Reserves (3P) scenarios. For other resource categories, terms like low estimate, best estimate, and high estimate are recommended.

Two distinct approaches to applying a deterministic method to reserves estimation have emerged: the risk-based approach and the uncertainty-based approach. Risk refers to the probability of a specific event occurring, while uncertainty encompasses the range of possible outcomes in an estimate. Estimating the recoverable volume of petroleum involves both technical and commercial uncertainties, making exact estimation impossible. This distinction between risk and uncertainty is well understood in exploration.

Undrilled prospects are evaluated based on the risk of a dry hole and the uncertainty of recoverable volumes if a discovery is made. While early recognition of uncertainty issues in reserve estimation is acknowledged, North American practices have often relied on risk as the basis for classifying reserves as proved, probable, or possible. Conversely, many other regions have used uncertainty-based systems for years. This divergence reflects the industry's development variations, including differences between onshore and offshore environments, property sizes, reporting regulations, and other factors. Both approaches are valid and serve user needs differently, with uncertainty-based systems offering significant advantages due to their compatibility with both deterministic and probabilistic methods, and their ability to track uncertainty over time.

3.4.2 Risk-based Philosophy

In the risk-based approach, reserve quantities for each category (proved, probable, and possible) are estimated deterministically as fixed volumes. The uncertainty inherent in these volumes is not accounted for in the reserve quantities themselves. Instead, the risk (for probable and possible categories) pertains to the possibility of not having that volume present or recoverable. According to the SPE/WPC definitions, probable reserves must have a greater than 50% chance of being recoverable; thus, there is a risk of less than 50% that these reserves will not be produced. Possible Reserves, having a lower likelihood of recovery than probable reserves, face a higher risk of not being produced. This risk may relate to the presence of petroleum, the feasibility of development, or both.

For instance, if a large gas accumulation exists in an area with no current gas market, the chance of future development may be minimal. In such cases, some countries or companies might classify the entire accumulation as possible reserves, reflecting the high risk of development not occurring. Since all quantities are classified within a single category (e.g., possible reserves), this method does not account for uncertainties in the recoverable volume, which is substantial due to the undeveloped nature of the accumulation.

The risk-based approach attempts to capture both volumetric uncertainty and project maturity risk in a unified system. Industry practice often advises against aggregating quantities from different reserve categories (e.g., proved plus probable reserves) or combining them without risk adjustments. Consequently, this approach does not facilitate tracking of uncertainty over time or easily aligns with probabilistic methods. Despite these limitations, it remains a valid interpretation and application of the SPE/WPC petroleum reserves definitions.

3.4.3 Uncertainty-based Philosophy

The probabilistic evaluation method is inherently an uncertainty-based approach but can also be applied deterministically. In deterministic cases, a more suitable term is the "scenario approach," which involves evaluating three specific scenarios rather than relying on a single deterministic estimate. Project maturity is considered separately.

In the uncertainty-based philosophy, the goal is to first establish a "best estimate" of the recoverable volume, which represents the quantity expected to be most accurately recovered from the accumulation. Subsequently, estimates are made for an upside and a downside case, which reflect the range of uncertainty around the best estimate. This approach generates three scenarios for the accumulation: a downside case consistent with the proved category, a best estimate consistent with the proved plus probable category, and an upside case consistent with the proved plus probable plus possible category. Each scenario is crafted to represent a realistic combination of parameters in line with the definitions and considers potential dependencies between parameters. The scenarios account for both technical and commercial uncertainties.

The scenario approach is comparable to the probabilistic method but differs in that it estimates three representative reserve values rather than the complete range of possible outcomes. While it may not be as mathematically rigorous as the probabilistic method, it has the advantage of clearly identifying each assumption (e.g., average porosity values used for the proved scenario), facilitating reserve auditing. A crucial aspect is the relationship between the development plan (and recovery factor) and the geological model (and petroleum-initially-in-place). This dependency is often overlooked in probabilistic methods but is more easily addressed with the scenario approach.

When applying an uncertainty-based philosophy, whether deterministic or probabilistic, resource uncertainty categories help track estimates over the life of an accumulation. In deterministic methods, the confidence associated with low, best, and high estimates should align with the technical criteria for corresponding reserve categories. For instance, if an accumulation is noncommercial solely due to the lack of a market, a developed market, without new technical data, should make the proved reserve estimate close to the previous low estimate. Probabilistic methods should use probabilities consistent with reserve categories: at least a 90% probability that the potentially recoverable quantity will meet or exceed the low estimate, 50% for the best estimate, and 10% for the high estimate. These resource uncertainty categories—low estimate, best estimate, and high estimate—are related to reserve categories as depicted in Table 3.1.

A consistent method for estimating volumetric uncertainty, as described, enables the use of a powerful tool for tracking uncertainty in estimated ultimate recovery (EUR) over time (see Figure 3.5). Ideally, with increased information, uncertainty should decrease. Discrepancies from this trend often indicate that previous estimates were flawed, providing a basis for reviewing and refining estimation methodologies.

3.4.4 Project Status Categories

To create a detailed resource reporting system for portfolio management, many countries and companies further subdivide resource categories based on project status or maturity. As an accumulation matures, the probability of achieving commercial production increases (risk decreases). Project status categories are

Table 3.1 Resource uncertainty categories.

Resource uncertainty category	Reserves	Other resources	
	Scenario	Probabilistic	Scenario
Low estimate	Proved (1P)	P90	Low
Best estimate	Proved + Probable (2P)	P50	Most likely
High estimate	Proved + Probable + Possible (3P)	P10	High

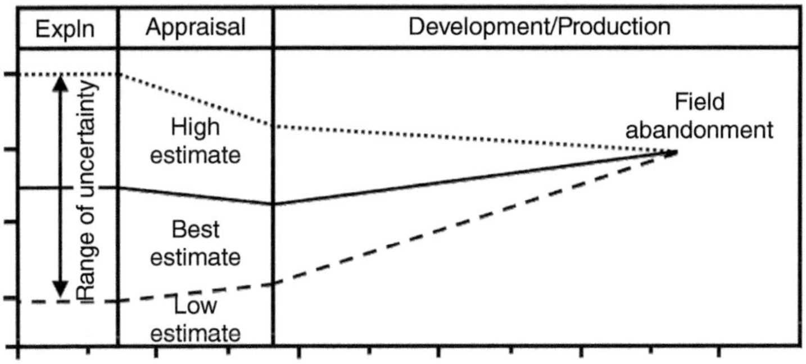

Figure 3.5 Uncertainty in resource estimation.

independent of the uncertainty associated with potentially recoverable volumes; however, uncertainty generally decreases as maturity increases.

The following project status categories are illustrative. SPE, WPC, and AAPG do not endorse specific subdivisions of reserves, contingent resources, or prospective resources, recognizing that countries and companies will establish their own categories consistent with their classification objectives. In the classification system shown in Table 3.2, each accumulation is categorized according to its project status/maturity, reflecting the actions (business/budget decisions) required to move it toward commercial production.

The text below provides an explanation of the terms used in the table above:

1. **Reserves**
   - *On production*: The project is actively producing and selling petroleum to the market.
   - *Under development*: All necessary approvals have been obtained, and the development of the project is in progress.
   - *Planned for development*: Meets all criteria for reserves with a firm intent to develop, but detailed planning and/or necessary approvals/contracts are still pending.
2. **Contingent Resources**
   - *Development pending*: Requires further data acquisition and/or evaluation to confirm commercial viability.
   - *Development on hold*: Of significant size but awaiting market development or the removal of other constraints (technical, environmental, political, etc.) to proceed.
   - *Development not viable*: No current plans to develop or acquire additional data due to limited production potential.

3.4.5 Prospective Resources

1. *Prospect*: Potential accumulation is well-defined enough to be considered a viable drilling target.

2. *Lead*: Potential accumulation is poorly defined and requires more data acquisition and/or evaluation to be classified as a prospect.
3. *Play*: Recognized prospective trends of potential prospects requiring more data acquisition and/or evaluation to define specific leads or prospects.

Table 3.2 Classification according to project status/maturity.

Total	Discovered/Undiscovered	Commerciality				Project status	Risk/maturity
TOTAL PETROLEUM INITIALLY-IN-PLACE	DISCOVERED PETROLEUM INITIALLY-IN-PLACE	Commercial	PRODUCTION			PROJECT STATUS	
			RESERVES			On Production	LOWER RISK
			PROVED	PROVED plus PROBABLE	PROVED plus PROBABLE plus POSSIBLE	Under Development	
						Planned for Development	↑
		Sub-Commercial	CONTINGENT RESOURCES			Development Pending	PROJECT MATURITY
			LOW ESTIMATE	BEST ESTIMATE	HIGH ESTIMATE	Development on Hold	
						Development not Viable	
			UNRECOVERABLE				↑
	UNDISCOVERED PETROLEUM INITIALLY-IN-PLACE		PROSPECTIVE RESOURCES			Prospect	
			LOW ESTIMATE	BEST ESTIMATE	HIGH ESTIMATE	Lead	
						Play	HIGHER RISK
			UNRECOVERABLE				
			← RANGE OF UNCERTAINTY →				

3.5 Russian Classification of Reserves

In Russia, the estimation of reserves for natural gas, gas condensate, and crude oil involves two main components (Figure 3.6):

1. *Geological reserves*: The total quantity of hydrocarbons contained in the subsoil.
2. *Extractable reserves*: The portion of geological reserves that can be economically extracted under current market conditions using modern technology, while adhering to subsoil and environmental protection regulations.

The Russian reserves classification system is based exclusively on geological attributes and is divided into several categories (Figure 3.7).

3.5.1 Explored Reserves

- *Category A*: These reserves are calculated for portions of a deposit that have been drilled in accordance with an approved development project for the field. This category includes reserves that have been extensively analyzed to define the deposit's type, shape, size, hydrocarbon saturation, reservoir type, reservoir characteristics, and major development features. These include well productivity, reservoir pressure, and other relevant parameters.
- *Category B*: Reserves in this category are determined based on commercial oil or natural gas flows obtained from wells at various depths. The deposit's type, shape, size, effective saturation, reservoir characteristics, and the composition and properties of hydrocarbons are studied in detail. Category B reserves are computed for deposits drilled in accordance with either a trial industrial development project for gas fields or an approved technological development scheme for oil fields.

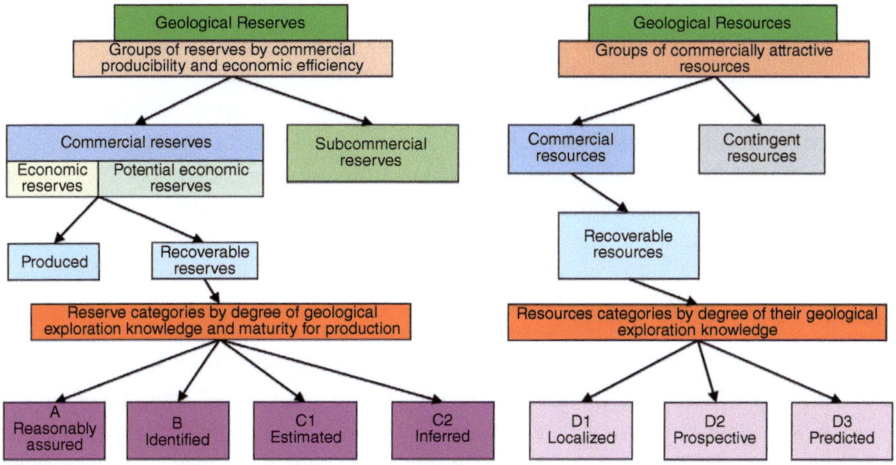

Figure 3.6 Russian classification of reserves.

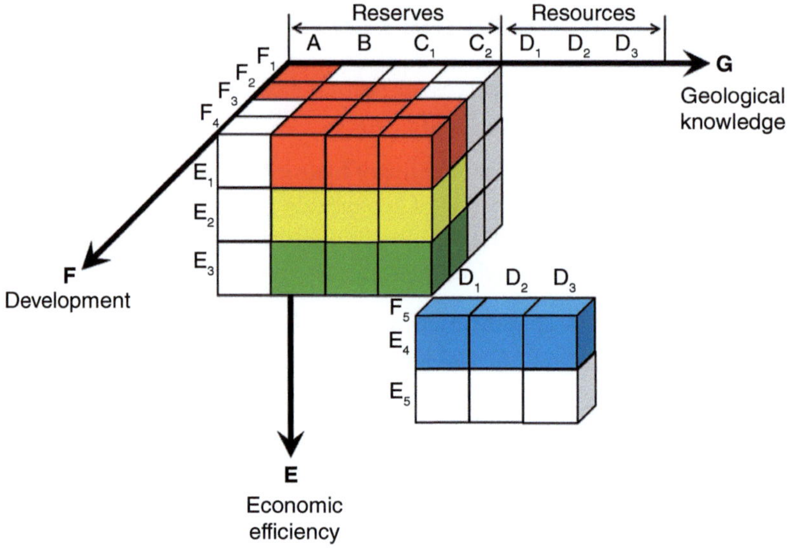

Figure 3.7 Principles of Russian classification of reserves and resources.

- *Category C1*: Reserves are assessed based on commercial flows of oil or natural gas obtained from wells, with additional data from geological and geophysical exploration. This category includes reserves where the deposit's type, shape, size, formation structure, and reservoir characteristics have been studied in detail through drilling and exploration. The data includes well productivity, reservoir pressure, and hydrocarbon characteristics, and is used to prepare a trial industrial development project for gas fields or a technological development scheme for oil fields.

3.5.2 Preliminary Estimated Reserves

- *Category C2*: Preliminary reserves in this category are estimated based on geological and geophysical research of unexplored sections adjoining known reserves or within partially explored fields. These reserves provide a general understanding of the deposit's shape, size, and characteristics, helping in planning further geological and exploration activities. Category C2 reserves are used to assess the development potential of a field and to plan additional exploration work.

3.5.3 Potential Resources

- *Category C3*: Prospective resources in this category are identified for potential drilling within the oil-and-gas-bearing area or unexplored sections of known fields. These resources are estimated based on geological and geophysical research and analogies with similar explored fields. Category C3 resources are used to plan prospecting and exploration activities in areas that may contain new reserves.

3.5.4 Forecasted Resources

- *Category D1*: Resources in this category are calculated based on regional geological, geophysical, and geochemical research and analogies with explored fields within the region. These reserves are located within large regional structures confirmed to contain commercial reserves of oil and natural gas.
- *Category D2*: Resources in this category are estimated using general geological concepts and analogies with better-studied regions. They are located within large regional structures not yet confirmed to contain commercial reserves. The potential for these resources to be oil-and-gas-bearing is evaluated based on geological, geophysical, and geochemical research.

3.5.5 Evaluation Methods

- *Volume method*: This method is used for estimating reserves in both new and existing fields. It calculates reserves by examining the deposit's area, the effective depth of hydrocarbon saturation, porosity, and level of hydrocarbon saturation, taking into account thermobaric conditions.
- *Material balance method*: This method is applied to existing fields, considering temporal changes in reservoir pressure due to hydrocarbon extraction and water influx. It helps in assessing the reserves by analyzing changes in reservoir conditions over time.

3.5.6 Regulatory Framework

- *State examination*: According to the Law on Subsoil, reserves in Russia must undergo a mandatory state examination conducted by organizations under the Federal Agency on Subsoil Use. Only reserves deemed commercially feasible are entered into the State Balance of Mineral Products.
- *Licensing and reporting*: Subsoil users must obtain exploration, development, or production licenses following state examination and submit annual statistical reports reflecting changes in reserves. Significant changes require approval from the State Reserve Commission or its regional organizations.

This classification system provides a detailed and structured approach to evaluating and managing hydrocarbon reserves, ensuring accurate reporting and effective resource utilization.

3.6 The United Nations Framework Classification for Resources

The United Nations Framework Classification for Resources (UNFC) is a comprehensive, project-based classification system designed to assess the environmental, socio-economic, and technical viability of resource projects. It provides a standardized approach to evaluating and reporting on resource projects, ensuring consistency and clarity across different types of resources and projects.

3.6.1 Key Components of UNFC

1. Resources and products:
   - *Resources*: Include various sources, such as solar, wind, geothermal, hydro-marine, bioenergy, hydrocarbons, minerals, nuclear fuels, and water. These resources may be in their natural state or derived from anthropogenic sources like tailings.
   - *Products*: The outputs of resource projects, which can be sold, used, or bought, include electricity, heat, hydrocarbons, hydrogen, minerals, and water. In some cases, such as renewable energy projects, the products (like electricity or hydrogen) differ from the sources (like wind or solar irradiation). In other projects, like petroleum, the sources and products are similar (oil and gas), though their state and properties may change from reservoir to surface conditions.
2. Project definition:
   - A *Project* is defined as a development or operation that serves as the basis for environmental, social, economic, and technical evaluation and decision-making. The project plan may range from detailed to conceptual, depending on the project's maturity and the needs of stakeholders.
3. Application areas:
   - UNFC is designed to address the needs of various applications, including:
     - Policy formulation based on resource studies.
     - Resource management functions.
     - Corporate business processes.
     - Financial capital allocation.

The (UNFC) evaluates resource projects based on three key criteria: environmental-socio-economic viability (E), technical feasibility and maturity (F), and degree of confidence (G). Each criterion is divided into categories and subcategories that reflect the project's status and potential.

UNFC operates across three key dimensions (Figure 3.8), which are:

1. Environmental-socio-economic viability (E Axis)
2. Technical feasibility and maturity (F Axis)
3. Degree of confidence in estimates (G Axis)

Each dimension plays a critical role in classifying resources and projects, and they combine to form different "Classes" within the system.

3.6.1.1 E Axis—Environmental-Socio-Economic Viability

The E Axis evaluates whether a project is viable based on environmental, social, and economic factors. This assessment determines if a project can be developed and operated successfully given current and projected conditions.

1. *E1*: This category signifies that the project is confirmed to be environmentally, socially, and economically viable. All necessary conditions, such as permits and contracts, have either been met or are expected to be met within a reasonable timeframe. This includes scenarios where short-term adverse conditions do not undermine the long-term viability of the project.

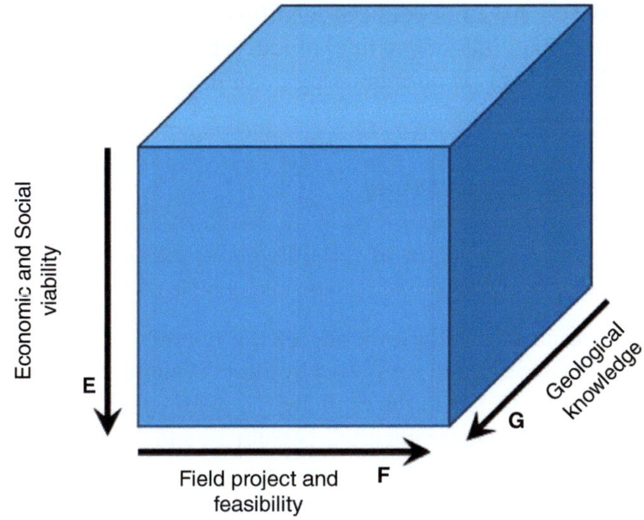

Figure 3.8 UNFC classification criteria.

2. *E2*: Projects in this category are anticipated to become viable in the foreseeable future based on realistic assumptions. Although they are not yet confirmed to be viable, there are reasonable prospects that they will meet the viability criteria as conditions improve or become clearer.
3. *E3*: Projects classified under E3 are those that are currently not expected to become viable in the foreseeable future. This could be due to insufficient information or unfavorable conditions. Additionally, it includes estimates for projects that are forecasted to be developed but will be unused or consumed during operations.

3.6.1.2 F Axis—Technical Feasibility and Maturity

The F Axis assesses the technical feasibility and maturity of a project. This dimension evaluates how ready and mature the technology and project plans are, which is crucial for determining whether a project can be successfully implemented.

1. *F1*: Projects in this category have confirmed technical feasibility. This means that development or operation is either ongoing or detailed studies have demonstrated that the project can be technically executed. All necessary approvals and commitments from stakeholders, including governments, should be in place.
2. *F2*: Technical feasibility is still under evaluation for projects in this category. Preliminary studies provide enough evidence to justify further investigation. Additional data and studies are required to confirm the feasibility of development.
3. *F3*: This category includes projects where technical feasibility cannot yet be fully evaluated due to limited data. Initial studies suggest potential, but more data and analysis are needed to assess the feasibility.

4. *F4*: Projects here have not yet been identified for development. This category represents the quantities of potential products that have not yet been developed or planned. The focus is on remaining quantities that, if developed, could be bought, sold, or used.

3.6.1.3 G Axis—Degree of Confidence

The G Axis reflects the confidence level in the estimates of resource quantities. This dimension categorizes the degree of certainty about the amount of resources that can be produced.

1. *G1*: High confidence in the estimate of the quantities. This means that the resource estimates are well supported by direct evidence.
2. *G2*: Moderate confidence in the estimate. The estimates are supported by some evidence, but there is more uncertainty compared to G1.
3. *G3*: Low confidence in the estimate. The estimates are less certain and are based on limited evidence.
4. *G4*: Used for prospective projects based on indirect evidence. This means that the existence of a developable resource is inferred from indirect data, and further validation is needed to confirm the estimates.

Confidence levels can also be expressed through scenarios or probabilistic analyses:

- *P90*: There is a 90% chance that the actual outcome will be at least as much as this estimate.
- *P50*: There is a 50% chance that the actual outcome will be at least as much as this estimate.
- *P10*: There is a 10% chance that the actual outcome will be at least as much as this estimate.

3.6.2 Classes and Subclasses

The combination of categories from each axis creates different Classes, which help in categorizing projects more specifically. Here's how they are structured:

1. *Viable Projects*: Projects that have been confirmed as viable in terms of environmental, social, and economic conditions and technical feasibility. These include:
   - *On production*: Projects that are currently producing or operational.
   - *Approved for development*: Projects that have been approved and are in the process of development.
   - *Justified for development*: Projects that are justified based on current information and are planned for development.
2. *Potentially viable projects*: Projects that have potential to become viable in the future, but are not yet confirmed. These include:
   - *Development pending*: Projects awaiting development decisions.
   - *Development on hold*: Projects where development has been paused or delayed.

3. *Nonviable projects*: Projects that are not expected to become viable in the near future. These include:
   - *Development unclarified*: Projects with unclear development status.
   - *Development not viable*: Projects considered unlikely to become viable due to various constraints.
4. *Prospective projects*: Projects based on indirect evidence, with potential for future viability. These include:
   - *Remaining products not developed*: Quantities from identified projects that remain undeveloped.
   - *Remaining products not developed from prospective projects*: Potential products from prospective projects that may never be developed.

3.6.3 Detailed Explanation of Subcategories

3.6.3.1 E Axis Subcategories

1. *E1.1*: Development is viable based on current conditions and future assumptions.
2. *E1.2*: Development is made viable through external support such as government subsidies.
3. *E2*: No subcategories defined. This includes projects that might become viable but are not yet confirmed.
4. *E3.1*: A forecasted product that will be consumed or unused in operations.
5. *E3.2*: Viability cannot be determined due to lack of information.
6. *E3.3*: No reasonable prospects for viability in the foreseeable future.

3.6.3.2 F Axis Subcategories

1. *F1.1*: Production is currently taking place.
2. *F1.2*: Funds are committed, and development is underway.
3. *F1.3*: Technical feasibility has been demonstrated, and necessary approvals are expected.
4. *F2.1*: Ongoing activities to justify future development.
5. *F2.2*: Activities on hold or delayed significantly.
6. *F2.3*: No plans for further development or data collection due to limited potential.
7. *F3.1*: Potential development identified with confidence but needs further testing.
8. *F3.2*: Potential development identified but requires more data for confirmation.
9. *F3.3*: Very preliminary stage, based on regional studies.

3.6.3.3 G Axis Subcategories

1. *G4.1*: Low estimate of quantities.
2. *G4.2*: Incremental amount to provide a best estimate.
3. *G4.3*: High estimate combining previous estimates.

A comparative analysis of the aforementioned classifications is presented in Table 3.3.

Table 3.3 Comparison of different classifications.

Object	Classification			
	RF		UNFC	SPE/WPC/AAPG
	current	new		
Field reserves				
Oil		Commercial reserves	E2—potential economic	Undeveloped
		Potential economic reserves	F2—Contingent development project	Justified for development
	C1 + C2	B + C1 + C2	G1 + G2 + G3	Proved + probable reserves
Gas				
Horizon 1		Commercial reserves	E2—potential economic	
		Potential economic reserves	F2—Contingent development project	Development unclarified or on hold
	C1 + C2	B + C1 + C2	G1 + G2 + G3	Contingent resources
Horizon 2		Commercial reserves	E2—potential economic	
		Potential economic reserves	F2—Contingent development project	Development unclarified or on hold
	C1 + C2	B + C1 + C2	G1 + G2 + G3	Contingent resources
Horizon 3		Commercial reserves	E2—potential economic	
		Potential economic reserves	F2—Contingent development project	Development unclarified or on hold
	C1 + C2	C1 + C2	G2 + G3	Contingent resources
Resources				
Oil		Commercial resources	E3—intrinsically economic	
			F3—project undefined	Prospect
	C3	D1	G4	Prospective resources
Gas		Commercial resources	E3—intrinsically economic	
			F3—project undefined	Prospect
	C3	D1	G4	Prospective resources

Bibliography

Aladasani, A., & Bai, B. (2010, June 8–10). Recent developments and updated screening criteria of enhanced oil recovery techniques. *Paper SPE-130726-MS Presented at the CPS/SPE International Oil & Gas Conference and Exhibition. Beijing, China.*

Arps, J. J. (1962). *Petroleum Production Handbook* (T. S. Frick, Ed.). Society of Petroleum Engineers, AIME.

Brashear, J. R., Becker, A. V., & Biglarbigi, K. N. (1988, March 23–25). Incentives, technology and EOR: Potential for increased oil recovery at lower oil prices. *Paper SPE-17454-MS Presented at the California Regional Meeting. Long Beach, SA.*

Carcoana, A. (1992). *Applied Enhanced Oil Recovery*. Prentice Hall.

Hughes, D., Larsen, S., & Wright, R. (2010). Review of low salinity water flooding. Document A10DEC015A. *Synergy Report conducted for DECC (British Department of Energy and Climate Change).*

Latil, M., Bardon, C., Burger, J., & Sourieau, P. (1980). *Enhanced Oil Recovery*. Technip.

Lovejoy, W. F., & Homan, P. T. (1965). *Methods of Estimating Reserves of Crude Oil, Natural Gas and Natural Gas Liquids*. Johns Hopkins University Press.

Society Adopts Proved Reserves Definitions. (1981, November). *Journal of Petroleum Technology*. 2113–2114. https://www.spe.org/industry/docs/Proved-Reserves-Definitions-1981.pdf

Taber, J. J. (1983, October 5–8). Technical screening guides for the enhanced recovery of oil. *Paper SPE-12069-MS Presented at the SPE Annual Technical Conference and Exhibition, San Francisco, California.*

Taber, J. J., Martin, F. D., & Seright, R. S. (1997). EOR screening criteria revisited. Part 1: Introduction to screening criteria and enhanced recovery field projects. Part 2: Applications and impact of oil prices. *SPE Reservoir Engineering*, *12*, 189–198; 199–205.

US Department of Energy, National Petroleum Council. (1984). *Enhanced Oil Recovery*. Washington, DC.

4

Oil Recovery Methods

4.1 Introduction

Hydrocarbon recovery occurs through two main processes: primary and additional, or secondary, recovery. Primary recovery is accomplished by the natural energy of the reservoir, while secondary hydrocarbon recovery is accomplished by energy brought into the reservoir from outside, by injecting fluids to maintain or increase the original energy in the reservoir.

4.2 Primary Recovery

During initial recovery, driving energy can be obtained through the expansion of rock and fluid, release and expansion of gas dissolved in oil as reservoir pressure decreases (depletion drive), expansion of the gas cap or active aquifer, gravity, or a combination of these factors (see Figure 4.1).

4.3 Secondary Recovery

Secondary hydrocarbon recovery involves introducing external energy into a reservoir through one well and extracting oil and gas from others. Typically, secondary hydrocarbon recovery involves immiscible displacement by injecting water, gas, or water–gas mixtures into injection wells (see Figure 4.2). However, the most common fluid introduced into a reservoir to maintain its energy is water due to its availability and low cost.

Typically, implementing secondary recovery early in reservoir development before the reservoir's natural energy is depleted resulting in a higher oil recovery factor than that achieved by natural mechanisms during primary oil recovery.

Methods for Enhanced Oil Recovery: Fundamentals and Practice, First Edition.
Baghir A. Suleimanov and Elchin F. Veliyev.

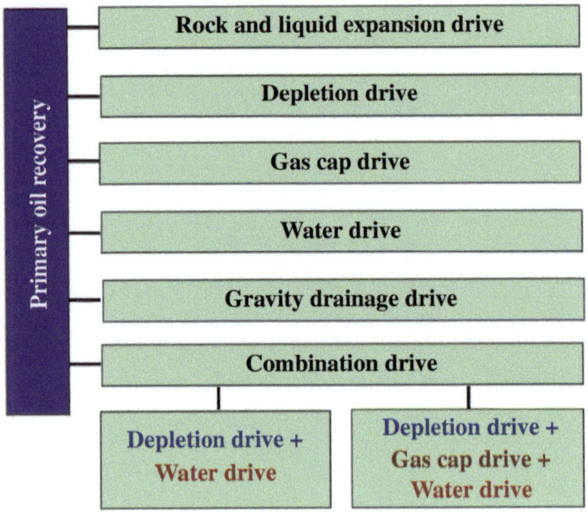

Figure 4.1 Primary oil recovery.

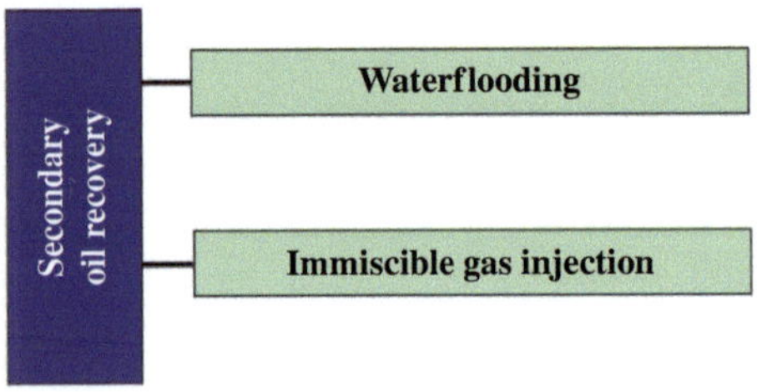

Figure 4.2 Secondary hydrocarbon recovery.

4.3.1 Water Injection

Waterflooding is carried out by pumping water into a series of injection wells and withdrawing hydrocarbons from surrounding wells. Waterflooding is carried out to achieve any of the following goals or combinations thereof:

- maintaining reservoir pressure,
- disposal of produced water, and
- creation of a water pressure regime to displace hydrocarbons from injection wells to production wells.

Water injection is usually carried out using various flooding systems, such as pattern flooding, peripheral flooding, and crystal flooding. Areal flooding is used in reservoirs with a small depth but a large area. Figure 4.3 shows possible options for the location of wells during areal flooding. Economic factors are the main criteria for choosing a specific well location geometry; these factors include the cost of drilling new wells and the cost of converting production wells to injection. The most commonly used are direct-line drive and staggered-line drive rows of injection wells (see Figure 4.3) because they require minimal investment. However, if it is necessary to increase the injection rate, a seven- or nine-point flooding system is usually used (see Figure 4.3).

With contour flooding, injection wells are located around the perimeter of the reservoir. Figure 4.4 shows two cases of contour flooding in the presence of underlying aquifers. In the anticlinal reservoir shown in Figure 4.4a, injection

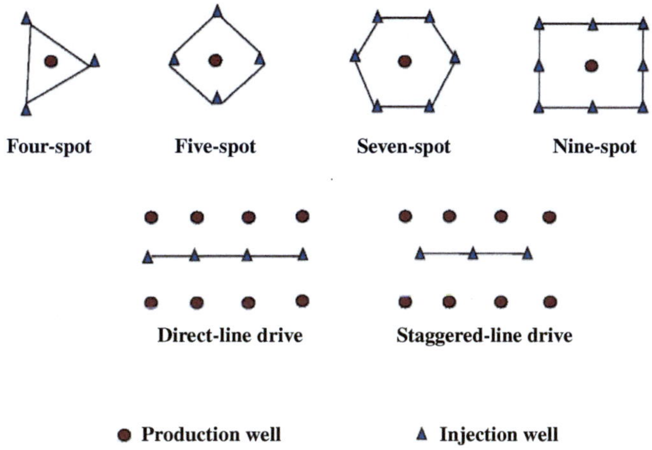

Figure 4.3 Schemes for areal waterflooding.

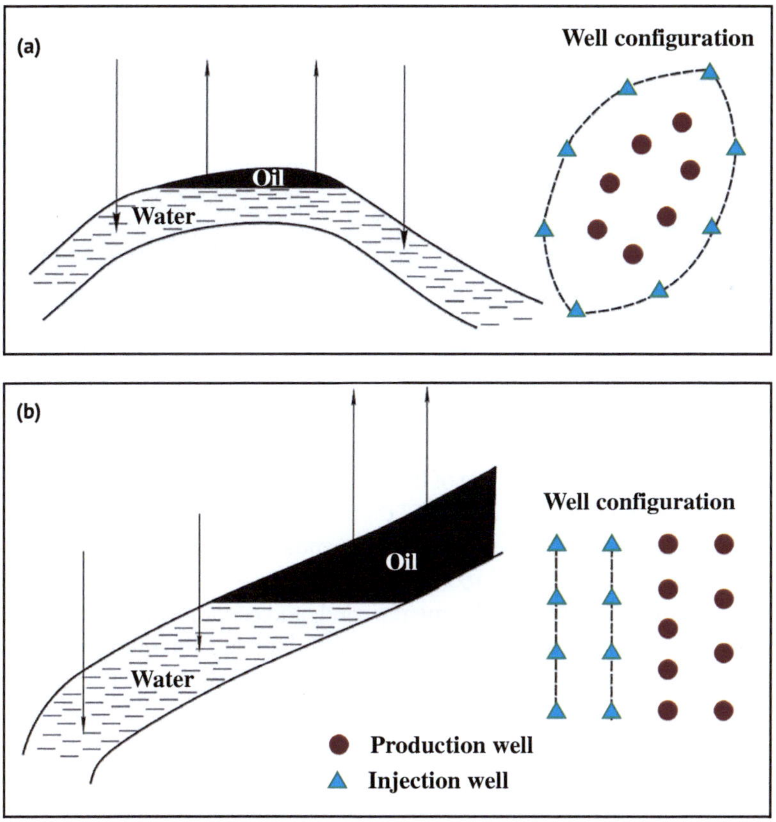

Figure 4.4 Well location for contour flooding.

wells are placed in such a way that water enters the aquifer or the zone near the water–oil contact, displacing oil to production wells located in the upper part of the reservoir. For the monoclinal reservoir shown in Figure 4.4b, injection wells are located below the oil–water contact to take advantage of gravity.

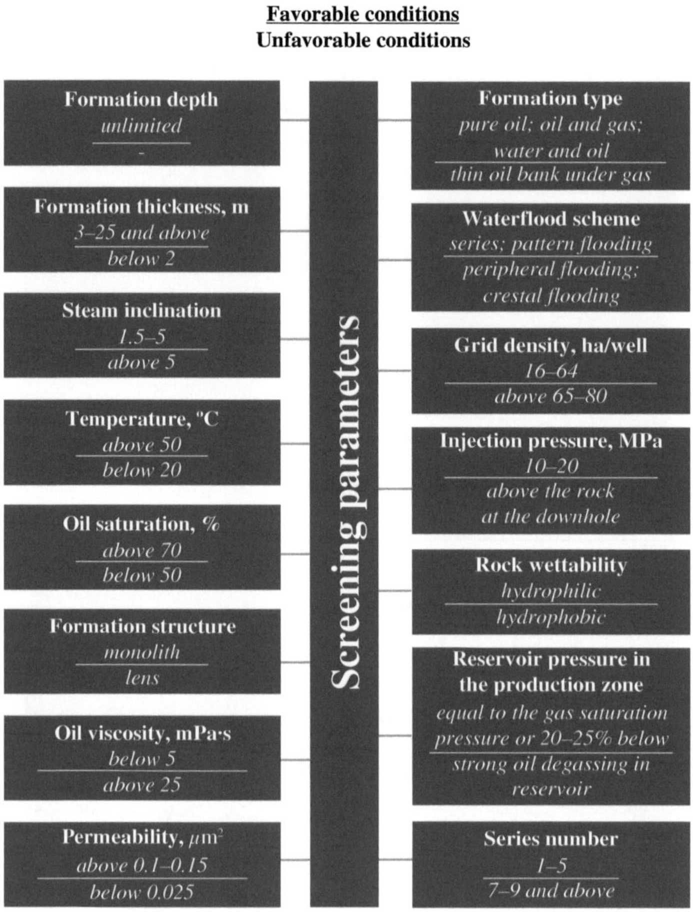

Figure 4.5 Waterflooding applicability criteria.

In reservoirs with an acute angle of incidence, injection wells are located in the roof part to effectively displace the oil located there. This system is known as roof flooding. In any case, the location of injection wells depends on the following factors: physical and chemical characteristics of rocks and presenting fluids, reservoir heterogeneity, injection pressure, planned development period, and economic indicators.

The use of waterflooding makes it possible to increase the oil recovery factor by 10–30%. The applicability criteria for waterflooding are shown in Figure 4.5.

4.3.2 Gas Injection

Immiscible gas injection is carried out to maintain reservoir pressure, slow down the rate of decline in the natural conditions of the reservoir, and sometimes support the gravity regime. The immiscible gas is usually introduced in slugs alternating with water. For immiscible displacement, associated petroleum gas, nitrogen, or flue gases are used. In this case, the gas injected into the well behaves in the same way as in the gas cap mode: the gas expands, displacing oil to the production wells. To carry out gas injection, the use of high-pressure compressors is required.

Injection of immiscible gases is inferior in efficiency to flooding, but in some cases, it is the only possible method of secondary oil recovery. This is especially evident in low-permeability reservoirs and rocks containing swelling clays.

4.4 Tertiary Recovery

Despite the increase in the oil recovery factor as a result of the use of secondary oil recovery methods, in particular, the most effective—waterflooding, a significant amount of residual oil remains in the reservoir. According to the work of Aurel Carcoana (*Applied Enhanced Oil Recovery*), for reservoirs with low permeability (tight oil reservoirs) or containing heavy oils, the final oil recovery when using primary and secondary oil recovery is 5–10%, developed in dissolved gas mode—10–25%, developed under partially water-pressure mode, gas injection or gravity mode—25–40%, developed by conventional waterflooding—40–55%. To increase the oil recovery factor above the specified values, tertiary methods for extracting hydrocarbons are used. The purpose of using tertiary EOR methods is to increase final oil recovery by:

- increasing the reservoir sweep efficiency due to:
  - reducing the ratio of the mobility coefficient of the injected and displaced liquids,
  - blocking of washed high-permeability water-saturated zones and diversion of injected liquid into low-permeability oil-saturated zones of the reservoir;
- regulation of molecular surface forces in the reservoir due to:
  - reducing the interphase surface tension between oil and displacing liquid,
  - reducing the influence of capillary forces,
  - changes in reservoir rock wettability,
  - changes in disjoining pressure;
- combination of these processes.

Tertiary hydrocarbon recovery processes are carried out after primary and secondary recovery processes have already been carried out. In this case, fluids are introduced into the formation that are different from those used during secondary recovery (ordinary water and immiscible gas; see Figure 4.6). EOR is a broader concept since EOR can be applied at any stage of field development, including primary, secondary, and tertiary recovery of hydrocarbons. Thus, EOR can be implemented as a tertiary process if it follows flooding or immiscible gas injection, but it can also be a secondary process if it directly follows primary hydrocarbon recovery. However, most EOR projects are implemented after waterflooding. At this stage, it is important to establish the difference between EOR and improved oil recovery (IOR) to avoid misunderstandings. The term "IOR" refers to the application of any EOR operation or any other advanced oil recovery technology that is implemented during any ongoing oil recovery process. Examples of IOR are any production enhancement techniques that are used during primary, secondary, or tertiary recovery of hydrocarbons. Other examples of IOR applications include hydraulic fracturing, scale control, acidizing, infiltration drilling, and horizontal wells (see Figure 4.7).

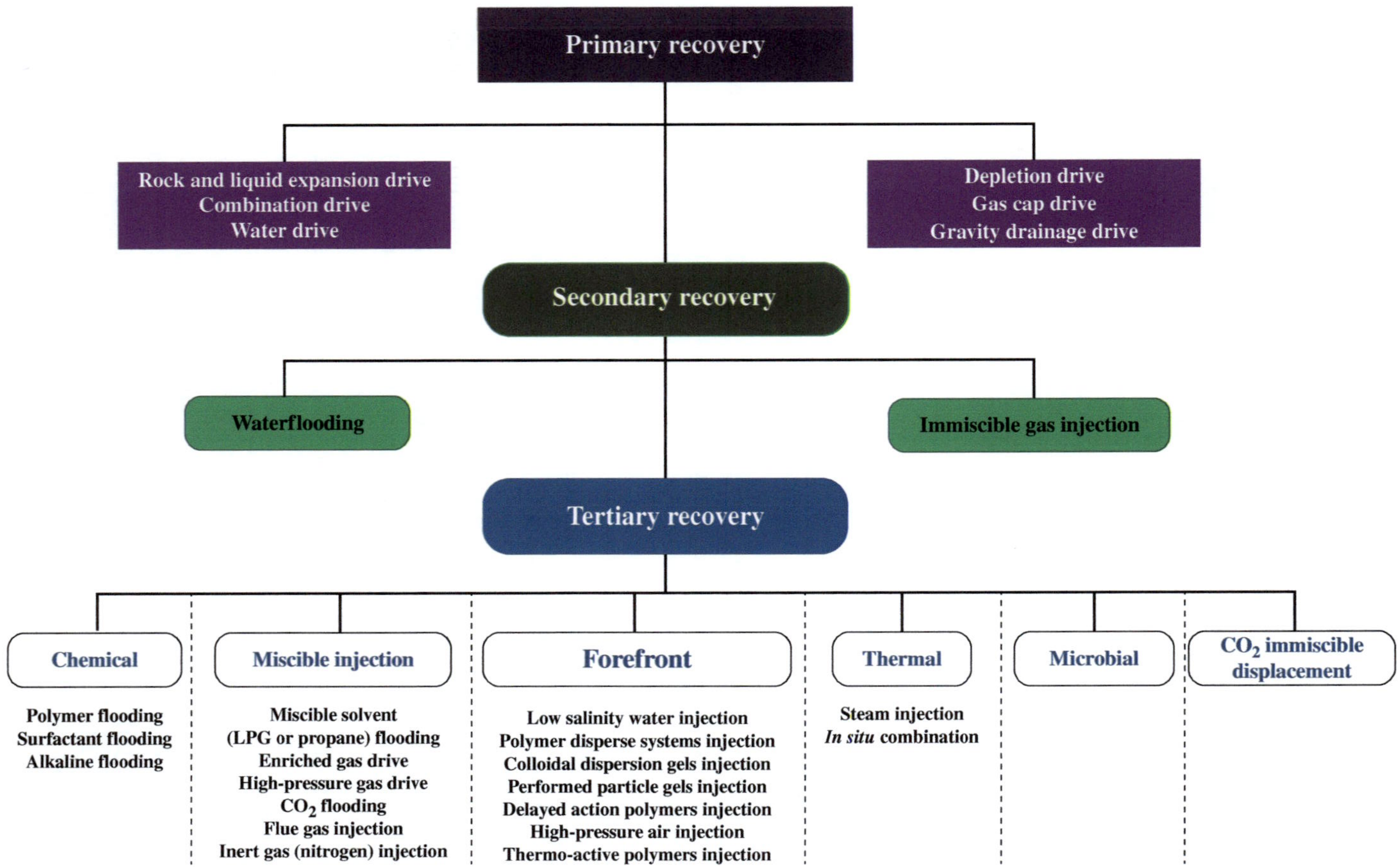

Figure 4.6 Oil recovery stages and applied extraction technologies.

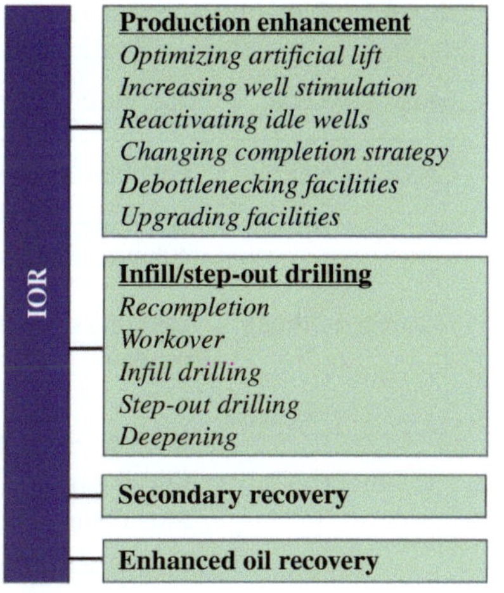

Figure 4.7 EOR processes.

Figure 4.8 shows the currently used EOR. Table 4.1 shows the EOR mechanism for the most commonly used EOR methods.

Chemical methods of EOR are characterized by the addition of chemical reagents to the injected water in order to change the physicochemical properties of the fluid or interfacial tension, which are more favorable for displacing oil. Polymer flooding uses polyacrylamides or polysaccharides—it is a conceptually simple and inexpensive solution. Although the technological efficiency of the method is not so high, it is used for commercial purposes. Surfactant flooding is a complex and expensive technology that requires detailed laboratory testing when drawing up a field development project. Alkaline flooding is used in reservoirs containing certain types of oil with a high acid number.

Mixing methods have the greatest potential for enhancing oil recovery for fields containing low-viscosity oils. Among these methods, miscible CO_2 flooding is the most promising.

Thermal methods are characterized by the supply of coolant to the reservoir to reduce the viscosity of oil, ensuring its higher mobility and increasing the efficiency

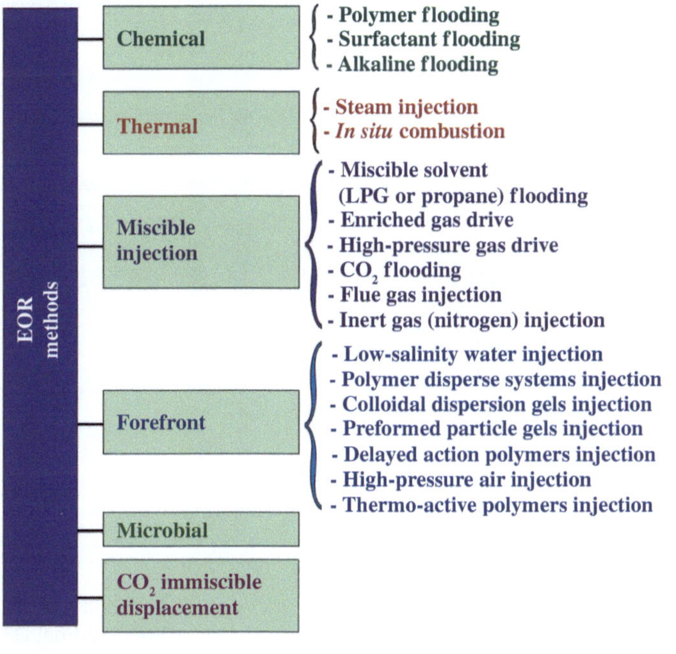

Figure 4.8 Methods for EOR.

Table 4.1 EOR mechanism for the most commonly used EOR methods.

Generalized EOR	Specific oil recovery methods	Basic mechanism
Chemical methods	Polymer flooding Surfactant flooding Alkaline flooding	Improvement of sweep efficiency; improvement of displacement efficiency
Miscible methods	Miscible solvent flooding (LPG or propane) Enriched gas drive High-pressure gas drive CO_2 flooding Flue gas injection Injection of inert gas (nitrogen)	Improvement of displacement efficiency
Thermal methods	Steam injection *In situ* combustion	Improvement of both sweep efficiency and displacement efficiency

of displacement to production wells. Steam injection, which has been used since the 1960s, is the most advanced of all EOR technologies in terms of practical implementation in field conditions. The effectiveness of steam injection can be estimated with less uncertainty than any other EOR. *In situ* combustion is typically used for tanks containing heavy oils. Quite a lot of projects were implemented, but only a few of them were commercially successful.

Figure 4.9 shows EOR and ultimate oil recovery for the most commonly used EOR methods.

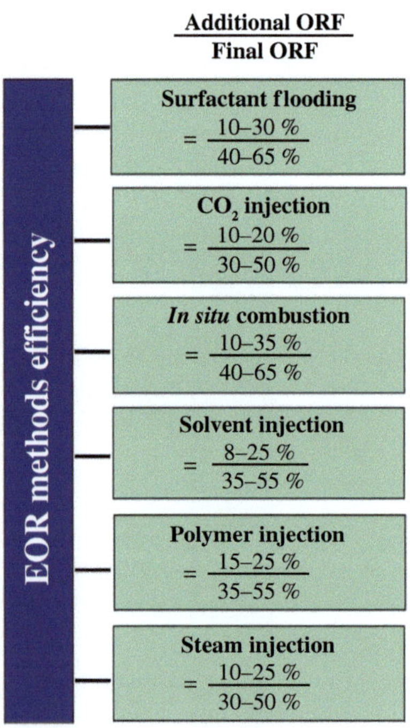

Figure 4.9 Additional oil recovery factors (ORF) and final ORF figures for the most commonly used EOR methods.

4.5 Sweep Efficiency

The ultimate goal of EOR processes is to increase the overall oil displacement efficiency, which is determined by microscopic and macroscopic efficiency. Microscopic efficiency is realized on the scale of formation pores and is determined by the interaction of the displacing fluid with the rock and the displaced fluid. For example, microscopic efficiency can be increased by reducing capillary forces, interfacial tension, wettability changes, and disjoining pressure.

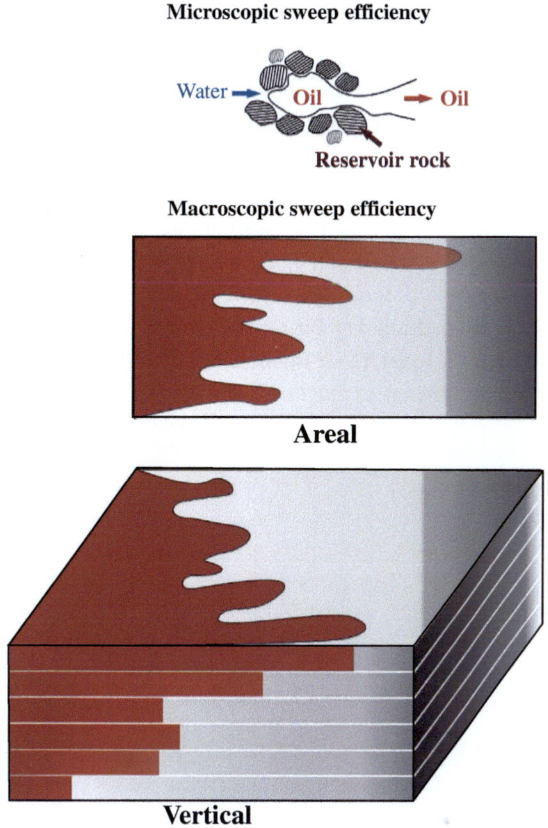

Figure 4.10 Representation of microscopic and macroscopic sweep efficiencies.

Macroscopic, or volumetric, efficiency is realized on the scale of the entire tank. The efficiency of volumetric displacement is determined by the efficiency of displacement both across the reservoir area and vertically, as well as by how effectively the injected fluid displaces oil to the production wells. Figure 4.10 shows the sweep efficiency diagram: microscopic and macroscopic (areal and vertical sweep).

The overall displacement efficiency of any oil displacement process can be increased by improving the mobility ratio of the displacing and displaced fluids. The mobility coefficient is defined as the ratio of the mobility of the displacing fluid divided by the mobility of the displaced fluid.

The mobility coefficient is determined as follows:

$$M = \frac{k_1/\mu_1}{k_2/\mu_2} = \frac{k_1\mu_2}{k_2\mu_1} \tag{4.1}$$

where index 1 is the displacing fluid, index 2 is the displaced fluid, k_1 and k_2 are the permeability of the displacing and displaced fluids, respectively, μ_1 and μ_2 are the viscosity of the displacing and displaced fluids, respectively.

Bibliography

Carcoana, A. (1992). *Applied Enhanced Oil Recovery*. Prentice Hall.

Craft, B.C. & Hawkins, M.F. (1991). *Applied Petroleum Reservoir Engineering*. Prentice Hall.

Lyons, W.C. & Plisga, G.J. (2005). *Standard Handbook of Petroleum and Natural Gas Engineering*. Gulf Professional Publishing.

Romero-Zerón, L. (2012). Advances in enhanced oil recovery processes. In L. Romero-Zerón (Ed.), *Introduction to Enhanced Oil Recovery (EOR) Processes and Bioremediation of Oil-Contaminated Sites*. IntechOpen.

Satter, A., Iqbal, G., & Buchwalter, J.L. (2008). *Practical Enhanced Reservoir Engineering: Assisted with Simulation Software*. Penn Well.

5

Thermal Enhanced Oil Recovery (EOR)

5.1 Introduction

Due to the fact that industrial oil production has been going on for more than 150 years, most of the fields currently being developed contain hard-to-recover oil reserves. This type also includes fields with heavy, highly viscous oils. Low mobility of highly viscous oils in a porous environment significantly reduces the efficiency of field development. Oil viscosity depends significantly on temperature. Therefore, to effectively extract oil with a viscosity of more than 25–50 mPa·s, thermal effects on the formations are required to reduce its viscosity.

Thermal recovery processes use thermal energy in one form or another to both increase the temperature of the reservoir and reduce the viscosity of the oil, thereby effectively displacing oil to production wells. Thermal oil recovery processes around the world are the most advanced EOR processes, which are classified into two main technologies as follows:

1. Steam injection
2. *In situ* combustion

5.2 Steam Injection

The first experimental studies on steam displacement of oil were carried out by S. L. Stovall in the 1930s. Using a reservoir model, he obtained a steam oil displacement coefficient of 98%. The results of laboratory studies provided the basis for the pilot industrial implementation of steam injection at one of the fields in Texas. In 1939, Lapuk published a review of early work on thermal effects. After a short break in 1951, Winkler A.K. conducted industrial experiments on the use of superheated steam for the production of high-viscosity oil from the Leoprechting formation (Austria). The temperature of the injected steam at the bottom was brought to 200–220 °C. Field experience in using steam in this field has shown results that are within the profitability range. Since 1960, the USSR and the USA began to widely introduce steam injection into oil reservoirs. Steam injection is the most common method of increasing oil recovery from formations containing highly viscous oils.

Methods for Enhanced Oil Recovery: Fundamentals and Practice, First Edition.
Baghir A. Suleimanov and Elchin F. Veliyev.

Continuous steam injection involves producing 80% steam at the wellhead and injecting it into injection wells. When steam enters the tank, it heats the oil and reduces its viscosity. Moving through the formation, the steam cools and condenses. Heat from steam and hot water vaporizes lighter hydrocarbons or turns them into gases. These gases move ahead of the steam, cool, condense, and dissolve in the oil. Thus, gases and steam provide additional energy, helping to effectively displace oil. Hot water also forces lower-viscosity oil toward production wells.

In addition to continuous, **cyclic steam injection is also used**, which is also called *puff and huff.* Cyclic steam injection is carried out by periodically injecting steam into the oil reservoir through production wells, keeping them closed for some time, and resuming their operation to extract oil of low viscosity. The purpose of the technology is to heat the near-wellbore zone of the formation and on the fluids saturating it, a concomitant decrease in oil viscosity, an increase in pressure, and a subsequent increase in oil flow to the wells.

5.2.1 Process Mechanism

Continuous injection of steam is carried out in formations with low temperatures and high viscosity of oil through steam injection wells inside the oil-bearing contour. Steam, which has a high heat capacity (more than 5,000 kJ/kg at 230 °C), introduces a significant amount of thermal energy into the formation, which heats the formation and reduces the viscosity of the oil. In this case, three zones are formed in the formation, differing in temperature and saturating phases (Figure 5.1).

1. *Zone 1*: Characterized by high temperatures, varying from steam temperature to the temperature at which condensation begins (200–400 °C). The formation temperature increases and becomes almost equal to the steam temperature. In this zone, oil is distilled, light fractions are released, and they are displaced by steam through the formation. Thus, joint filtration of light fractions of oil and steam occurs.
2. *Zone 2*: In this zone, the steam cools and condenses, and the temperature changes from the temperature at which condensation begins to the formation temperature. The hot water formed as a result of steam condensation displaces oil under non-isothermal conditions.

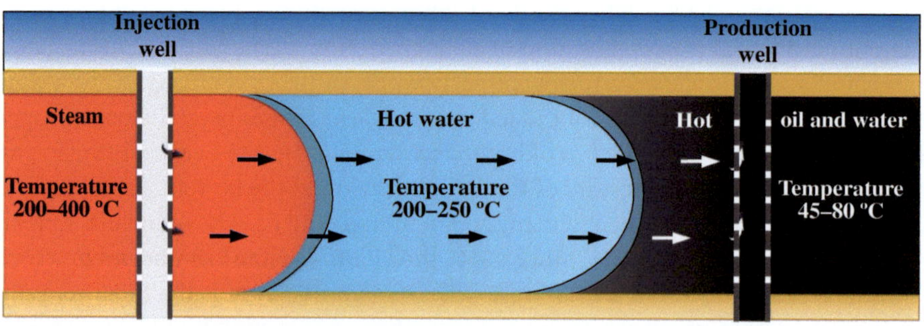

Figure 5.1 The mechanism of the steam injection process into the reservoir.

3. *Zone 3*: In this zone, the temperature practically corresponds to the reservoir temperature. In this zone, oil is displaced by the formation of water. However, since the specific volume of steam is much greater than that of water, all other things are equal, and the rate of filtration of water is higher.

Increase as the process continues, and the third zone decreases. As a result, water and steam break through into production wells, which precludes the further continuation of the process of steam injection into the formation.

When the formation is heated, oil distills, a decrease in viscosity and volumetric expansion of all formation agents, a change in phase permeabilities, formation wettability, and the mobility of oil and water occur.

In the process of continuous steam injection, the effect is achieved primarily by reducing the viscosity and distillation of oil, which leads to an increase in the mobility of the displacement front. Oil expansion and formation wettability have little effect on the efficiency of the process.

The target of continuous steam injection in formations is to heat the formation and oil in the bottomhole zones of production wells, reduce the viscosity of oil, increase pressure, facilitate filtration conditions, and increase the flow of oil to the wells.

The mechanism of processes occurring in the formation during cyclic steam injection is, in principle, the same as with continuous injection, with the difference being that during the period of keeping the well closed, reverse capillary impregnation occurs, ensuring the displacement of oil from small to large pores. Indeed, when steam is injected into a formation, it first penetrates into highly permeable interlayers. Therefore, during the holding period in the heated zone of the formation, an active redistribution of saturation occurs due to capillary forces: hot water displaces low-viscosity oil from small capillaries and weakly permeable interlayers into large pores and highly permeable layers.

In the process of cyclic injection of steam into the formation, the effect is ensured primarily by increasing the permeability of the bottomhole zone of the formation due to a decrease in oil viscosity and a concomitant increase in well production. In addition, there is an increase in the drainage zone of the near-wellbore formation zone (i.e., impact coverage).

Since the mid-1980s, with the development of horizontal drilling technology in Canada, **steam-assisted gravity drainage (SAGD) technology**, or SAGD, was developed and became the industry standard for bitumen extraction. Using this technology, an injection well is drilled parallel to and above the production well. When steam is injected into an injection well, it heats the bitumen or extra-heavy oil, reducing its viscosity. Gravity promotes the flow of bitumen or extra-heavy oil into the production well.

5.2.2 Applicability Criteria

The main disadvantage of the steam injection method is the use of high-purity water, with a content of mechanical impurities of less than 0.005 mg/L; it must be free of organic substances, dissolved gas, as well as magnesium and calcium cations. Otherwise, it will be impossible to obtain steam with high saturation (80%) and a heat capacity of 5,000 kJ/kg. Water purification for the purpose of using it in steam generators requires the use of expensive chemical reagents, the cost of which reaches 30–35% of the total cost of steam production.

At the low effective thickness of the formation, the relative heat exchange surface increases, as a result of which heat losses greatly increase, so the formation is heated less uniformly and to a lower temperature. The same happens with a large distance between wells; therefore, the density of the well network should be at least 1 hectare/well.

Injecting steam into fields with a depth of no more than 900 m and an oil viscosity of at least 1,000 mPa·s can ensure final oil recovery of up to 55% against a possible 15–18% during waterflooding.

Under the most favorable conditions, 2.5–3.5 tonnes of steam are consumed per tonne of additional oil. At the same time, for 1 tonne of oil spent on producing steam in a steam generator, 4–5 tonnes of oil are produced. This is a completely satisfactory specific efficiency of the process of steam injection into the reservoir.

In weakly cemented sand formations, the processes of displacement of oil by steam are accompanied by the removal of sand, which complicates their exploitation.

Reservoir coverage during oil displacement by steam is significantly lower than during flooding since the viscosity of steam is significantly lower than the viscosity of water, which leads to a violation of the stability of the displacement front. This is especially evident when the oil viscosity is more than 1,000 mPa·s.

The most important problem when pumping steam into a reservoir is heat loss in the casing strings of injection wells, which can reach 4% per 100 m depth. For wells more than 900 m deep, heat losses in injection wells can reach 40%. Thermal insulation of injection pipes reduces these losses, but despite this, they remain quite high, which limits the use of steam injection in wells deeper than 900 m.

It should be noted that the most important indicator of the injection process is the injection rate, with an increase in which heat losses are greatly reduced.

With cyclic steam injection, the final oil recovery factor can reach 10–30%. The main limitations for cyclic steam injection are the porosity of the formation (it must be at least 25%), the thickness of the formation (at least 6 m), and the depth of the deposit (no more than 900 m).

With the combined use of continuous and cyclic steam injection, the efficiency of field development increases significantly, and the increase in recoverable reserves increases to 40%.

Figure 5.2 shows the main criteria for the applicability of steam injection into oil reservoirs.

SAGD technology, or steam-injected gravity drainage, is used in fields with unconsolidated rocks with high vertical permeability for the production of bitumen and super-heavy oils.

5.2.3 Field Implementation

Steam injection has become widespread in countries with high-viscosity oil fields, mainly in Canada, the former Soviet Union, the USA, and Venezuela. Several projects have also been implemented in Brazil and China. It should be noted that SAGD technology is used to extract bitumen, which is being implemented in Alberta fields.

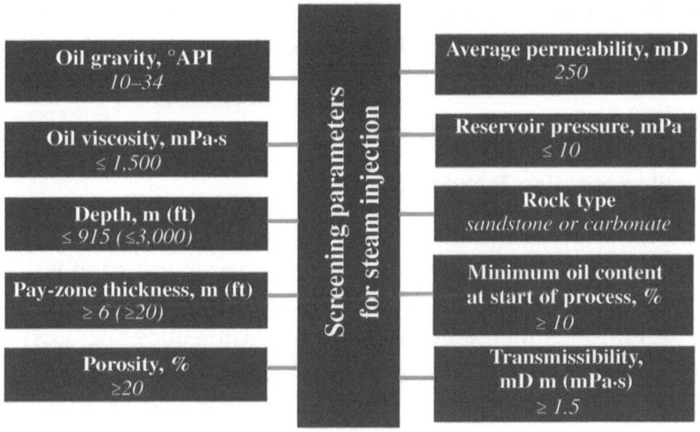

Figure 5.2 Steam injection applicability criteria.

Injecting steam into the reservoir to increase the oil recovery factor began about 60 years ago. In the last 50 years, steam injection has been used in the Mene Grande and Tia Juana fields in Venezuela and the Yorba Linda and Kern River fields in California. The continuous steam injection was later used in the Crud E fields in Trinidad, Schoonebeek in the Netherlands, and Alto do Rodrigues in Brazil.

In the former USSR, steam injection into formations began to be carried out in the 1960s in Azerbaijan in the Khorasani area of the Balakhani-Sabunchu-Ramany field, the Binagadi-Kirmaki area, the Binagadi-North field of the Binagadi field, and the Pirallahi field; in the Krasnodar region and Ukraine. A technology was proposed and implemented at the Okha field (Sakhalin Island) to displace oil by injecting a thermal fringe, which is subsequently moved by pumping water. The technology showed high efficiency and was also implemented in the Yaregskoye and Kenkiyakskoye fields. Since 1982, steam injection began at the Karazhanbas field (Kazakhstan). Steam was pumped into 27 injection wells. The volume of steam injection exceeded 400 thousand tonnes/year, and oil production using the method amounted to more than 150 thousand tonnes/year.

At the pilot site of the Usinsky field (Russia), for the first time in world practice, equipment, and technology were mastered that made it possible to pump high-temperature steam (320–330 °C) into carbonate reservoirs located at a depth of 1,400 m.

Modifications of the steam injection method are used, in particular a combination of continuous and cyclic steam injection into formations. It was used in the Kern River, San Ardo, and Waig Wolf fields (California, USA). The deposit depth is 200–600 m, the reservoir thickness is 25–70 m, and the oil viscosity is more than 3,000 mPa·s. Since the 1960s, these fields have used a combination of continuous and cyclic steam injection in more than 2,500 wells per year. As a result of implementation, recoverable oil reserves reached 35–37% of geological ones.

SAGD technology is being implemented in bitumen and extra-heavy oil fields, especially in Canada and Venezuela. Industrial testing of the technology was carried out in China, the USA, and Venezuela. However, on an industrial scale, this technology is used mainly in Canada. Since 1999, gravity drainage with steam injection began to be used for the extraction of bitumen in Tatarstan (Russia) at the Ashalchinskoye field; up to 320 tonnes of bitumen are produced per day at this field.

5.2.4 Implementation Technology

The technology of steam injection into the reservoir is carried out in two versions. In the traditional version (see Figure 5.3), steam with a saturation of 80% is produced in a steam generator (see Figure 5.4) installed at the wellhead and supplied to the injection well through thermally insulated pipes (to reduce heat loss). The well is also equipped with a heat-resistant packer.

The second option for steam injection involves installing a downhole steam generator (see Figure 5.5).

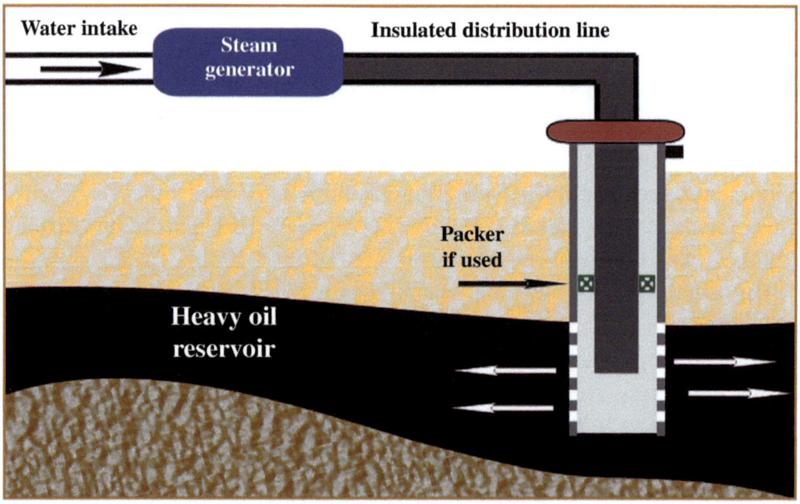

Figure 5.3 Technology of steam injection from the wellhead.

Figure 5.4 Steam generator.

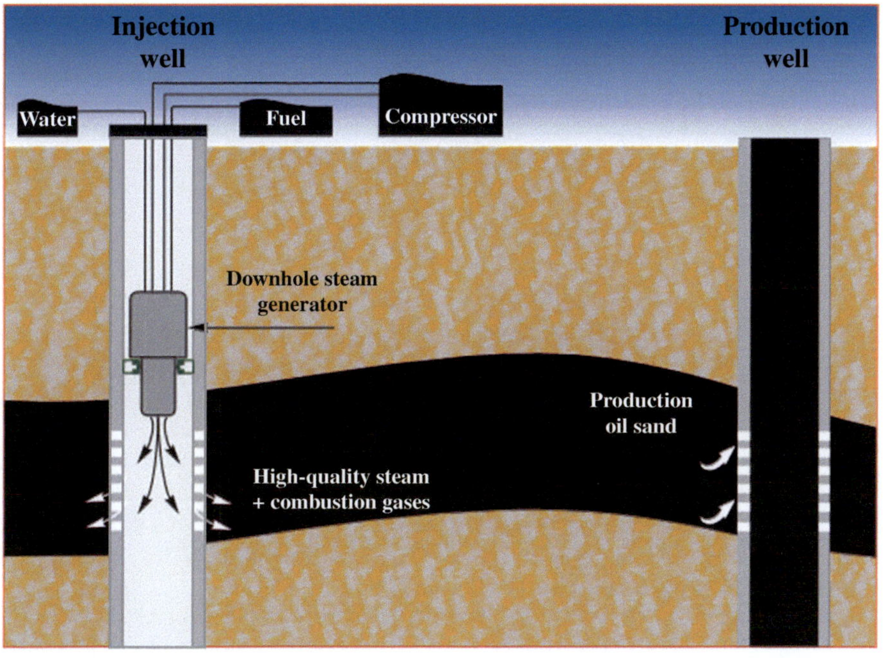

Figure 5.5 Steam injection technology using a downhole steam generator.

Excessive air pollution and heat losses of up to 32% when using wellhead steam generators led to the development of direct combustion downhole steam generators. In this case, the combustion process occurs at high pressure, then water and a mixture of carbon dioxide and nitrogen are added, as a result of which the resulting steam goes directly into the tank.

The technology of **steam-cyclic stimulation** of formations is carried out as follows:

1. *Stage 1*: Steam is pumped into the production well (see Figure 5.6) for several days or weeks (up to 1 month) in a volume of 30–100 tonnes per meter of formation thickness. The volume of injected steam strongly depends on the viscosity of the oil and reservoir energy: the higher the viscosity and the lower the reservoir energy, the more steam is injected, other things being equal.
2. *Stage 2*: Steam injection is stopped, and the well is closed for one to two weeks to complete the process of heat and mass transfer and capillary impregnation. The holding period of a closed well depends on the amount of reservoir energy: the lower the reservoir energy, the shorter the holding period since it is important to use steam pressure to displace oil. During the soaking period, the steam in the formation condenses, forming a zone of hot water and low-viscosity oil.
3. *Stage 3*: The production well is put into operation and operated to the maximum cost-effective oil flow rate within 8–12 weeks.

Following the first, the second and subsequent cycles are carried out with a longer exposure time. This cyclical process of steam injection, soaking, and operation can be repeated until oil production stops.

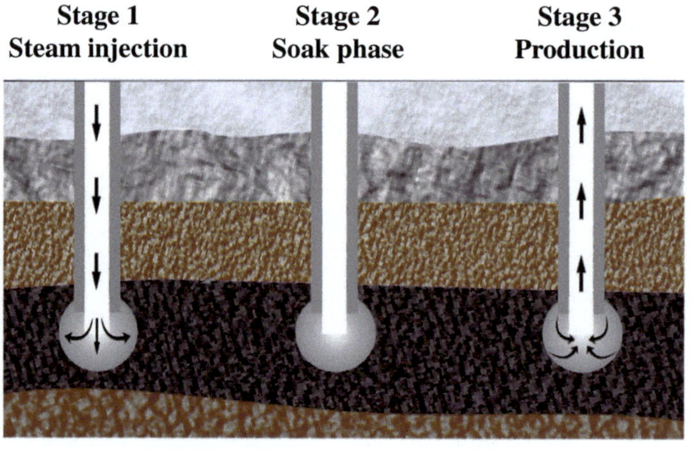

Figure 5.6 Cyclic steam injection stages.

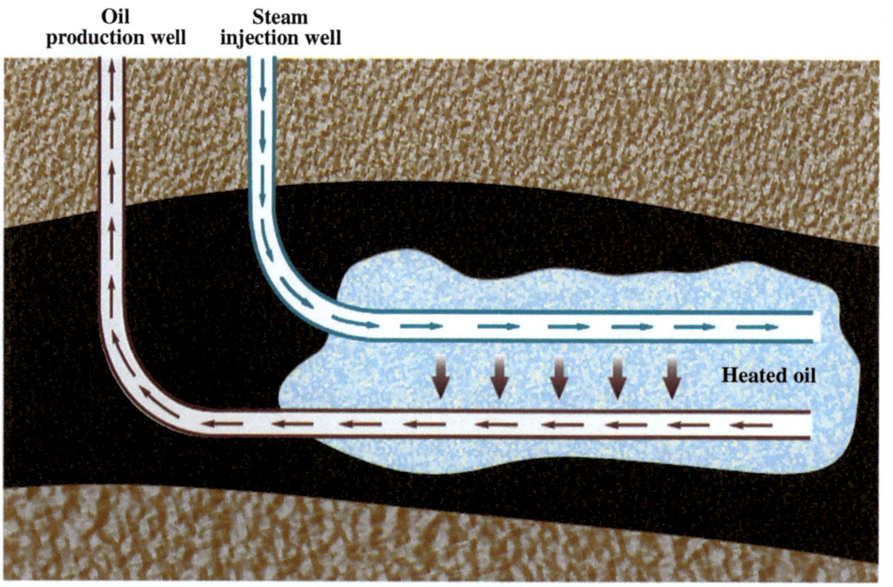

Figure 5.7 Well arrangement and schematic of oil extraction process during SAGD.

SAGD technology, two horizontal wells are drilled, located parallel to one another, with a distance between them of 5 m and a horizontal well length of up to 1,000 m (see Figure 5.7). After a preparatory period, during which steam is circulated in both wells, steam is injected into the upper well, and bitumen is taken from the lower well. As a result of steam injection into the upper well, the formation warms up, the viscosity of bitumen is reduced, and it flows into the production well, along with condensing steam. Due to the fact that bitumen is constantly in contact with steam, heat losses are negligible.

5.3 *In situ* Combustion

In situ combustion is a process in which oxygen-containing gas is introduced into a reservoir as an oxidizer, where it reacts with oil to create a high-temperature, self-sustaining combustion front (200–650 °C) that propagates through the formation. Ignition can be caused by electrical or gas igniters and can also occur spontaneously (autoignition) if the crude oil is sufficiently reactive. In most cases, the air is injected into the reservoir as an oxidizer, and the fuel that supports the combustion of oil is a residual product obtained by a complex process of cracking, coking, and steam distillation that occurs before the combustion front. The heat from combustion helps reduce the viscosity of the oil, release gas, and evaporate water to form steam. Steam, hot water, and gas displace oil into production wells (see Figure 5.8). *In situ* combustion is possible if the interaction of crude oil and rock produces a sufficient amount of fuel (or it is additionally introduced into the formation) to maintain the combustion front. There are two main types of *in situ* combustion: direct-flow combustion (forward combustion)—in this case, the combustion front moves in the direction of movement of the oxidizer; reverse combustion (reverse combustion)—the combustion front moves in the direction opposite to the movement of the oxidizer.

Soviet scientists A. B. Sheinman, K. K. Dubrovay, and others were the first to propose the use of *in situ* combustion for oil extraction. In 1934, for the first time in world practice, they carried out the first successful industrial test of *in situ* combustion at the Neftyano-Shirvanskoye field in the Maikop oil-bearing region, and in 1935 they repeated the success at the Chusovskie Gorodoki field in the Perm Territory. Subsequently, this method of increasing oil recovery became widespread in the former USSR.

Since the 1960s, *in situ* combustion has been used in the USA and Venezuela. In 1964, the implementation of *in situ* combustion began at the Suplacu de Barcau field, Romania). The successful implementation of the technology in this field made it possible to increase the oil recovery factor to 56%. In the last 20 years, *in situ* combustion technology has also been successfully applied in India and Canada.

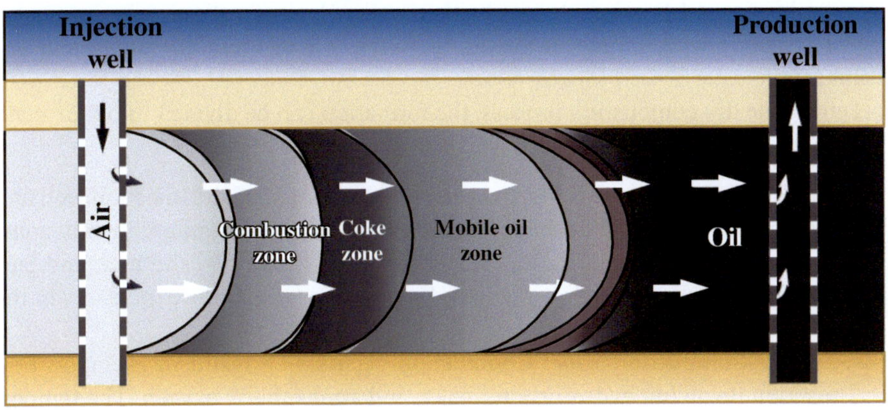

Figure 5.8 *In situ* combustion process.

5.3.1 Process Mechanism

The essence of *in situ* combustion is the combustion of oil in the reservoir, and by burning part of the oil, the remaining part is displaced. Oil combustion is carried out near the bottom of the injection well by heating and pumping air (in this case, the oxidizing agent is atmospheric oxygen). A necessary condition for oil combustion in a reservoir is preheating of the rock (on average up to 65–95 °C). Therefore, downhole electric heaters, gas burners (fire heaters), as well as various chemical reactions are used to ensure combustion. The creation of a combustion source, subsequent injection of air, and removal of its products (N_2, CO_2, etc.) from the combustion source ensure the maintenance of *in situ* combustion. The heat released during combustion is sufficient to increase the temperature of the rock to a sufficiently high value (200–650 °C), ensuring the movement of the combustion front through the formation. During the process of *in situ* combustion, part of the oil (6–25%) is burned. The amount of oil burned depends significantly on the properties of oil and rock, and it is known that the volume of burned oil increases with increasing its density and viscosity and decreases with increasing permeability of the rock.

There are three main options for *in situ* combustion: forward flow with air injection (also called dry forward combustion); forward flow with injection of air and water (also called wet forward combustion); and reverse flow with air injection. With forward combustion, ignition and air injection are carried out in the same well. With wet forward combustion, water is additionally pumped in, either together with air or sequentially. Reverse combustion is carried out through two wells. In the first well, ignition is carried out, and through the injection well, air is pumped into the unheated part of the formation, which moves toward the combustion source. Thus, oil moves toward the ignition well, which becomes a production well. Reverse combustion is carried out in deposits with super-heavy (immobile) oil or bitumen, i.e., where direct-flow combustion is difficult to carry out.

5.3.1.1 Dry Forward Combustion

In this version of the technology, the combustion front acts as a piston, displacing noncombustion oil fractions to production wells. Heavy fractions turn into coke, which, reacting with oxygen in the air, plays the role of fuel that supports *in situ* combustion.

During the dry combustion process, the formation can be divided into four main zones (see Figure 5.9).

1. *Zone 1*: The burnt zone is located between the bottom of the injection well and the combustion front. Ideally, dry formation material remains in this zone, without any impurities. Here, the injected air is heated by the rock and part of the combustion energy is recovered so that the temperature increases in the direction of the flow.
2. *Zone 2*: Combustion zone: Combustion occurs as a result of a combustion reaction involving hydrocarbons, coke, and atmospheric oxygen. The temperature reached in this zone mainly depends on the properties of the rock and the liquids and gases saturating it.

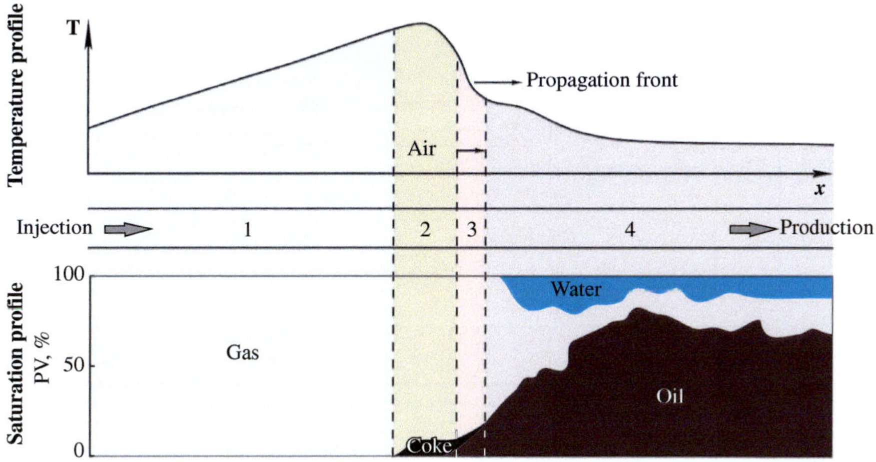

Figure 5.9 Zoning for dry forward burning.

3. *Zone 3*: Coke formation zone: Heavy fractions of oil that have not been displaced or evaporated undergo pyrolysis.
4. *Zone 4*: Characterized by a significant drop in temperature and no further noticeable chemical changes occur. The zone is engulfed by gaseous combustion products and displaced liquids, and the following phenomena occur:
   - in the area adjacent to the reaction zone, sequential evaporation of light fractions of oil and condensation of coherent (interstitial) water formed during the combustion process occurs;
   - in the area where the temperature is below the condensation temperature of the water, there is a zone with water saturation higher than the initial water saturation (water bank), while water is introduced into the zone with oil saturation higher than the initial one (oil bank) and displaces oil. If the oil is very viscous, this can lead to formation blockage. Beyond the oil-saturated zone, reservoir characteristics (pressure, temperature, etc.) correspond to the initial conditions.

5.3.1.2 Wet Forward Combustion

During the dry combustion process, due to the low heat capacity of air compared to the rock, the formation heating front lags behind the moving combustion front, and therefore more than 80% of the generated heat remains behind the combustion front and is dissipated into the surrounding rocks. In addition, up to 40% of operating costs are due to air injection. Therefore, the cost of oil when implementing this method of increasing oil recovery is quite high. The transfer of heat to the combustion front is possible by improving heat transfer with the addition of a working fluid with high heat capacity, for example, water, to the air. In this regard, a wet forward combustion technology was developed, which involves either joint or sequential injection of air and water. In this case, water evaporates upon contact with combustion-heated rock. Water vapor transfers heat to the region ahead of the combustion front.

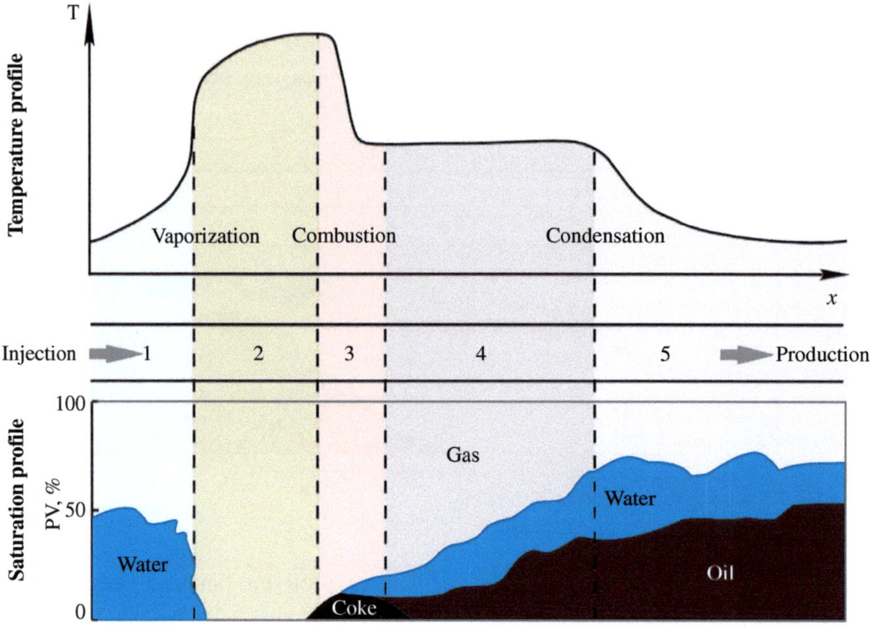

Figure 5.10 Zoning for wet forward burning.

During the wet combustion process, the formation can be divided into five main zones (see Figure 5.10).

1. *Zone 1*: Since the zone was already covered by the combustion front, it practically does not contain hydrocarbons. However, since the temperature is below the boiling point of water, the formation is saturated with water and injected air.
2. *Zone 2*: Water is in the vapor phase; the formation is saturated with a mixture of injected air and steam. The evaporation front of the injected water is located at the boundary between zones 1 and 2.
3. *Zone 3*: Combustion zone: Combustion occurs as a result of a combustion reaction involving hydrocarbons, precipitated coke, and atmospheric oxygen.
4. *Zone 4*: Evaporation/condensation zone: The temperature in this zone is close to the evaporation temperature of water, and progressive condensation of steam and water formed during the combustion process occurs. In addition, some light and intermediate fractions of oil evaporate. If the temperature is high enough, certain chemical reactions may occur.
5. *Zone 5*: The zone is characterized by high back pressure due to the formation of water and oil swell areas. Behind the oil bank, reservoir characteristics (pressure, temperature, etc.) correspond to the original conditions.

5.3.1.3 Reverse Combustion

After ignition, the combustion front moves from the production well to the injection well.

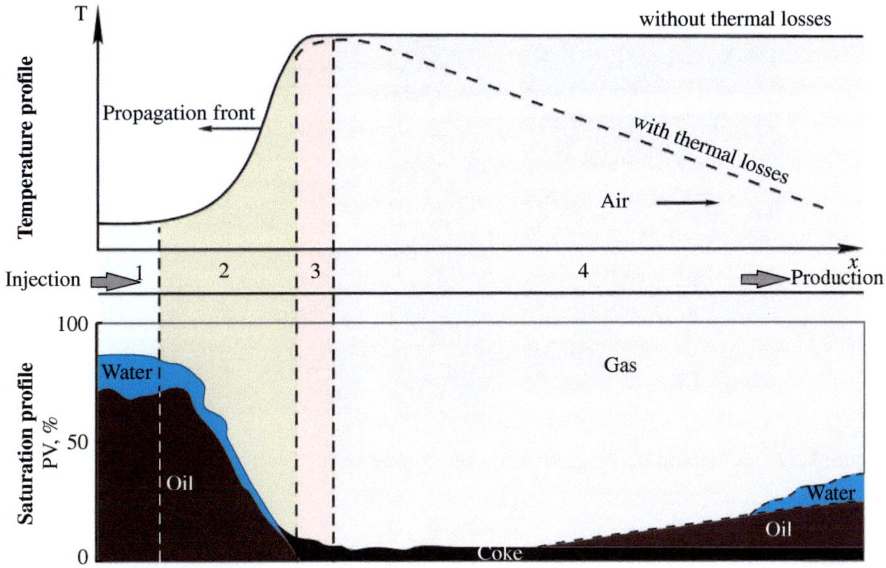

Figure 5.11 Zoning for reverse combustion.

During the reverse combustion process, the formation can be divided into four main zones (see Figure 5.11).

1. *Zone 1*: The reservoir is in initial conditions. However, due to air injection, if the reservoir temperature and oxidability of the oil are high, certain oxidation reactions may occur.
2. *Zone 2*: The temperature rises due to heat entering the hot zone. The onset of oxidation also contributes to an increase in temperature. Evaporation of produced water, distillation of light oil fractions, and cracking of certain hydrocarbons in the presence of oxygen occur. Liquid and vapor fractions are displaced, and some components form coke.
3. *Zone 3*: Combustion zone: The temperature reaches its maximum value. Oxidation and combustion reactions involving the most reactive hydrocarbon molecules absorb all the oxygen not used by reactions in the preceding zones.
4. *Zone 4*: Unburned coke is deposited, while steam and the liquid phase continue to move. If there were no heat loss, the temperature along the flow would remain the same as for the combustion front. In reality, the temperature decreases with distance from the combustion zone. Thus, condensation of distilled fractions of oil and, possibly, steam occurs.

5.3.2 Applicability Criteria

In situ combustion is applicable to a wide range of reservoirs (see Figure 5.12). As can be seen from the figure, many criteria are identical to SAGD.

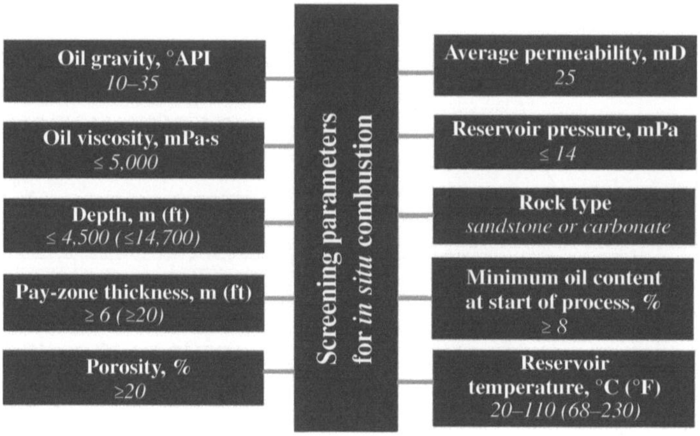

Figure 5.12 Screening parameters for *in situ* combustion.

The first and most important criterion of applicability, which is not identical for steam injection and *in situ* combustion, is the depth of occurrence. Although the depth limit for internal combustion is supposed to be 3,500 m is not a limit if air can be introduced at existing reservoir pressure and crude oil releases sufficient fuel to maintain the combustion front.

Reservoir pressure is the second criterion by which the two thermal methods of EOR differ. Although the reservoir pressure limit for *in situ* combustion is 14 MPa, this is not the limit. In principle, the level of reservoir pressure depends on the cost of air compression and pumping speed. Currently, projects have been implemented where the reservoir pressure significantly exceeds 14 MPa.

Rock permeability is the third criterion by which steam injection and *in situ* combustion differ. *In situ* combustion can be used in tanks that have lower permeability than steam combustion because the air injection rate is sufficient to maintain the combustion front. Permeability during *in situ* combustion must provide the volume of air necessary for combustion, which is calculated based on the properties of the oil and rock. Therefore, permeability should not be lower than the minimum since it affects the injectivity of injection wells and the air injection pressure into the formation.

The properties of oil are of great importance since deposits containing heavy (density from 0.85 to 0.99 g/cm^3) highly viscous oils are suitable for the *in situ* combustion process because it does not require the introduction of additional fuel into the formation.

The lower limit of density is determined by the fact that the main components of light oil evaporate before the combustion front, and the coke (fuel) necessary to maintain *in situ* combustion is not formed. Super-heavy oils with a density of more than 0.99 g/cm^3 during direct-flow *in situ* combustion can clog the pore space, which reduces the efficiency of the process. For such oils, the reverse combustion method is more effective.

An important indicator of oil is the resin content since it provides the formation of fuel to maintain combustion. The minimum required content of sulfuric acid resins in oil must be at least 20%.

The most favorable geological structure of the deposit where *in situ* combustion is planned is an alternation of reservoir rocks and clay layers. Clay layers, preventing heat loss, provide higher thickness coverage of the formation with increased temperature.

When the thickness of the formation is below the minimum (about 6 m), high heat losses occur in the surrounding rocks through the roof and base, which causes attenuation of the combustion front. With a layer thickness of 6–10 m, heat loss is already about 20–30%.

Since in the process of *in situ* combustion the combustion front tends to move toward the roof of the formation (due to the effect of gravitational separation of oil, air, and combustion gases), then with a sufficiently high thickness of the formation, 20 m and above, there is a significant decrease in the coverage of the formation by combustion.

When choosing fields, you must first pay attention to indicators such as oil density, resin content, depth, thickness, and permeability of the formation. For selected deposits, it is necessary to conduct laboratory studies to determine such parameters as fuel content, specific air consumption, optimal water-air ratio for wet combustion, etc.

5.3.3 Field Implementation

After the first experimental implementation of *in situ* combustion in the 1930s in the former USSR in the 1960s successful *in situ* combustion projects were carried out in the USA (Sloss, Delhi, Bellevue, Caddo Parish, and Carlide fields), Romania (Suplacu de Barqueu field), and Venezuela (East Tia-Juan). Of greatest interest are the results in fields with high formation depths and low oil viscosity.

Sloss light oil field, after complete watering in May 1963, wet combustion began. As a result of the use of wet combustion, by 1968, 19.1% of the remaining oil reserves after waterflooding were produced. The experience of implementing *in situ* combustion at the Sloss field shows that the use of wet combustion is effective for the additional development of low-viscosity oil deposits after they have been flooded.

At the Suplacu de Barqueu field (Romania), the process of dry *in situ* combustion began in 1966, and the flow rate of individual wells increased by 20–40 times. The average specific air consumption was about 2200 m^3/t. In June 1973, along with air, pumping of water began, and the airflow rate decreased to 1700 m^3/t. Oil recovery behind the combustion front was about 50%.

In 1971, at the Bellevue field (USA), the implementation of wet combustion began at a pilot site. During the development period of 50 years, oil recovery from the site was only 5%. During the first 2 years of testing, it was possible to achieve a cumulative oil production of 65% of oil reserves at the beginning of the process. Specific air consumption was 2 768 m^3/m^3.

In the former USSR, the first wet *in situ* combustion project was implemented in Azerbaijan in the Khorosany area of the Balakhani-Sabunchi-Ramany field starting in 1973. As a result of the implementation of the event, about 700 thousand tonnes of oil were additionally produced over 15 years, and the current oil recovery increased by 25%. The average specific air consumption was 1840 m^3/t.

In the 1980s, more than 50 *in situ* combustion projects were carried out in various countries. Basically, a combination of *in situ* combustion with waterflooding was used. Oil production through the use of *in situ* combustion during this period in the USA exceeded 600 thousand tonnes/year, and in Romania-430 thousand tonnes/year.

According to Alvarado and Manrique, despite its long history and commercial success, *in situ* combustion technology is viewed with a high degree of doubt by operating companies due to the high failure rate and uncertainty of *in situ* combustion implementations. However, it should be noted that unsuccessful implementations in most cases are due to the use of deposits that are not entirely suitable for *in situ* combustion.

Currently, *in situ* combustion is used only in a few heavy oil deposits. These are the Veteran fields in Canada; Suplacu de Barqueu in Romania, Balol, Bechraj, Lanva, and San Tai in India, and Bellevue in the USA.

5.3.4 Implementation Technology

The currently traditional scheme for carrying out the *in situ* combustion process is as follows: a heater is lowered into the well, and a high-temperature zone is created, which moves into the formation through air injection. Then, water is added to the pumped air. When most of the formation warms up, the injection of the mixture stops and the injection of cold water begins.

To implement the specified *in situ* combustion technology, the following equipment is required: heaters of various designs and compressors of various types. Of great importance are the piping of the wellhead, the design of the well bottom, and the separation of well products.

1. **Ignition of oil in a reservoir** occurs as a result of spontaneous ignition or forced ignition.

   Spontaneous ignition can occur in highly reactive oils, with oxidation at normal reservoir temperature leading to the release of heat sufficient to warm the reservoir and initiate ignition. Spontaneous combustion occurred in a number of fields, such as South Belridge (USA), Midway-Sunset (USA), California C, D, E, F, G, H (USA), and Tia Juana (Venezuela).

   Forced ignition is carried out using special deep fire and electric heaters of various designs, as well as chemical initiation (for example, using easily oxidized linseed oil). Heaters, depending on the design of the gas burner device, are lowered to the bottom of the ignition well using pipes, cables, or rope. Figure 5.13 shows a deep gas-air heater.

   The heater is lowered into the tubing on a rope, and the gas intake chamber is installed at the bottom of the pipe string, which also serves to supply gas. From the gas intake chamber, gas is directed to the ejector, where it is mixed with air pumped into the annular space. The gas-air mixture is ignited in an ignition device (electrical or chemical). The heater provides heating of the well bottom during the day up to 260 °C.

The most common method of initiating the combustion of reservoir oil is using downhole electric heaters. To lower electric heaters into the well, a heat-resistant, high-strength, reinforced cable is used. Electric heaters are powered from the field network, but it is possible to use mobile electric generators.

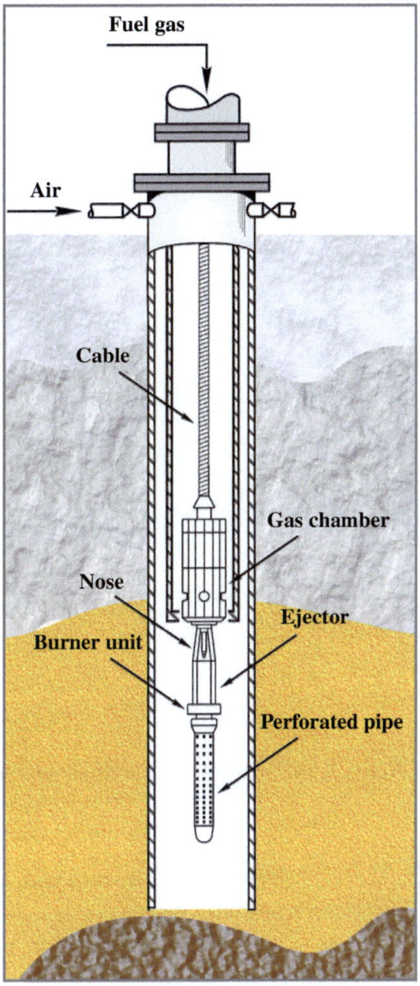

Figure 5.13 Deep gas-air heater.

2. To compress the design pressure and pump air into the design volume, special **compressor equipment is required**. Field experience in implementing *in situ* combustion shows that to implement the process, it is necessary to pump in more than 20 and no less than 250 thousand m^3/day of air at a pressure of more than 20 and no less than 90 atm. The choice of a compressor is based on the calculation of the required air volume and injection pressure. There are three types of compressors: reciprocating, centrifugal, and axial. Axial compressors are chosen when there is a need for high air flow rates, while if it is necessary to increase the outlet pressure, centrifugal compressors are installed.
3. **The wellhead of the ignition well** is tied according to Figure 5.14, when using a deep electric heater to ignite reservoir oil. When reservoir oil is ignited by a deep fire heater, the wellhead of the ignition well is tied according to Figure 5.15. As can be seen from the figures, when implementing the *in situ* combustion process, serious reconstruction of the wellhead is not required.
4. **Borehole design:** The main requirement for the design of both injection (ignition) and production wells is to prevent rapid wear of the casing material as a result of corrosion and high temperature. In this regard, casing material must be made of special heat-resistant materials.

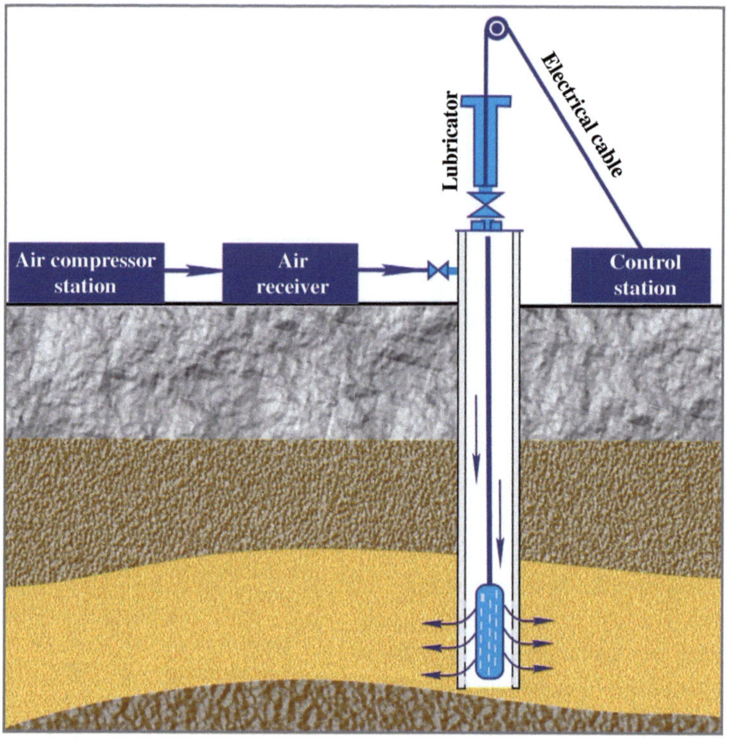

Figure 5.14 Ignition well arrangement for ignition by the downhole electrical ignitor.

Completion of ignition wells, where possible, should be completed with an open hole, with casing freely suspended to provide thermal expansion.

In production wells with weakly cemented rocks, where there is a risk of destruction of the bottomhole zone and sand removal, it is possible to use special anti-sand heat-resistant ceramic filters.

5. **Separation of well products:** When implementing *in situ* combustion, well products, namely gas, contain substances that cause corrosion. In this case, the main part of the gases during mechanized operation is removed through the annular space of the wells. It should be taken into account that untreated combustion gases, when released into the atmosphere, lead to serious environmental problems. In this regard, it is necessary to dispose of untreated gaseous products.

Of great importance is monitoring the composition of the produced gas, which makes it possible to assess the location of the combustion front and regulate its progress. In addition, analysis of the gas composition makes it possible to assess the degree of atmospheric pollution during their release and outline ways for its disposal. It is also important to monitor the properties of reservoir fluids in order to determine the water content in the product and estimate the acid number of oil.

Figure 5.15 Ignition well arrangement for ignition by the submerged fired heater.

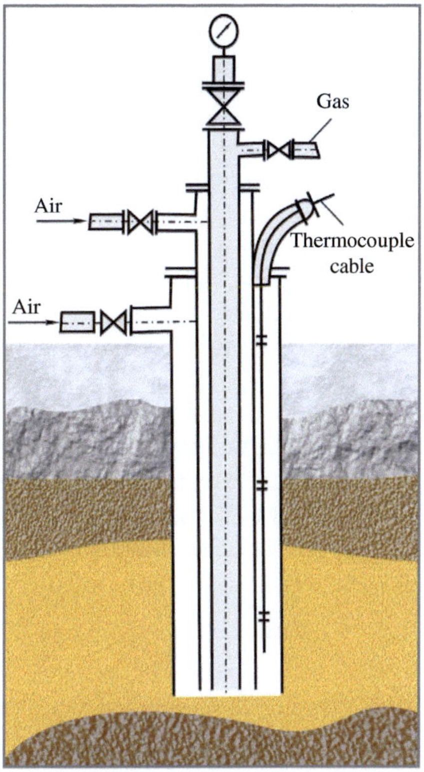

Bibliography

Alvarado, V. & Manrik, E. (2011). *Methods for Enhancing Oil Recovery: Planning and Application Strategies*. Premium Engineering LLC.

Amelin, I.D. (1980). *In situ Combustion*. Nedra.

Baibakov, N.K. & Garushchev, A.R. (1981). *Thermal Methods for Developing Oil Fields*. Nedra.

Burger, J., Sourieau, P., & Combarnous, M. (1984). *Recuperation assistee du petrole: Les methodes thermiques*. Editions Technip.

Carcoana, A. (1992). *Applied Enhanced Oil Recovery*. Prentice Hall.

De Haan, H.J. & Van Lookeren, J. (1969). Early results of the first large-scale steam soak project in the Tia Juana field, Western Venezuela. *Journal of Petroleum Technology*, *21*(January), 101–110.

Eson, R.L. (1982, March 24–26). Downhole steam generator: Field tests. *Paper SPE-10745-MS Presented at the SPE California Regional Meeting. San Francisco, California*.

Green, D.W. & Willhite, G.P. (1998). *Enhanced Oil Recovery*. Society of Petroleum Engineers.

Ismailov, F.S., Mehdiev, U.S., & Gasimly, A.M. (2011). *Experience in Applying Thermal Methods of Influence in Oil Fields of Azerbaijan*. Mars Print.

Kuhn, C.S. & Koch, R.L. (1953). *In situ* combustion: Newest methods of increasing oil recovery. *Oil and Gas Journal, 52*(14), 92–96, 113–114.

Latil, M., Bardon, C., Burger, J., et al. (1980). *Enhanced Oil Recovery*. Technip.

Ramey, H.J. (1967, April 2–9). A current review of oil recovery by steam injection. *Paper WPC-12247 Presented at the 7th World Petroleum Congress. Mexico City, Mexico.*

Romero-Zerón, L. (2012). Advances in enhanced oil recovery processes. In L. Romero-Zerón (Ed.). *Introduction to Enhanced Oil Recovery (EOR) Processes and Bioremediation of Oil-contaminated Sites*. IntechOpen.

Stovall, S.L. (1934, August 13). Recovery of oil from depleted sands by means of dry steam. *Oil Weekly, 73*, 17–24.

Surguchev, M.L. (1985). *Secondary and Tertiary Methods for Increasing Oil Recovery*. Nedra.

Tatar-inform.ru. (2013, April 27). [News article]. http://www.tatar-inform.ru/news/2013/04/27/358128/

Winkler, A.K. (1951, May 28–June 6). The exploitation of oil-fields of extremely high oil-viscosity by wells, under the application of thermal energy. *3rd World Petroleum Congress. The Hague, Netherlands.*

6

Gas Flooding

6.1 Introduction

Gas injection is by far one of the oldest methods of EOR, and its use is increasing. At the same time, most of the field implementation of the method relates to the use of nonhydrocarbon gases—carbon dioxide, nitrogen, and flue gases.

Miscible hydrocarbon gas displacement can be divided into three distinct methods (liquefied gas or solvent injection, enriched gas injection, and high-pressure gas injection), all of which have undergone field testing or full-scale application. In this case, for the injection of a slug of liquefied gas or solvent, enriched gas, and high-pressure gas, a sufficiently high pressure is required to achieve miscibility. If the reservoir characteristics are unfavorable, then an early breakthrough of the working agent occurs, and, accordingly, most of the oil reserves remain untapped. In addition, the hydrocarbons needed to carry out the injection processes are valuable, consumer-demanded products, and there is increasing reluctance to inject them back into the reservoir. Therefore, in recent years, especially in the USA, the emphasis has shifted to less valuable nonhydrocarbon gases such as CO_2, nitrogen, and flue gases. Although nitrogen and flue gases displace oil significantly worse than hydrocarbon gases, the overall economic efficiency may be slightly more favorable.

A significant influence on the process of the displacement of oil by gas injected into the reservoir is its miscibility with oil. As a rule, at normal temperatures and pressures, dry natural gas (methane) and oil do not mix. To ensure the mixing of oil and gas, a transition zone between them must exist. In this case, the hydrocarbon mixture forming the transition zone must be dissolved in both gas and oil. This is usually achieved by enriching the injected gas.

Formation fluid: miscibility at first contact (first contact or direct miscibility) and miscibility at multiple contacts, or dynamic miscibility (multiple-contact or dynamic miscibility).

The miscibility of the first contact is achieved by injecting liquefied hydrocarbon gases into the formation, which are mixed with reservoir oil in any proportion, and the solution always remains single-phase.

Methods for Enhanced Oil Recovery: Fundamentals and Practice, First Edition.
Baghir A. Suleimanov and Elchin F. Veliyev.

Dynamic miscibility is observed when high-pressure natural gas, enriched natural gas, nitrogen, flue gases, and carbon dioxide are injected into the reservoir. These fluids do not mix with formation oil upon first contact and form two-phase systems with formation fluids. Miscibility is achieved through the mass transfer of components resulting from repeated contact with oil during flow in the formation.

There are two modes by which dynamic miscible displacement in the formation can be achieved, namely, the condensing gas drive process or the vaporizing gas drive process.

In the first case, miscibility is achieved due to the dissolution of intermediate hydrocarbons present in the injected enriched gas and in the reservoir oil.

In the second case, high-pressure dry natural gas, nitrogen, flue gases, and carbon dioxide can be used as a propellant. In this case, miscibility is achieved at a pressure above the minimum miscibility pressure due to the evaporation of intermediate hydrocarbon components from oil and their dissolution in the injected dry gas. (These processes are discussed in detail in Figures 6.1 and 6.2.)

The displacement of oil under conditions in which complete mixing with gas does not occur is called immiscible displacement. It is obvious that the best results in increasing oil recovery are observed with miscible displacement of oil by gas.

An analysis of the use of gas methods in the USA shows that there are two time intervals during which interest in this method of EOR has increased significantly (see Figures 6.1 and 6.2). The first time interval was 1980–1986, when, due to rising oil prices, the number of implemented projects increased. During this period, the number of projects increased from 35 to 105, and oil

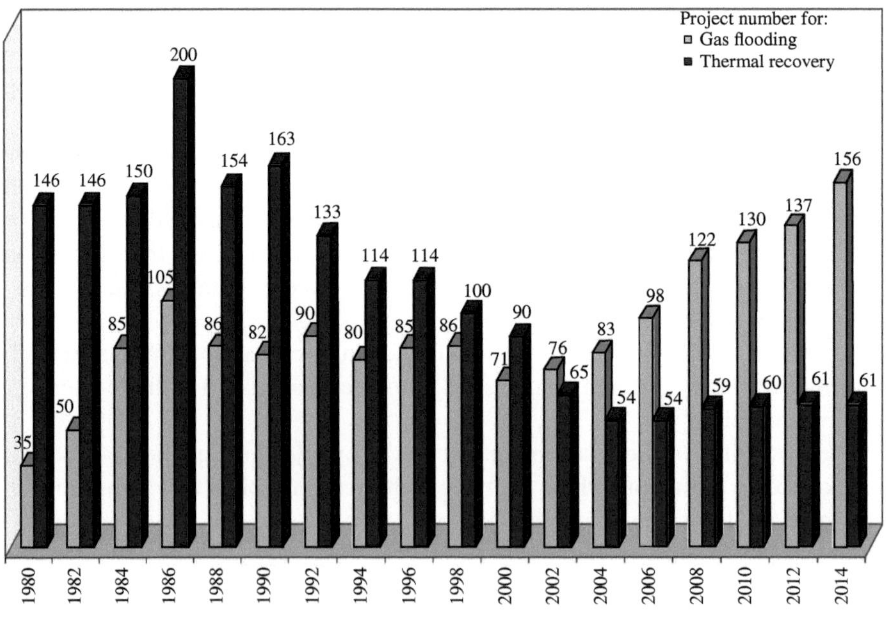

Figure 6.1 Number of thermal and gas EOR projects implemented in the USA.

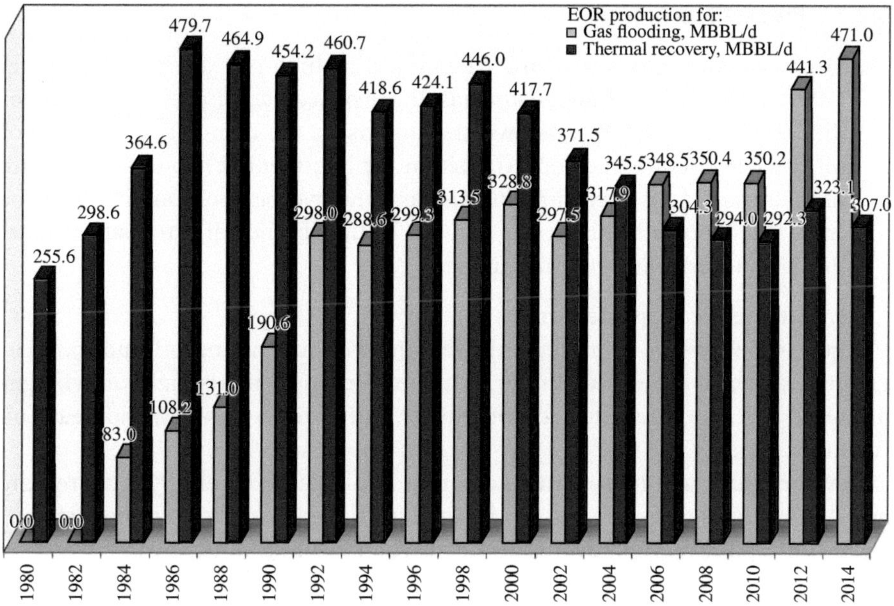

Figure 6.2 EOR production of thermal recovery and gas flooding in the USA.

production reached 108.2 thousand barrels per day. The second time interval began in 2002, also due to a significant increase in oil prices. During this period, the number of projects increased from 76 (2002) to 156 (2014), and production increased from 297.5 to 471 thousand barrels per day. Moreover, in 2006, for the first time, oil production obtained by gas methods exceeded oil production obtained by thermal (thermal) methods of EOR (see Figure 6.2).

6.2 Injection of Hydrocarbon Gases

As noted above, when hydrocarbon gases are injected into a reservoir, oil can be extracted both without mixing and with mixing. Dry natural gas does not mix with oil at normal pressures and temperatures, even in deep-lying formations.

To ensure mixing, a transition zone is required in which both oil and gas dissolve. However, when dry hydrocarbon gases are injected, no transition zone is formed, and oil is extracted without mixing. To achieve more efficient miscible displacement, three different methods are used: high-pressure gas injection, enriched gas injection, and liquefied gas injection.

Mechanism of mixing displacement of oil by hydrocarbon gas:

- creating miscibility when injecting high-pressure gas and enriched gas (this ensures zero interfacial tension between the oil and the working agent, as well as a capillary number equal to infinity),
- increase in oil volume (swelling), and
- reduction of oil viscosity.

6.2.1 Process Mechanism

1. **High-pressure gas injection.** When injecting high-pressure gas, depending on the composition of the oil and gas, the displacement process occurs under different conditions. It is known that to ensure the complete miscibility of oil and gas, the presence of a transition (or intermediate) zone of a mixture of hydrocarbons is necessary. In this case, the hydrocarbon mixture must be mutually dissolved in both gas and oil. In the absence of a transition zone, displacement occurs without mixing.

1.1. Immiscible displacement

Figure 6.3 shows a displacement diagram without mixing oil and gas. For greater clarity, oil is depicted in the form of three components: light oil with gas dissolved in it, intermediate components (from ethane to hexane), and heavy oil fractions.

Low-pressure gas injection: A diagram of oil displacement by low-pressure gas is shown in Figure 6.3a. As can be seen from the figure, behind the displacement front, oil saturation decreases to a certain limiting value. At low injection pressure, there is practically no dissolution of gas in oil. In this regard, the composition of residual oil is practically no different from the composition of the original oil.

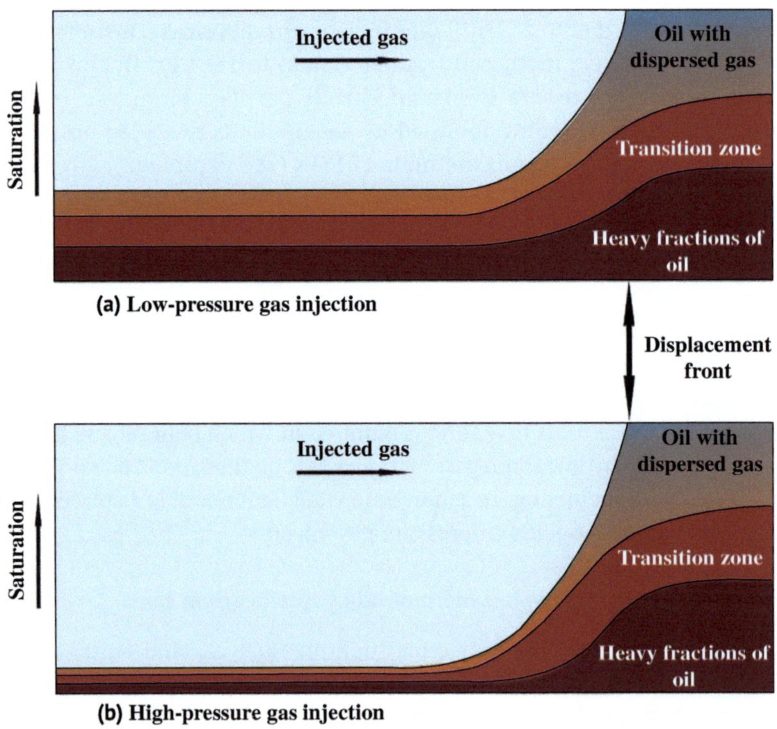

Figure 6.3 Oil displacement scheme for immiscible hydrocarbon gas injection.

High-pressure gas injection: A diagram of oil displacement by high-pressure gas is shown in Figure 6.3b. As can be seen from the figure, behind the displacement front, the oil saturation is significantly less than when injecting low-pressure gas. A significant decrease in residual oil saturation, compared to low-pressure gas injection, is due to three main reasons:

- a decrease in the viscosity of oil when gas is dissolved in it, an increase in the viscosity of gas at high pressure, and a resulting decrease in the ratio of the mobility of the displacing agent and the displacer;
- an increase in the volume (swelling) of oil as a result of enrichment with intermediate components; and
- enrichment of gas with intermediate components and subsequent shrinkage of oil.

It should be noted that the saturation pressure of oil with gas has a significant impact on the efficiency of the displacement process. Moreover, if the pressure of the injected gas is higher than the saturation pressure, the oil recovery factor is significantly higher than in the opposite case. This is due to the fact that the injected gas, dissolving in the oil, promotes its swelling (increase in volume).

1.2. Miscible displacement

Figure 6.4 shows a diagram illustrating the mechanism of oil displacement with mixing during high-pressure gas injection. As can be seen from the figure, four zones are formed during the displacement process: the first zone is occupied by injected dry gas; in the second zone, gas partially enriched during filtration through the oil-saturated formation moves; the third zone is occupied by already enriched gas with a high content of intermediate hydrocarbons compared to partially enriched gas; the fourth zone is occupied by displaced oil. At the boundary of the third and fourth zones, mixing displacement of oil by gas is already taking place.

Figure 6.4 also shows the change in the concentration of intermediate hydrocarbons until a concentration is reached that ensures complete mixing of oil and gas (the boundary of zones 3 and 4).

Let us consider the nature of the displacement of oil by gas using triple-phase diagrams. A complex hydrocarbon system is assumed to consist of three components: C_1, $C_2 - C_6$, C_{7+}. The phase diagram is shown in Figure 6.5.

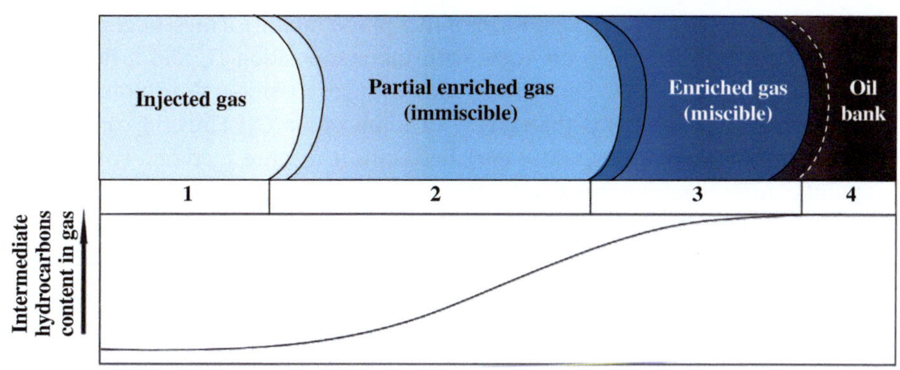

Figure 6.4 Oil displacement scheme for miscible hydrocarbon gas injection.

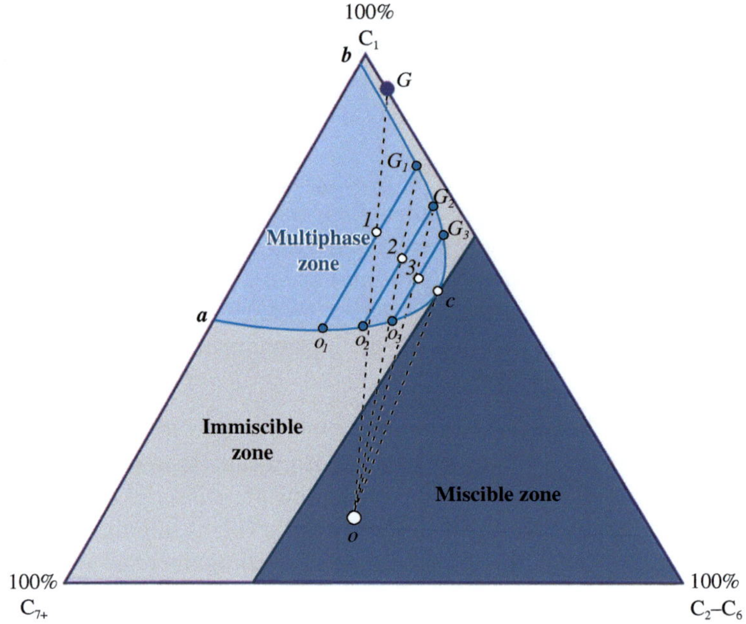

Figure 6.5 Ternary diagram (*acb* – boundary phase curve, *ac* – curve of the beginning of evaporation or degassing, *cb* – curve of the beginning of condensation).

Let us consider the process of the displacement of oil with composition O by gas with composition G. As can be seen from the figure, gas consists almost exclusively of methane, and oil contains a sufficient amount of intermediate components. When gas is injected into a reservoir, it mixes with oil to form a mixture. The composition of the mixture corresponds to point *1* on a straight-line OG. This mixture is in the two-phase region and therefore splits into a gas with composition G_1 (located on the curve of the beginning of condensation) and a liquid with the composition O_1 (located on the curve of the beginning of evaporation or degassing). We get the connection O_1 G_1. Obviously, a gas with a composition of G_1 contains a larger amount of intermediate hydrocarbons than a gas with a composition of G. Since gas with a composition of G_1 moves faster than liquid with composition of O_1, when mixed with oil with composition of O, it forms a two-phase mixture *2* on straight line OG_1. This mixture breaks down into a gas with the composition G_2 and a liquid with the composition O_2 (O_2 G_2 connection). Gas with composition G_2 is more enriched in intermediate components than gas with composition G_1. Thus, in contact with reservoir oil, the gas is constantly enriched until it reaches a critical composition at point c. After reaching this point, complete miscibility of gas with oil is achieved (i.e., all mixtures will be single-phase).

Figure 6.6 shows a ternary phase diagram with boundary phase curves for three pressures. As can be seen from the figure, with decreasing pressure, the miscibility zone decreases, and the two-phase region, on the contrary, increases. Considering the process of displacement of oil with composition O by gas with composition G, we will find that at 21 and 24.5 MPa, miscibility of oil and gas is

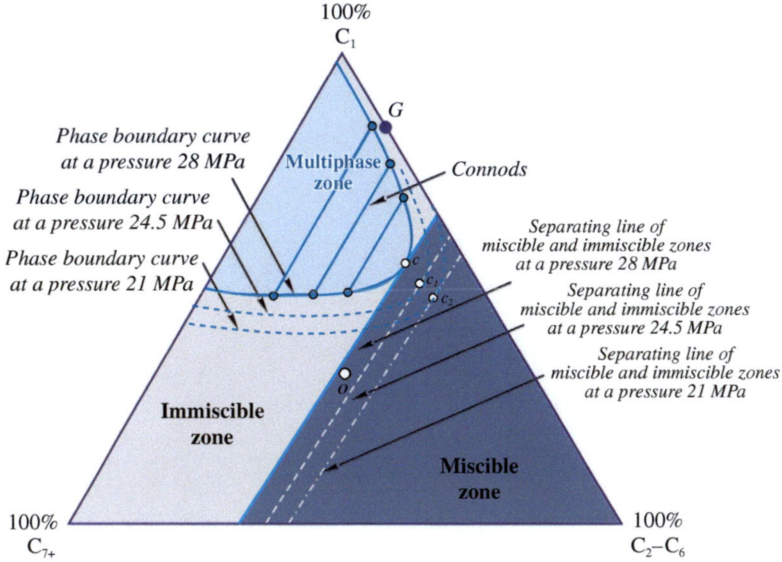

Figure 6.6 Ternary diagram with different injection pressure.

not achieved. With a given composition of oil and gas, miscibility occurs only at a pressure of 28 MPa. Studies have shown that to achieve miscibility at a pressure of 28 MPa, it is necessary that the concentration of intermediate components in oil be at least 35%; otherwise, higher pressure will be required. Thus, the most important factor ensuring miscibility is the composition of the oil, namely the presence of a sufficient amount of intermediate hydrocarbons.

Let us now consider the process of displacement under conditions of immiscibility (see Figure 6.7). As can be seen from the figure, in the process of displacement of oil by gas, the gas composition, although enriched, does not reach the critical composition mixtures.

In this case, the volume of oil does not increase, since the reservoir pressure does not exceed the saturation pressure, and gas does not dissolve in oil. In this regard, an increase in oil recovery occurs due to an increase in gas viscosity as a result of enrichment with intermediate hydrocarbons and a resulting decrease in the ratio of oil and gas mobilities.

2. **Injection of enriched gas:** Gas enriched with intermediate hydrocarbons ($C_2 - C_6$), primarily propane, is injected into the formation. Moreover, due to the fact that the concentration of intermediate hydrocarbons in gas is higher than in oil, they begin to dissolve in oil, which leads to swelling (increase in volume) and a decrease in viscosity. As a result, oil recovery from the reservoir is significantly increased.

Let us consider the nature of the displacement of oil by enriched gas using ternary phase diagrams. The phase diagram is shown in Figure 6.8. Let us assume that gas with composition *G* is injected into the formation and displaces formation fluid with composition *L*. When pumping enriched gas into the reservoir, it mixes with

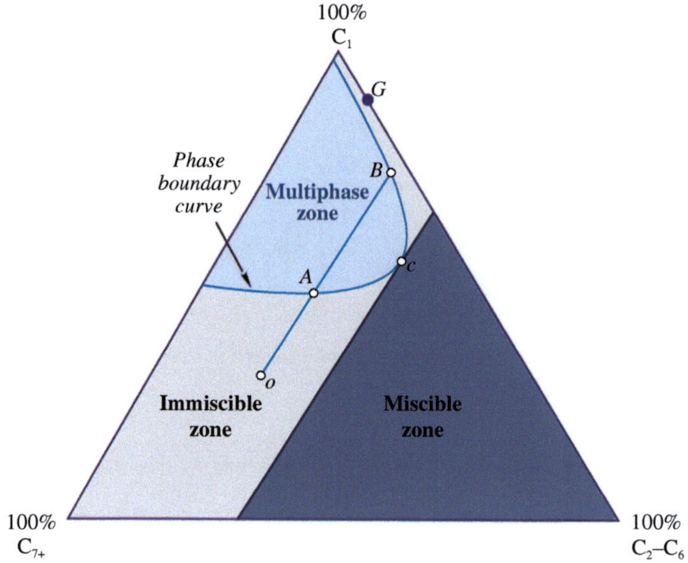

Figure 6.7 Ternary diagram for immiscible injected gas–oil combination.

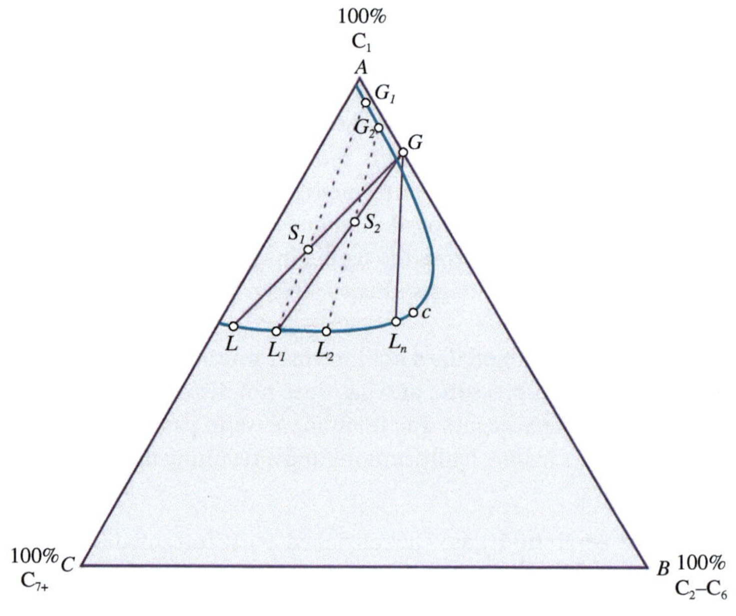

Figure 6.8 Ternary diagram.

residual oil, forming a mixture S_1. This mixture is in the two-phase region and therefore breaks down into a gas with composition G_1 and a liquid with composition L_1. The phases are in equilibrium, and their connection, L_1 G_1 passes through point S_1.

A fresh portion of gas entering the formation leads to the mixing of gas with composition G_1 and liquid with composition L_1, forming a mixture S_2. This mixture is also in the two-phase region and therefore breaks down into a gas with the composition G_2 and a liquid with the composition L_2. The process continues in the specified sequence, and the composition of the residual liquid will correspond to the point L_n. Thus, in contact with reservoir oil, the gas is constantly enriched until it reaches a critical composition at point *c*. After reaching this point, complete miscibility of gas with oil is achieved (i.e., all mixtures will be single-phase). As a result of an increase in the concentration of intermediate hydrocarbons, the initial composition of the liquid changed significantly, but due to the fact that direct GL_n is in the two-phase region, complete miscibility of oil and gas is not achieved.

Let's consider the case of achieving full miscibility when pumping enriched gas (Figure 6.9). In this case, the injected gas must contain a sufficiently large amount of intermediate hydrocarbons (35–40%).

Let us assume that gas with the composition G_3 is injected into the formation and displaces formation fluid with composition L. When enriched gas is injected into the reservoir, it mixes with residual oil, forming a mixture S_1.

In the first mixing cell, the mixture S_1 is separated into gas G_1 and liquid L_1. Gas G_1 moves to the next mixing cell, and liquid L_1 mixes with a fresh portion of gas, forming a mixture S_2. Liquid L_2 mixed with a fresh portion of gas, etc. Thus, in the first mixing cell, this process leads to the formation of a single-phase mixture. Since the gas phase has already passed the first mixing cell, miscibility develops at the trailing edge of the oil and gas mixing zone as a consequence of the enrichment of the liquid phase with intermediate hydrocarbons from the injected gas. Since the intermediate hydrocarbons condense from the gas into the liquid phase, the process in question is also called the condensing gas oil displacement process. Point G_3 shows the minimum concentration of intermediate

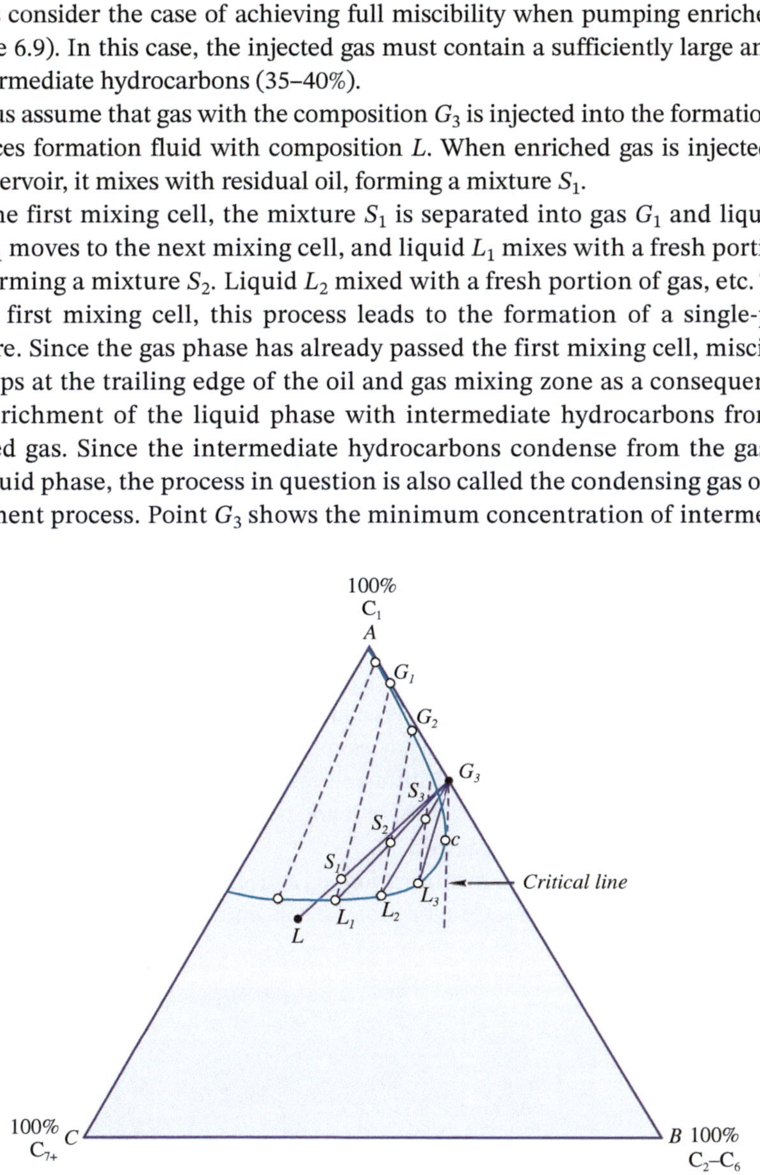

Figure 6.9 Ternary diagram.

hydrocarbons required to ensure complete miscibility of oil and gas. It should be noted that the minimum pressure of full miscibility depends significantly on the degree of gas enrichment with intermediate hydrocarbons. In this case, the higher the concentration of intermediate hydrocarbons, the lower the miscibility pressure.

3. **Injection of liquefied gas:** Beginning in the early 1950s, numerous field trials and several full-scale implementations were conducted in the USA using LPG (for miscible displacement). However, the need for excessively large quantities of LPG and the high market cost of liquid propane have made LPG injection economically unattractive. In this regard, the technology was modified to reduce cost. In this case, a small slug of liquefied petroleum gas is injected into the formation, which forms a transition zone between the dry gas that propels the slug and oil. Typically, propane or its mixtures with other intermediate hydrocarbons are used as a liquefied gas slug. The main advantage of this method is the absence of capillary forces, which prevent increased oil recovery when implementing immiscible methods, so the efficiency of oil displacement increases significantly. The injection pressure must ensure that the liquefied gas is maintained in a liquid state and that the liquefied gas is completely miscible with both oil and dry propellant gas. In this case, the required miscibility pressure is significantly lower than when injecting dry or enriched gas. The need to maintain liquefied gas in a liquid state limits the use of the method in fields with temperatures below the critical temperature for the gas used. The critical temperature for propane is 96.6 °C, so it cannot be used at higher temperatures.

Figure 6.10 shows a diagram illustrating the mechanism of oil displacement during the injection of liquefied gas (propane). As can be seen from the figure, during the displacement process, five sequentially moving zones are formed: the first zone is occupied by injected dry gas; in the second zone, a mixture of gas and propane moves; the third zone is occupied by pure propane; in the fourth zone, a mixture of propane and oil moves; the fifth zone is occupied by displaced oil. This distribution of zones is due to the fact that in the process of propane moving through the formation with dry gas, both simple and diffusion mixing occurs. Thus, propane moves behind the displacement front; in front of it and behind it there are transition zones containing its mixture of dry gas and oil.

As propane moves through the formation, its concentration in the slug decreases. Over time, a significant decrease in the propane concentration in the slug can lead to a

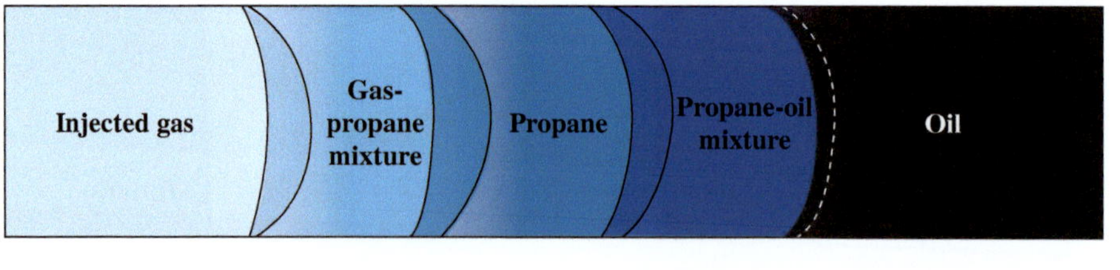

Figure 6.10 Oil displacement scheme for the liquid propane-supported flooding.

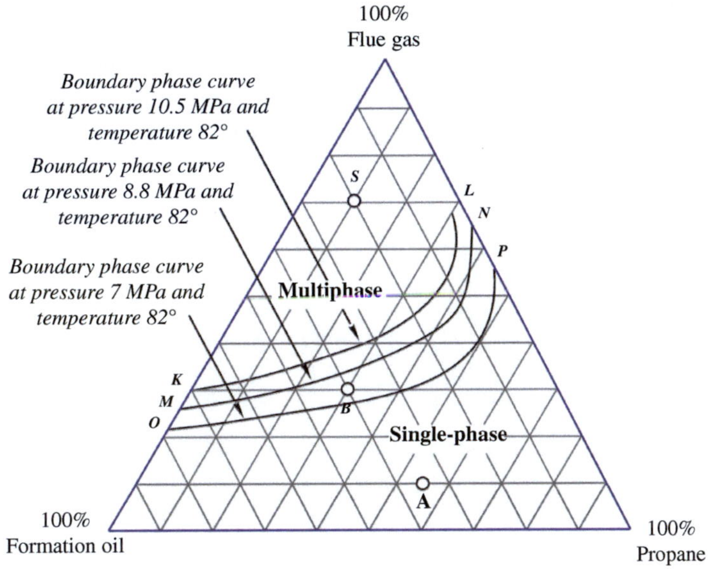

Figure 6.11 Ternary diagram.

significant decrease in process efficiency. To ensure the efficiency of the displacement process, it is necessary to calculate the optimal size of the propane slug. To solve this problem, phase diagrams are usually used. Based on them, based on the composition of the three-phase system, the permissible concentration of propane is determined at which the system is in a single-phase state until the end of the displacement process. Figure 6.11 shows the phase diagram for the "reservoir oil–flue gas–propane" system.

The figure shows that at a pressure of 7 MPa the permissible propane concentration is 50%, at 8.8 MPa – 42%, and at 10.5 MPa – 37%.

6.2.2 Applicability Criteria

Injection of hydrocarbon gases should be carried out in fairly deep deposits. The minimum depth is determined by the pressure required to maintain miscibility. The required pressure ranges from approximately 8 MPa when injecting liquefied gas to 20–35 MPa when injecting high-pressure gas, depending on the composition of the oil.

To carry out the process, steeply dipping formations are preferred, providing gravitational stabilization of the displacement front, which usually has an unfavorable mobility coefficient. The applicability criteria of the method are shown in Figure 6.12.

The disadvantages of the method include:

- viscous instability leads to low vertical and horizontal sweep efficiency;
- a large number of expensive, high-demand products are needed; and
- there are significant losses of expensive injected solvents, which cannot be completely recovered.

6.2.3 Field Implementation

As shown in Figure 6.12, the use of gas methods in the USA there have been two significant periods of interest in methods for enhancing oil recovery, including the injection of hydrocarbon gases (see Figure 6.13). The first period occurred in 1980

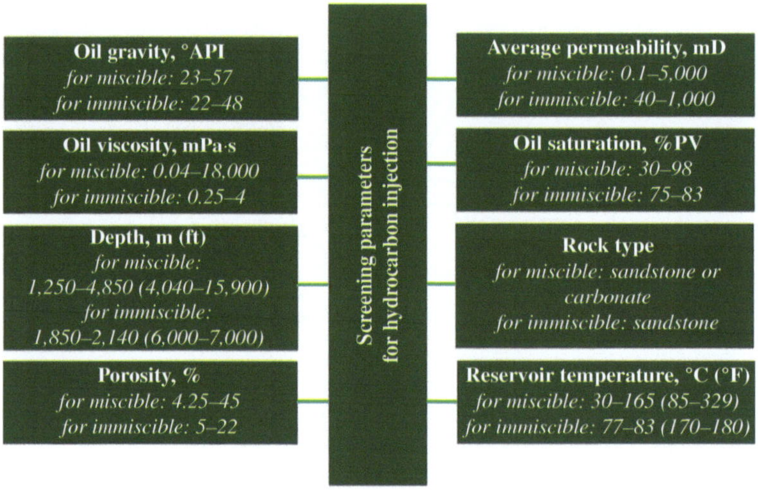

Figure 6.12 Applicability criteria for hydrocarbon gas injection.

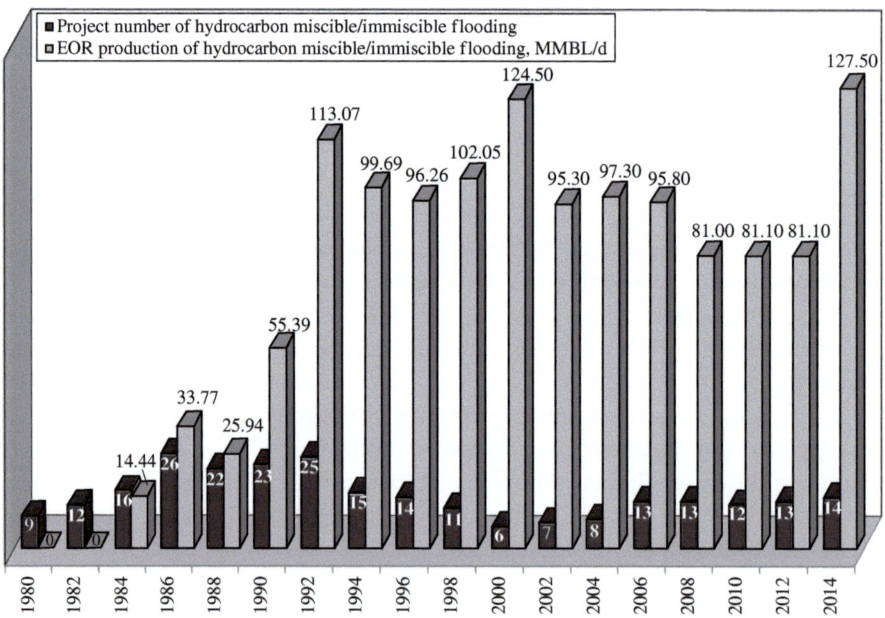

Figure 6.13 Hydrocarbon flooding EOR projects operating in the USA between 1980 and 2014.

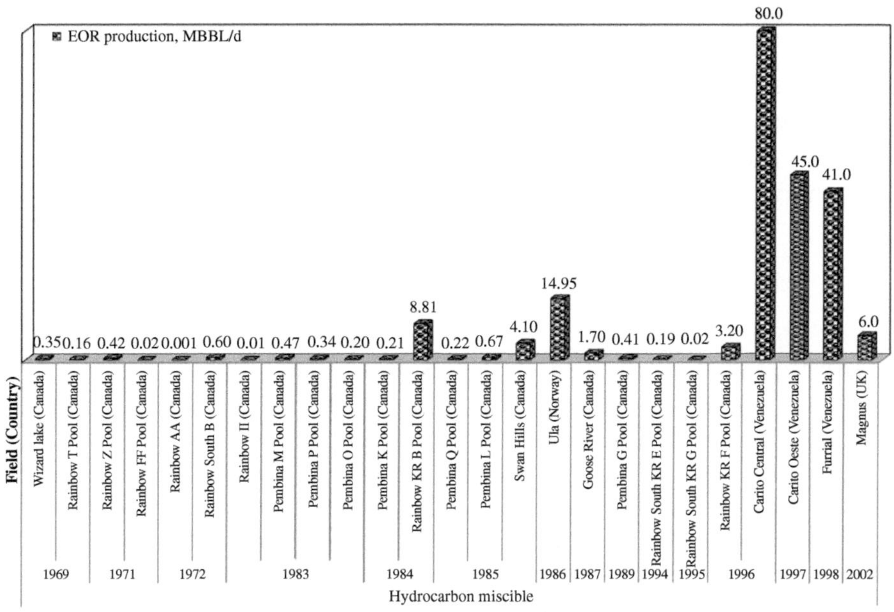

Figure 6.14 Implementation of the hydrocarbon gas injection method around the world between 1969 and 2014.

due to rising oil prices and continued until 1992. During this period, the number of hydrocarbon gas injection projects increased from 9 (1980) to 25 (1992), and oil production increased to 113 thousand barrels per day. The second period occurred after 2002, when oil prices soared to over \$100 per barrel, bringing significant profits from gas-based EOR techniques. During this period, the number of hydrocarbon gas injection projects increased from 7 (2002) to 14 (2014), and oil production increased from 95.3 to 127.5 thousand barrels per day. It should be noted that this is significantly lower than oil production from the use of carbon dioxide, which on the same date (2014) amounted to 335.53 thousand barrels per day.

Figure 6.14 shows the implementation of the method of injection of hydrocarbon gases around the world. As can be seen from the figure, in addition to the USA, this method has been implemented since 1969 in Canada, Norway, Venezuela, and the UK.

6.2.4 Implementation Technology

When implementing a hydrocarbon gas injection project, first of all, it is necessary to calculate its need (dry or enriched) to produce a unit of additional oil. In this case, the volume of injected gas is calculated based on reservoir pressure and temperature.

The volume of injected gas is determined as follows:

$$V_{ig} = V_o + V_w + V_g,$$

where V_{ig} is the volumetric flow rate of injected gas, reduced to reservoir conditions, m^3/day; V_o — volumetric flow rate of produced oil under reservoir conditions, m^3/day; V_w — volumetric flow rate of produced water under reservoir conditions, m^3/day; V_g — volumetric flow rate of free produced gas under reservoir conditions ($V_g > 0$ only in the case of production wells operating at bottomhole pressure above the saturation pressure or in the case of gas breakthrough into production wells from the gas cap), m^3/day.

The parameters included in the above equation are determined by the formulas:

$$V_o = V_{od} b_o,$$

$$V_w = V_{wd} b_w,$$

where V_{od} and V_{wd} are respectively the volumetric flow rates of oil and water after degassing (under standard conditions), m^3/day; b_o, b_w are the volumetric coefficients of oil and water under reservoir conditions, respectively.

If there is a sufficient amount of hydrocarbon gas in the field, the gas is prepared for injection. At the same time, the gas is purified from mechanical impurities and compressed. Gas purification can be carried out by oil or cyclone dust collectors, as well as gravity separators. To compress gas, it is necessary to create a compressor station, the performance of which is calculated from the required volume of hydrocarbon gas and injection pressure. If there is high-pressure gas in the field, the injection can be carried out without a compressor, which significantly reduces capital and operating costs.

Hydrocarbon gases are injected through the tubing, which is usually lowered to the upper holes of the filter. The annulus is closed with a drillable packer, which is lowered onto the tubing. This protects the casing from high gas injection pressure. The design of the injection well is shown in Figure 6.15.

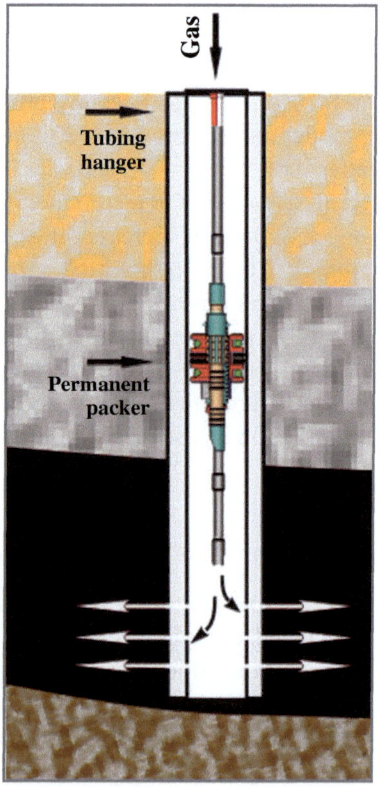

Figure 6.15 Gas injection well design.

Gas breakthroughs into production wells create a serious problem when implementing the technology. To prevent a breakthrough, it is necessary to control the gas injection rate, gas factor, and composition of the produced gas, and in case of a breakthrough, eliminate it. In this case, restriction of fluid withdrawal from wells, as well as isolation work, are applied.

Modifications of the method: To reduce the consumption of valuable hydrocarbons during the implementation of the method, after injection of a slug of enriched gas in a volume sufficient to effectively displace oil, they proceed to injection of cheaper inert gases. In particular, it is possible to use nitrogen and flue gases. It should be taken into account that the flue gas contains carbonic acid, which, upon contact with the formation and connected water, dissolves in it, and at the displacement front, there will be only nitrogen. Flue gas is known to consist of almost 90% nitrogen and about 10% carbonic acid. If the injection pressure is high enough and the oil contains a sufficient amount of intermediate hydrocarbons, then the nature of the displacing agent is not so important.

In some cases, it is possible to inject flue gases after hydrocarbon gases. In this case, hydrocarbon gases of 5% of the pore volume are sufficient. Note that an important advantage of flue gases is their high compressibility coefficient, which significantly reduces the financial costs of gas compression. The disadvantage of flue gases is that they must enter the field in compressed form.

When implementing the injection of enriched hydrocarbon gas, it should be taken into account that it is rare that sufficiently rich gas is produced at the field, and it is necessary to obtain enriched gas by mixing dry natural gas with liquefied gas. Therefore, in order to reduce the consumption of such a sought-after valuable liquefied gas, before the start of the process, it is necessary to determine the permissible dilution of liquefied gas with dry hydrocarbon gas at injection pressure and reservoir temperature.

Development systems: The field development system when using hydrocarbon gas injection is practically no different from the development system using conventional waterflooding. The main requirement for the development system is that contour or pattern flooding must be carried out. The placement of wells also does not differ significantly from conventional waterflooding. However, with high vertical permeability and the presence of high gravitational segregation of hydrocarbon gas and water, the well pattern should be compacted.

6.3 Nitrogen Injection

To replace expensive and high-demand hydrocarbon gases, relatively inexpensive nitrogen and flue gases are used as displacing agents. Moreover, depending on the pressure and composition of the oil, the displacement process can be miscible or immiscible. Due to their low cost, large volumes of these gases can be injected into the formation. Nitrogen or flue gases can also be used to propel a slug of hydrocarbon gases or carbon dioxide, which significantly reduces the cost of the project.

Mechanism of displacement of oil by nitrogen and flue gases:

- at a sufficiently high injection pressure, light fractions of oil evaporate, which ensures miscible displacement;
- ensuring a gas pressure regime in which a significant part of the reservoir volume is filled with relatively inexpensive nonhydrocarbon gases.

The advantages of nitrogen and flue gases include low cost and availability, low compressibility (3 times compared to carbon dioxide and 1.5 times compared to methane), which provides significantly lower compression costs (2–3 times less), and the absence of nitrogen (unlike flue gases) of corrosive activity.

The disadvantages include the low solubility of nitrogen in oil (35–45 m^3/m^3 for light oil, 15–25 m^3/m^3 for heavy oil). Flue gases, due to the fact that they are almost 90% nitrogen, and despite the fact that they contain about 10% carbon dioxide, differ little in solubility and displacement efficiency from nitrogen. In this regard, the minimum miscibility pressure for nitrogen exceeds 35 MPa, which is significantly higher than for natural gas (25 MPa) and carbon dioxide (8 MPa).

6.3.1 Process Mechanism

Along with determining the minimum miscibility pressure, there are two more parameters that must be determined experimentally. These are the oil swelling test, which determines the pressure at which first contact miscibility (FCM) is achieved, and the multicontact test.

The main mechanism for oil displacement during gas injection is the development of miscibility between gas and reservoir oil. There are three types of miscibility, which are determined by the pressure and composition of the oil and gas. The first type is FCM, in which oil and gas mix on first contact in any proportions. This type of miscibility is very difficult to achieve with the injection of any gaseous agent, but when it occurs (usually with liquefied petroleum gas injection, see Section 6.2.1), virtually all of the residual oil can be recovered. Oil swelling tests determine the minimum pressure at which FCM is achieved.

Miscibility can also develop in a reservoir through multiple contacts, in which gas and oil are mixed in repeated contacts at the forward and backward edges of the oil and gas miscibility zone or a combination of both. Miscibility by multiple contact is achieved by bringing the composition of the phases in each contact closer to a critical point (see Section 6.2.1).

The last type is immiscible. This type is more limited, but it achieves varying degrees of mass transfer between phases. In a strict sense, the term "immiscible" is actually not correct because gas always absorbs some of the components of the oil. Ideal immiscible gas injection is the limiting case in which the solubility of oil in the gas phase is very low or negligible (this occurs at low pressures). The closest analog of this displacement regime is the displacement of oil by water. Oil displacement efficiency increases as the gas stream becomes more miscible.

Let us consider the mechanism of oil displacement during nitrogen injection using ternary diagrams. Summarizing the results obtained from the analysis of ternary diagrams (Figures 6.5–6.9), we can construct the resulting ternary diagram shown in Figure 6.16. Figure 6.16 shows simplified phase behavior at constant temperature and pressure in the reservoir, when only three pseudocomponents are present (L – light, I – intermediate, H – heavy hydrocarbons). In the diagram, all possible compositions that are formed when gas and oil are mixed must lie on the segment drawn between them (O_n G_n). This straight line is called the mixing or dissolving line. From cutting O_2 G_3 does not pass through the two-phase region and characterizes miscible displacement (at first contact). The opposite case is represented by the segment O_1 G_1. It is completely in the two-phase region and characterizes immiscible displacement.

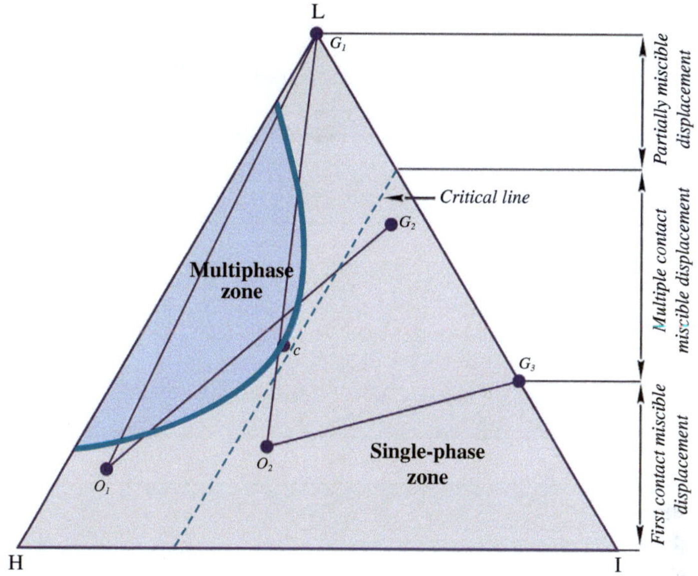

Figure 6.16 Ternary diagram.

When the points corresponding to the composition of the injected gas and formation fluid are located on opposite sides of the critical tie line, displacement is a process occurring either in the gas pressure mode with evaporation of O_2G_1, or in the mode of condensing gas O_1G_2.

Let's take a closer look at these two modes. When implementing the evaporating gas (O_2G_1) regime, intermediate oil hydrocarbons evaporate into an equilibrium gas phase injected into the formation. In this case, miscibility develops at the leading edge of displacement, because two-phase compositions resulting from contacts at the forward edge of the oil and gas mixing zone are constantly enriched, approaching the critical point, until complete mixing occurs. Indeed, when high-pressure gas is injected into the formation (Figure 6.17a), intermediate oil hydrocarbons dissolve in the injected dry gas, gradually enriching it until the complete mixing of oil and gas occurs at the displacement edge in front of it and the formation of a single-phase fluid. When implementing the condensing gas mode (O_1G_2), enriched gas is injected into the formation, and intermediate hydrocarbons from the condensing gas are dissolved in oil. In this case, miscibility now develops at the backward edge of the mixing zone as a consequence of the enrichment of the liquid phase with intermediate hydrocarbons condensing from the gas as a result of contacts at the backward edge of the mixing zone. Indeed, during the injection of enriched gas (Figure 6.17b), the liquid phase is enriched due to the depletion of the gaseous phase, so that miscibility is achieved at the rear edge of the mixing zone. The front of the mixing zone is an area of immiscible flow due to the continuous contact of new portions of the gas phase with oil (this is also true for the rear edge of the mixing zone in the vaporizing gas drive).

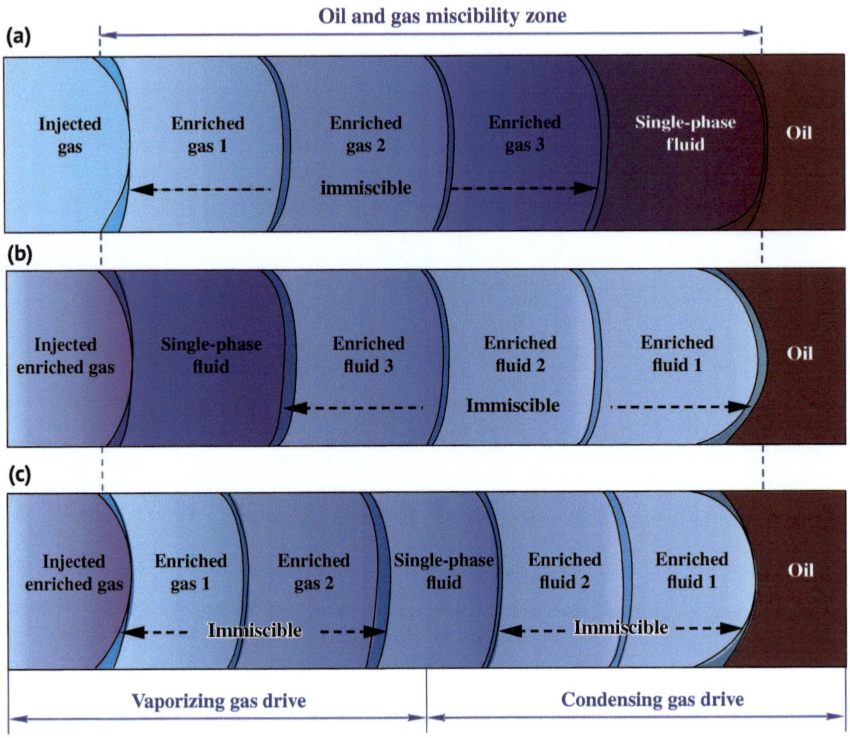

Figure 6.17 Vaporizing and condensing gas drive.

The described simplified model does not adequately reflect the process of displacement when the system contains more than three components. Most real displacement processes using CO_2 and enriched gas have elements of both drive mechanisms—vaporizing and condensing (Figure 6.17c). However, when injecting dry gases such as nitrogen, flue gases, and methane, only vaporizing gas drive occurs.

Figure 6.18 shows a ternary diagram for nitrogen.

As can be seen from the figure, almost the entire diagram is occupied by a two-phase region. Therefore, when nitrogen and flue gases are used, the displacement is generally immiscible. Miscible displacement for nitrogen and flue gases is only possible in a vaporizing drive.

6.3.2 Applicability Criteria

Injection of nitrogen and flue gases, providing miscible displacement, should be carried out only in deep-lying reservoirs since it is achieved only for light oils at high pressures.

As with the injection of hydrocarbon gases, ensuring gravitational stabilization of the displacement front, which usually has an unfavorable mobility coefficient, is possible only in steeply dipping formations. The applicability criteria of the method are shown in Figure 6.19.

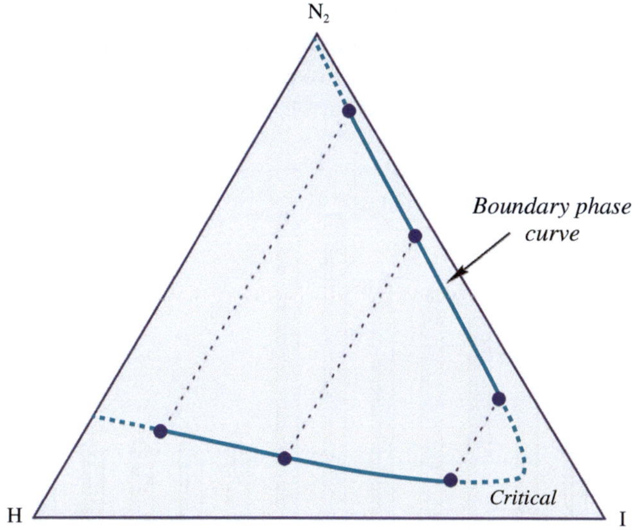

Figure 6.18 Ternary diagram for nitrogen.

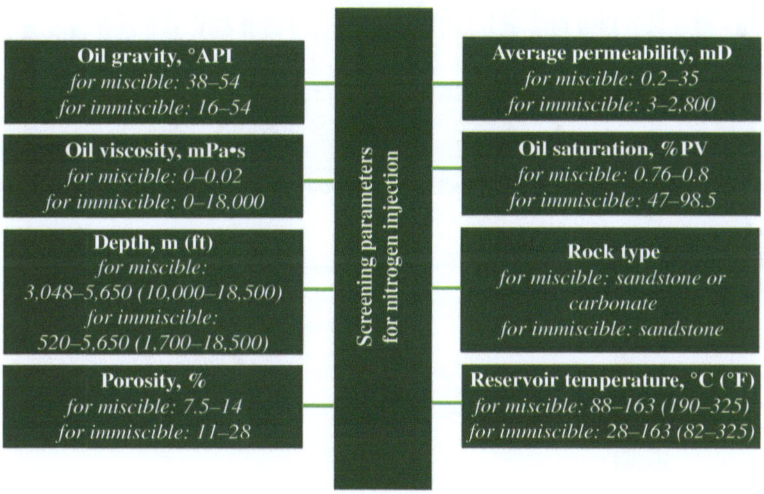

Figure 6.19 Applicability criteria for nitrogen and flue gas injection.

Disadvantages of the method:

- viscous instability leads to low vertical and horizontal sweep efficiency;
- corrosion can cause problems when injecting flue gases;
- oil and gas treatment costs increase, as nonhydrocarbon gases must be separated from the produced natural gas.

6.3.3 Field Implementation

The figures show nitrogen and flue gas injection projects implemented in the USA. As can be seen from the figures, the maximum number of nitrogen injection projects of 10 were implemented in the USA in 1998. At the same time, oil production amounted to 28.12 thousand barrels per day. After 1998, the number of projects implemented in the USA decreased significantly and reached 8 thousand barrels of oil per day in 2014. For flue gases, only two projects were implemented in 1992, and production amounted to 11 thousand barrels per day (see Figures 6.20 and 6.21).

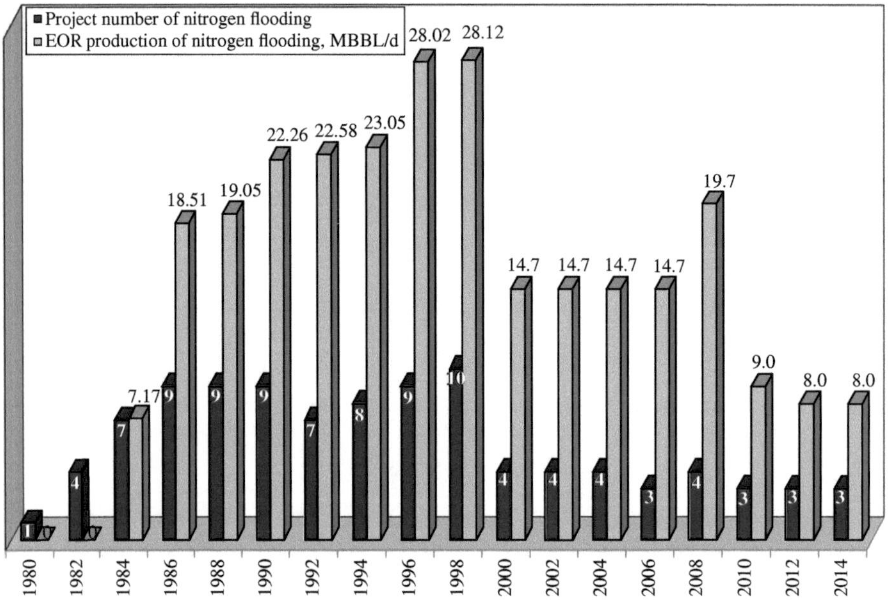

Figure 6.20 Nitrogen injection projects completed in the USA between 1980 and 2014.

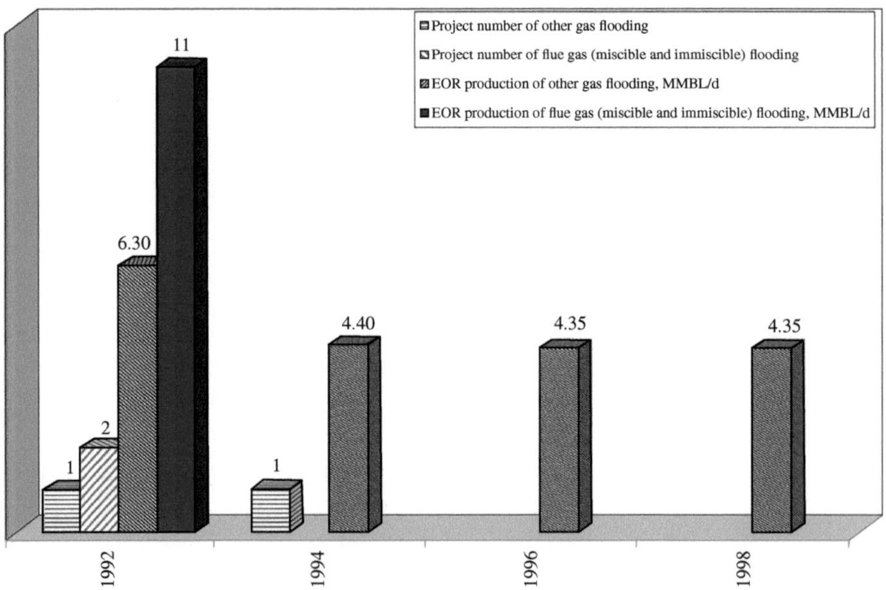

Figure 6.21 Flue gas and other gas injection projects implemented in the USA between 1992 and 2014.

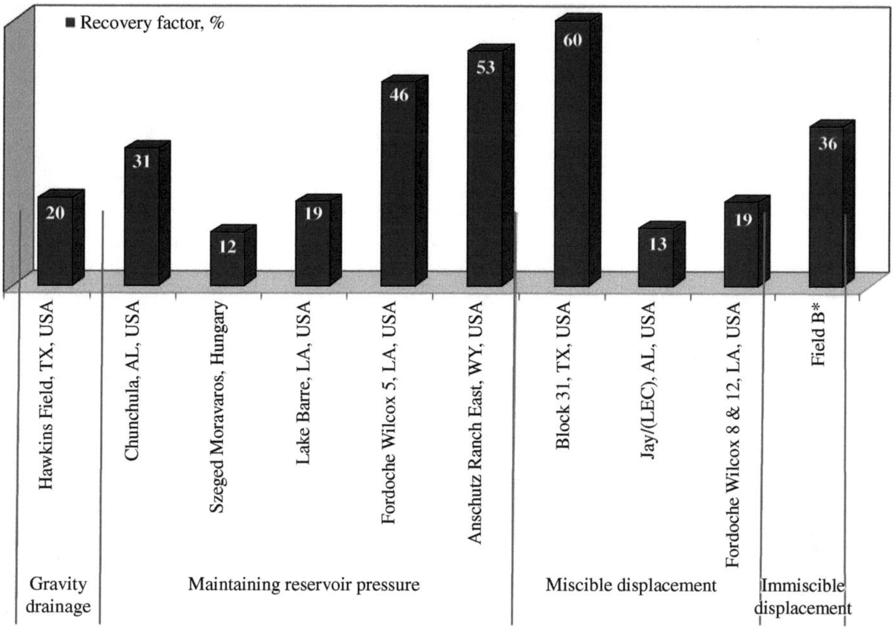

Figure 6.22 Nitrogen injection projects.

The figures show that oil production obtained from the injection of nitrogen and flue gases is significantly lower than for carbon dioxide and hydrocarbon gases.

Figure 6.22 shows the implementation of injection at various fields in the USA and Hungary.

As can be seen from the figure, the oil recovery factor reaches 60%, which is a fairly high figure. Figure 6.23 shows the results of using nitrogen injection at the Mexican Akal field, part of the Cantarell field complex located in the Gulf of Mexico. As can be seen from the figure, nitrogen injection began in 2000, and in 2004, oil production increased from 1,247 (2000) to 2,182 thousand barrels per day.

6.3.4 Implementation Technology

When implementing a project for injection of nitrogen and flue gases, as well as for other gases, first of all, it is necessary to calculate the need for the production of a unit of additional oil. In this case, the volume of injected gas is calculated based on reservoir pressure and temperature.

According to expert calculations, about 45–70 m^3 of nitrogen (1.6–2.5 Mscf) is needed to produce one barrel of additional light oil.

Nitrogen is obtained from air by cryogenic air separation or using noncryogenic membrane air separation technology. Cryogenic technology is the most common for producing nitrogen in the field. At the same time, a cryogenic installation is being built in the field. The productivity of the installations, depending on demand, is 0.3–1.5 million m^3/day. The potential output of a large plant can reach

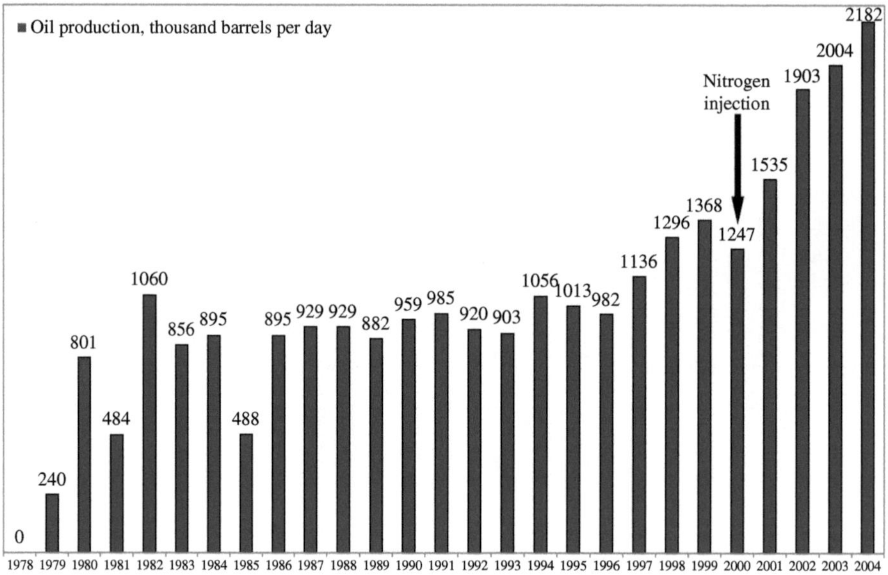

Figure 6.23 Nitrogen injection at the Akal field.

up to 6 million m^3/day. The largest cryogenic plant for producing nitrogen from air was launched in 2000 and expanded in 2004 to inject nitrogen into the reservoir at the Mexican complex of the Cantarell and Ku Maloob Zaap fields, located in the Gulf of Mexico. The capacity of this installation is about 43 million m^3/day (1.5 billion scf/day or 50,000 t/day). Typically, a cryogenic plant contains air compressors for supplying air, the cryogenic distillation plant itself, and compressors for delivering the resulting nitrogen.

The design of the injection well does not differ from the injection of hydrocarbon gases and is shown in Figure 6.15.

There are the following **modifications** of nitrogen injection.

Immiscible displacement of injection into the gas cap: Nitrogen is injected into formations that have a broad anticlinal structure with a significant gas cap (see Figure 6.24). In this case, nitrogen is injected into the roof of the reservoir to maintain reservoir pressure, extract hydrocarbons from the gas cap, and stabilize the gas-oil contact. This is ensured by the anticlinal shape of the structure, and the difference in the densities of nitrogen and displaced reservoir fluids (the density of nitrogen is higher than the density of dry natural gas but lower than the density of oil).

Mixing displacement: As noted above, mixed oil displacement during high-pressure nitrogen injection is achieved by evaporation of intermediate hydrocarbons from the oil and their dissolution in nitrogen. This creates a transition zone of mixing containing a mixture of nitrogen and intermediate hydrocarbons, which dissolve in the oil. This process allows oil to be effectively displaced to production wells (Figure 6.25).

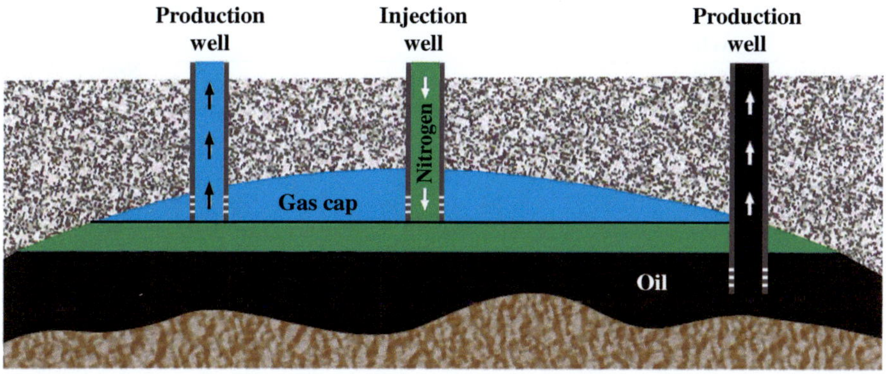

Figure 6.24 Injecting nitrogen into the gas cap.

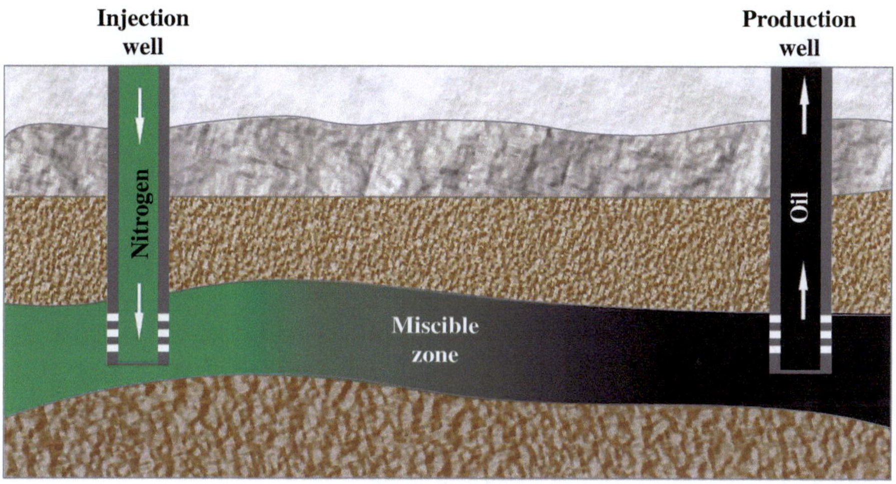

Figure 6.25 Miscible displacement by nitrogen injection.

In some cases, after nitrogen is injected, it moves through the formation with water. In this case, a rim of a mixture of water and nitrogen is formed, which ensures effective displacement of oil (Figure 6.26).

Gravity drainage: It is used in formations that have the potential for gravitational drainage due to the great depth or high thickness of the oil reservoir. Nitrogen, which has a lower density than reservoir oil, when introduced into the reservoir's cap, will displace the oil below it (see Figure 6.27). When implementing gravitational drainage, piston displacement is achieved, preventing viscous instability. In addition, the following is additionally provided: horizontal gas-oil contact; gravity dominates the gas flow; the maximum optimized time between gas injection and oil production; the greater the dip angle of the formation, the higher the injection and production rate without viscous instability, and the more effective the gravity drainage.

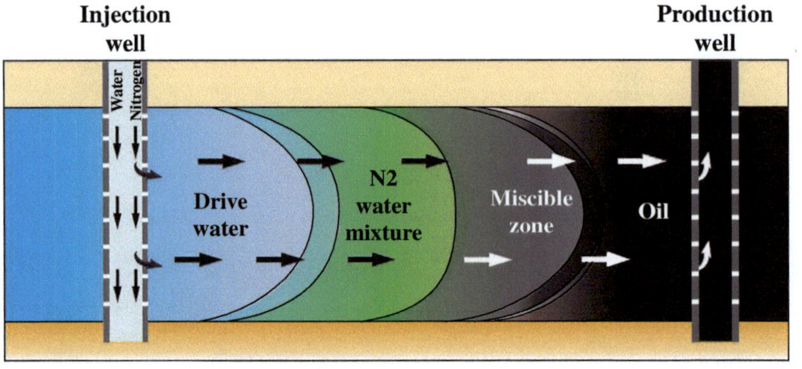

Figure 6.26 Nitrogen/water miscible displacement.

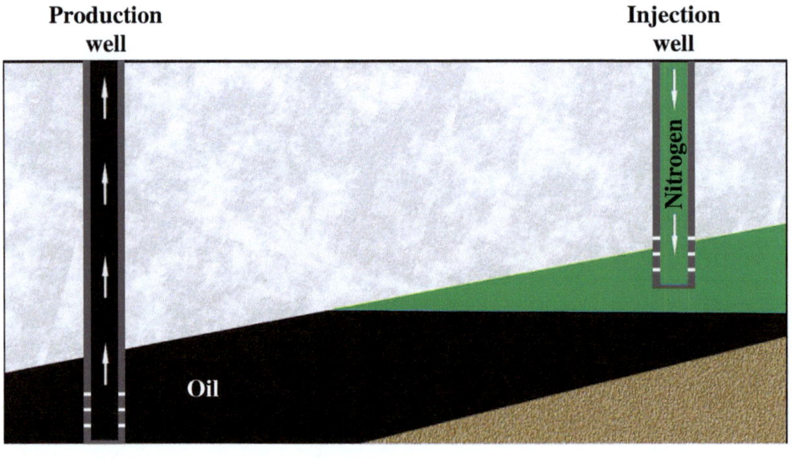

Figure 6.27 Gravity drainage with nitrogen injection.

Maintaining reservoir pressure: Nitrogen is injected into the gas condensate formation at high pressure to maintain reservoir pressure (see Figure 6.28). Maintaining reservoir pressure allows you to restore phase permeability and achieve high gas condensate recovery. The process is carried out in fields with reservoir pressure close to the pressure of the onset of condensation, a high rate of liquid precipitation during condensation (20–40% of the pore volume), with a significant yield of condensate, at least 10^{-3} m^3/m^3 (175 bbl/MMscf).

Development systems: The field development system when using nitrogen and flue gas injection is practically no different from the development system using conventional waterflooding. The main requirement for the development system is that the working agents must be injected via contour or pattern flooding. In this case, small-row development systems are preferable. The placement of wells also does not differ significantly from conventional waterflooding. However, with high vertical permeability and the presence of high gravitational segregation of nitrogen, flue gases, and water, the well pattern should be compacted.

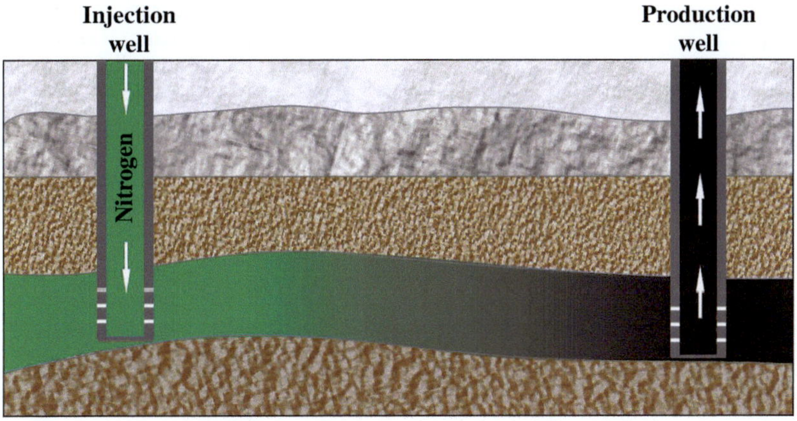

Figure 6.28 Maintaining reservoir pressure in a gas condensate field by injecting nitrogen.

6.4 CO_2 Injection

The utilization of carbon dioxide for enhancing oil recovery has proliferated since its inception in the 1950s. Extensive laboratory investigations and field trials have been meticulously conducted, refining our understanding and application of this innovative technique.

To understand why it has become an important oil injection agent, we must note its basic properties and the factors that make it a useful tool for EOR.

Basic properties of carbon dioxide: Carbon dioxide is a colorless, odorless, inert, and nonflammable gas. It has a molecular weight of 44.01, which is one and a half times higher than that of air. The phase diagram of carbon dioxide is shown in Figure 6.29.

Carbon dioxide is highly soluble in oil and less so in water. In this case, the following factors are observed that contribute to increased oil recovery:

- reducing the viscosity of crude oil and increasing the viscosity of water,
- swelling and decrease in oil density,
- acid attack on carbonate and clay rocks,
- miscibility effect.

As can be seen from Figure 6.29, at temperatures below 31.2 °C, carbon dioxide forms a liquid phase. When carbon dioxide contains hydrocarbons, the critical temperature can increase to 40 °C.

The critical pressure is 7.2 MPa, so at a pressure below the critical pressure, carbon dioxide turns into a gaseous state.

At a pressure of 8–25 MPa and a temperature of 20–100 °C, the density of carbon dioxide in the liquid state is 500–900 kg/m^3, and viscosity is 0.05–0.1 mPa·s. The density of carbon dioxide in the gaseous state under normal conditions is 1.98 kg/m^3, and the viscosity in the temperature range 20–100 °C and pressure 8–25 MPa is 0.2–0.08 mPa·s. If the pressure exceeds 15 MPa at a reservoir temperature below 40 °C, the density of carbon dioxide in liquid and gaseous states is almost the same and amounts to 600–800 kg/m^3.

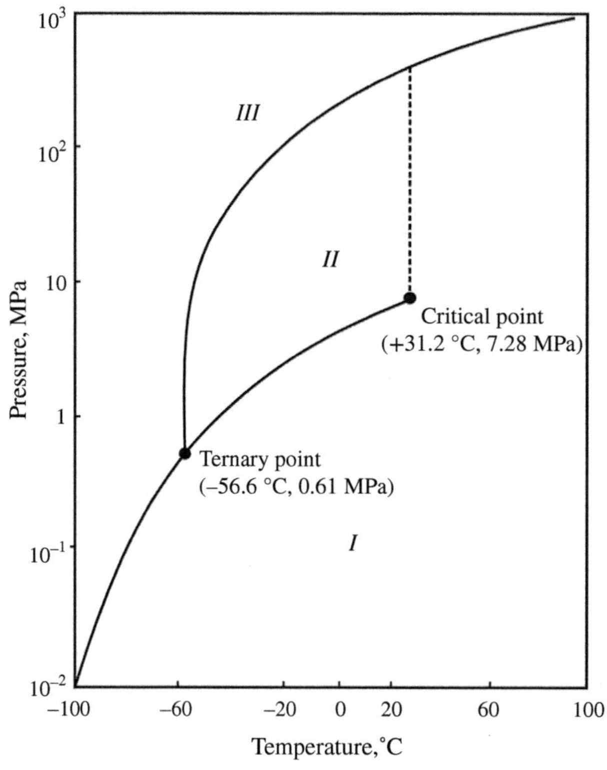

Figure 6.29 Phase diagram of carbon dioxide (I – gas, II – liquid, III – solid).

Dissolution in water: Carbon dioxide is much more soluble in water than hydrocarbon gases. At the same time, an increase in water salinity and temperature reduces the solubility of carbon dioxide; an increase in pressure, on the contrary, increases solubility. Under reservoir conditions, the solubility of carbon dioxide is 30–60 m^3/m^3 (3–5%).

Dissolving carbon dioxide in water increases its viscosity by 20–30%. When carbon dioxide dissolves in water, carbonic acid is formed, which has a significant effect on the dissolution of carbonates and clays. At the same time, the permeability of carbonate rocks increases by 6–75%, and of sandy rocks by 5–15%. Also, the swelling of clays in the presence of carbon dioxide is reduced, which has a significant effect on increasing the permeability of the formation.

Dissolution in oil: An important property of carbon dioxide is its good solubility in oil (4–10 times higher than in water), which ensures its transition into oil from an aqueous solution. In this case, the interfacial tension between them is significantly reduced, and the displacement becomes practically miscible. Unlimited miscibility of carbon dioxide and oil occurs when the pressure of full miscibility is exceeded, regardless of the percentage of CO_2. The indicated pressure significantly depends on the physical and chemical properties of the oil and can be 8–30 MPa. Moreover, with increasing oil density, temperature, and gas saturation pressure, the pressure of complete miscibility increases.

At pressures below miscibility pressure, carbon dioxide and oil separate to form gas and liquid phases. In this case, the gas phase is formed by carbon dioxide containing light fractions of oil, and liquid oil – without light fractions. Heavy hydrocarbons, in particular paraffin, may fall out of oil.

It should be noted that the viscosity of oil decreases significantly when carbon dioxide is dissolved in it. Separation of oil and carbon dioxide promotes the transition of light fractions of oil into carbon dioxide, which leads to an increase in its density, viscosity, and mobility behind the carbon dioxide front.

Complete mixing of carbon dioxide and oil does not occur from the beginning of the carbon dioxide injection process, due to the influence of these factors on miscibility. However, during the oil displacement process, carbon dioxide becomes enriched in hydrocarbons and the displacement becomes miscible. In this regard, the mixing pressure for carbon dioxide is significantly lower than for hydrocarbon gases, nitrogen, and flue gases. To displace light oil, the pressure required for miscible displacement by hydrocarbon gases is almost twice that of carbon dioxide.

It should be noted that oil swelling (increase in volume) with the dissolution of carbon dioxide has a significant effect on increasing oil recovery. In this case, a significant decrease in oil viscosity occurs. The volume of oil increases by 1.5–1.7 times, while the increase in density is insignificant (2–3%).

6.4.1 Process Mechanism

Miscible displacement: In miscible displacement, oil is displaced by carbon dioxide, similar to a normal solvent. In this case, three zones are formed in the formation in the direction of oil displacement: the carbon dioxide zone; the transition zone containing both CO_2 and oil; zone of residual oil. Laboratory experiments on natural cores show that the displacement efficiency of miscible displacement of oil by carbon dioxide can be as high as 0.95 (or 95%).

Determination of miscibility pressure: The pressure of complete miscibility is determined experimentally. The most common two methods for determining miscibility pressure.

Gravity-stable experiments use a vertical-fill test column, saturated with oil, into which carbon dioxide is injected from above, displacing the oil vertically downward at a rate slow enough to maintain a gravitationally stable flow, overcoming the tendency of the displacement front to break through. The experiment is carried out at different pressures and the oil displacement coefficient is determined. A graph of the displacement coefficient versus pressure is plotted (see Figure 6.30). As can be seen from Figure 6.30, the minimum miscibility pressure (MMP) is the pressure corresponding to the breakover point of the displacement ratio versus pressure.

The second method for determining the minimum miscibility pressure is a slim tube experiment. In this case, a spiral stainless steel slim tube with a diameter of 0.25 inches and a length of 1.2 m is filled with sand and saturated with oil at a given pressure and temperature. A sufficiently high ratio of the length of the tube to its diameter ensures a stable front for displacing oil with carbon dioxide without fingering. The results of the experiments are presented

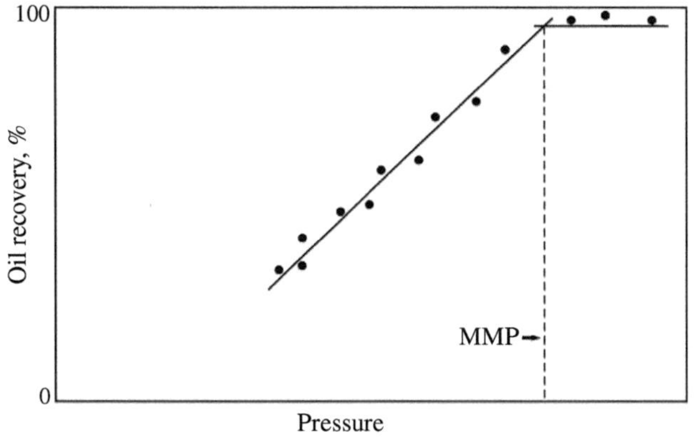

Figure 6.30 Minimal miscible pressure determination.

in the form of a dependence of the displacement coefficient on pressure (see Figure 6.30), from which the minimum miscibility pressure of carbon dioxide and oil is determined.

Immiscible displacement: In the case of immiscible displacement, light fractions of oil are converted to carbon dioxide, and part of the carbon dioxide is dissolved in oil. At the same time, due to the fact that carbon dioxide, enriched in light oil fractions, displaces oil saturated with CO_2, component separation of oil occurs, and hard-to-recover oil saturated with heavy components is formed in the washed zones.

At reservoir temperatures above critical, carbon dioxide is in a gaseous state, regardless of pressure. In this case, field development performance will be significantly worse than with mixing displacement. This is explained by the unfavorable ratio of the mobility of oil and gas, and the low coverage of heterogeneous formations by the influence. In this regard, to achieve technological and economic efficiency when injecting carbon dioxide, it is necessary to ensure that it is in a liquid state, since in this case, the efficiency of the oil displacement process is much higher. This is achieved by using the method in fields with temperatures close to critical (25–35 °C).

Volume effects: Along with leveling the mobility of the displacement front when injecting carbon dioxide by reducing the viscosity of oil and increasing the viscosity of water, an increase in oil volume is one of the important factors in increasing the efficiency of oil displacement by carbon dioxide. The content of light hydrocarbons, reservoir pressure, reservoir temperature, and the amount of dissolved gas have a significant impact on the volumetric expansion (swelling) of oil.

An artificial increase in oil saturation due to an increase in oil volume leads to an increase in pore pressure. As a result, stationary residual oil is additionally displaced. In this case, the oil recovery factor due to oil swelling can increase by 6–10%.

6.4.2 Applicability Criteria

Carbon dioxide injection should be carried out in reservoirs of moderately light oil (API ≥28), and the reservoir should be deep enough (≥1500 m) to provide high enough pressure to achieve miscibility. Carbon dioxide, when dissolved in water, can reduce the interfacial tension between oil and water. However, this process can also lead to corrosion problems. In this process, about 20–50% of the carbon dioxide slug is displaced by water. Water is usually injected together with CO_2 in water-gas mode to improve the mobility ratio of the displacing fluid and oil. The criteria for the applicability of the method are shown in Figure 6.31.

Carbon dioxide injection is the fastest-growing EOR method in the USA, with completed projects continuing to demonstrate good incremental oil production. Carbon dioxide injection is used as both a secondary and tertiary method for enhancing oil recovery. However, the largest CO_2 injection projects are tertiary projects in long-term fields that have been waterflooded for many years.

The disadvantages of the method include:

- early breakthrough of CO_2 due to its low viscosity,
- corrosion in production wells,
- the need to separate CO_2 from produced reservoir fluids,
- the need to pressurize CO_2 (for re-injection),
- high demand for CO_2 per unit of oil produced.

6.4.3 Field Implementation

Since the early 1970s, high oil prices have generated widespread interest in carbon dioxide injection. In this case, both miscible and immiscible displacement of oil with carbon dioxide was carried out.

Due to the presence of large reserves of carbon dioxide in the USA, which are located in the same geological basin with oil fields, this method of EOR is most widespread in this country.

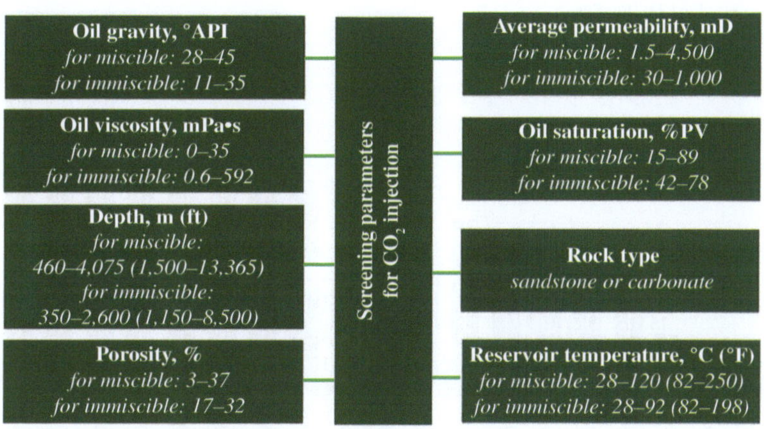

Figure 6.31 CO_2 injection applicability criteria.

To deal with the global oil crisis in 1972, the US government passed the Emergency Energy Security Act, which prompted a significant increase in US oil production and oil companies' exploration of promising EOR technologies.

Since 1980, oil and gas prices have increased significantly, leading to an increase in the number of carbon dioxide injection projects implemented, and in 2006, for the first time, oil production from this method exceeded oil production from thermal-EOR methods. At the same time, there were two surges of interest in the carbon dioxide injection method. The first surge occurred between 1980 and 1992 when laws and regulations were passed to encourage investment in the energy sector. At the same time, the number of projects increased from 17 to 54, and oil production reached 144.97 thousand barrels per day (see Figures 6.32 and 6.33). The second jump occurred in 2002 when the price of oil soared to \$100 per barrel, which led to a significant increase in profits from the sale of CO_2 injection.

During this period (after 2002), the number of projects increased from 67 (2002) to 137 (2014), and oil production increased from 187.41 to 292.74 thousand barrels per day.

As Figure 6.32 shows, there were 128 mixed displacement projects active in the USA in 2014, of which 104 were successful (81.25%). And only 9 projects were with immiscible displacement; they were less effective. At the same time, the main oil production (63%) comes from 24 largest projects.

Figure 6.34 shows the implementation of projects for miscible injection of carbon dioxide by oil-producing companies in 2014.

In addition to the USA, projects to inject carbon dioxide into the reservoir have been implemented in other countries (see Figures 6.35 and 6.36). As can be seen from Figure 6.35, carbon dioxide injection projects have been implemented in the following countries: Brazil, Hungary, Canada, China, Trinidad, and Turkey. From Figure 6.36, it can be seen that the best performance was obtained at the Weyburn Unit (Canada) and Batman (Turkey) fields.

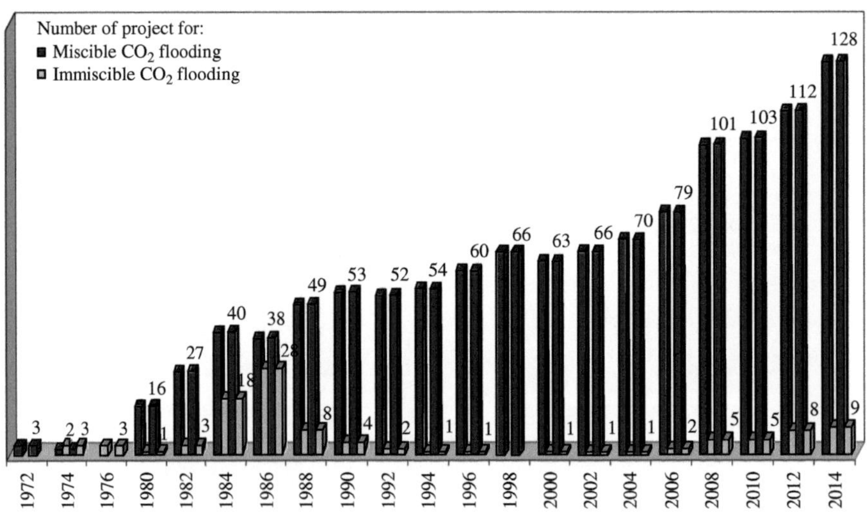

Figure 6.32 CO_2 injection projects implemented in the USA between 1972 and 2014.

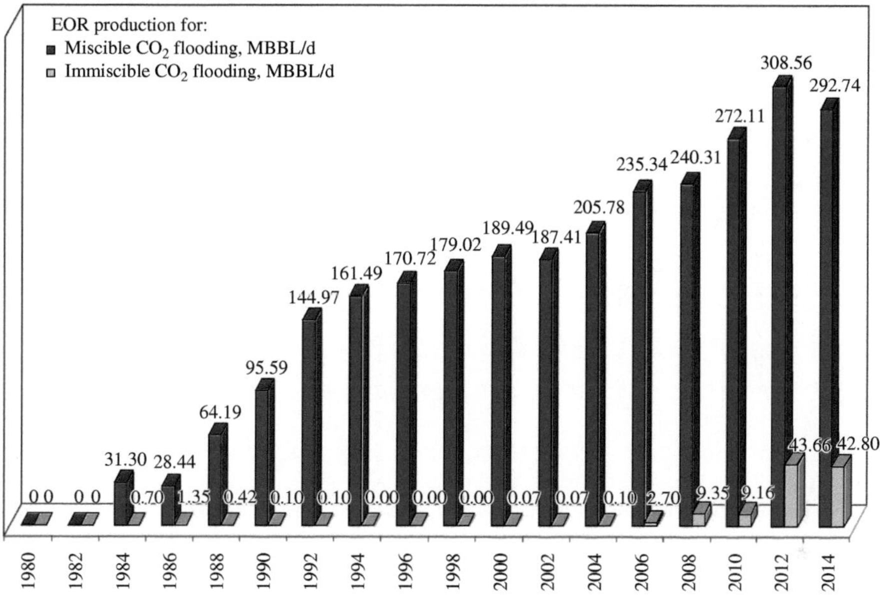

Figure 6.33 Oil production in the USA, obtained from the implementation of CO_2 injection into the reservoir.

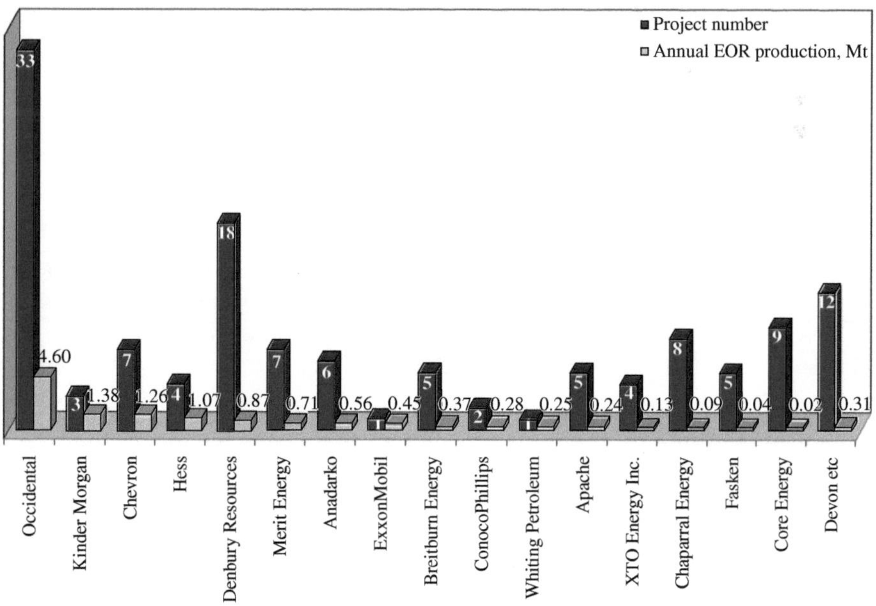

Figure 6.34 Number of projects and annual production in 2014 from the implementation of miscible CO_2 injection by operating companies.

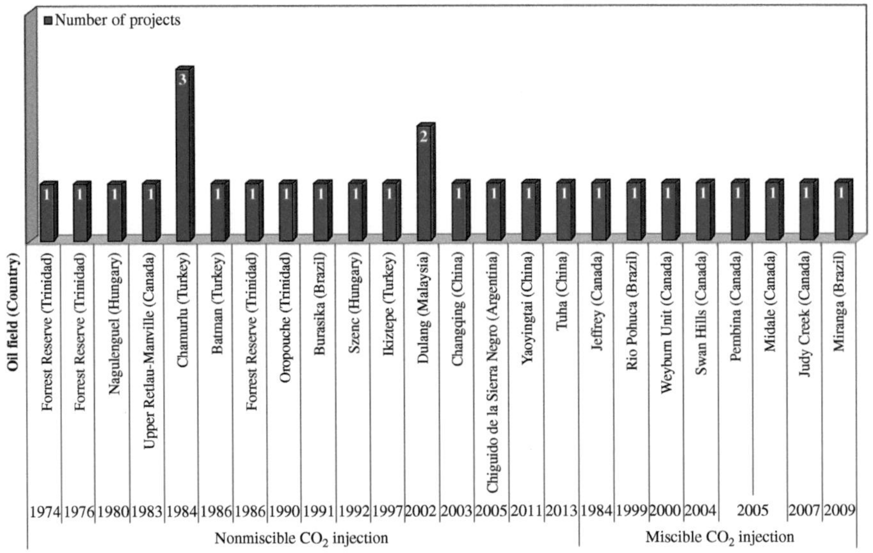

Figure 6.35 Carbon dioxide injection projects implemented in different countries.

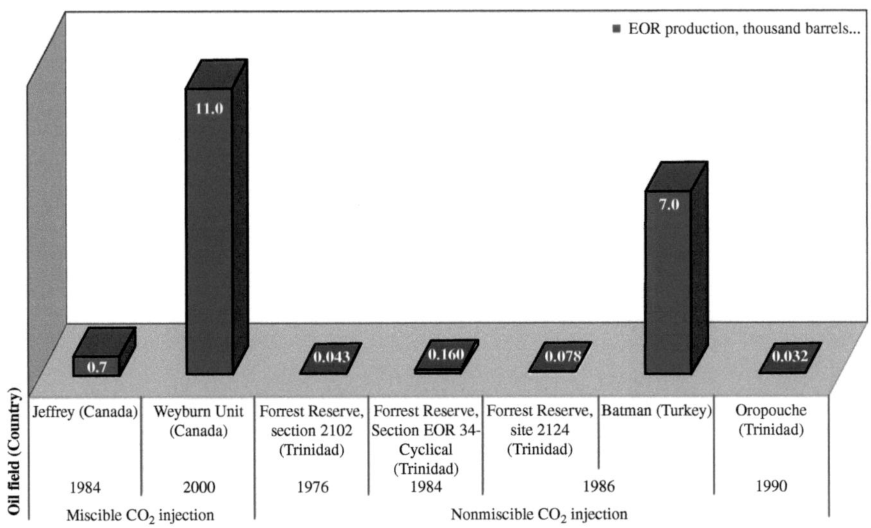

Figure 6.36 Oil production obtained from the implementation of CO_2 injection projects into formation in different countries.

Projects to inject carbon dioxide into the reservoir were also implemented in the former USSR (see Figure 6.37).

The first field experiment on carbon dioxide injection in the USSR was carried out at the Aleksandrovskaya area of the Tuymazinsky field. The event was carried out in 1967 with the start of the injection of carbonated water. Carbonated water was

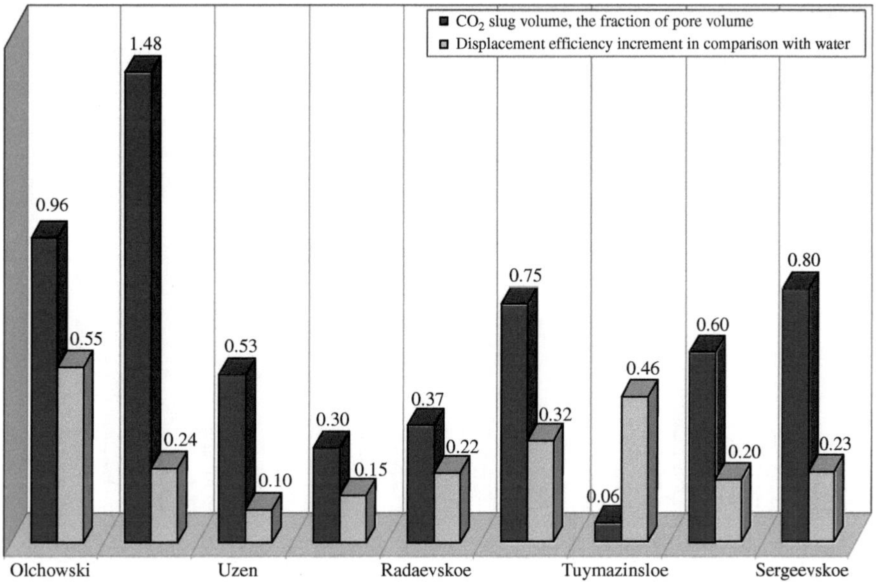

Figure 6.37 Carbon dioxide injection projects implemented in the former USSR.

produced at the bottom of the well by simultaneously injecting carbon dioxide into the tubing and water into the annulus. As a result of using the method, oil recovery increased by 15.6%.

Of interest is the analysis of carbon dioxide injection projects with immiscible displacement. Figures 6.38 and 6.39 show the number of projects and their distribution by country, respectively. As can be seen from Figure 6.38, in 1980, the number of projects increased sharply, which is associated with rising oil prices. The leader in the application of the immiscible displacement method using carbon dioxide is the USA.

Figure 6.40 shows the most successful immiscible displacement projects using carbon dioxide. As can be seen from the figure, the best results were achieved in the fields of the USA, Brazil, and Hungary.

6.4.4 Implementation Technology

When implementing a carbon dioxide injection project, first of all, it is necessary to calculate the need for carbon dioxide to produce a unit of additional oil, determine the source, and study the transport infrastructure for delivering carbon dioxide to the field.

CO_2 consumption per unit of production: According to calculations by experts, during secondary oil production using carbon dioxide, to extract one barrel of additional light oil, about 400 (13.5 Mscf) are needed, and for heavy oil – 800 (24.8 Mscf) m^3 of CO_2. In tertiary oil production, about 500 (16.4 Mscf) m^3 of CO_2 is required to produce one barrel of additional light oil. The calculation of the required amount of carbon dioxide can be calculated using the formula given in Section 6.2.4.

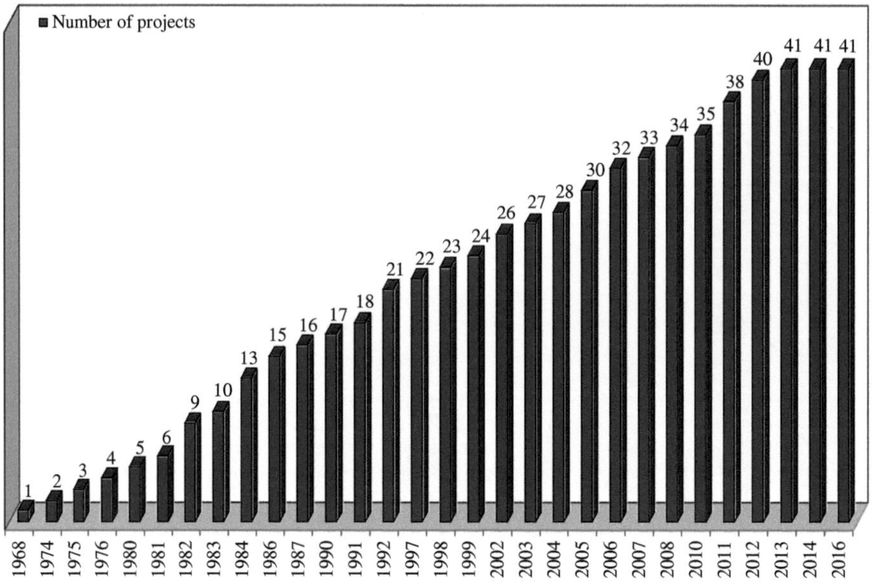

Figure 6.38 Number of immiscible displacement carbon dioxide injection projects.

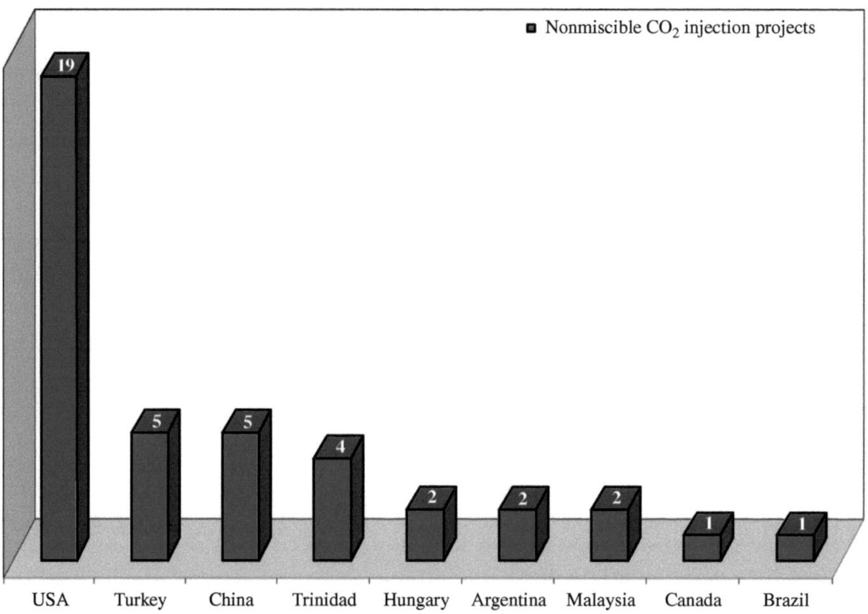

Figure 6.39 Distribution of the number of carbon dioxide injection projects with immiscible displacement by country.

A reliable source of supply is very important because gas must be supplied on a continuous basis in large volumes over a long period of time, up to 5 years or more. The gas used must be 90% pure or greater. If other gases such as methane or nitrogen are present, higher injection pressure will be required to provide miscible displacement.

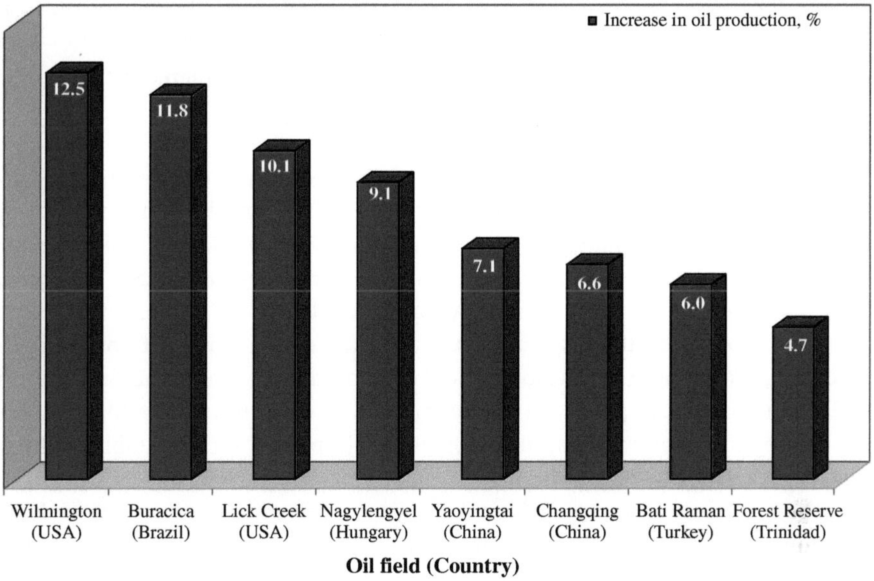

Figure 6.40 Increasing oil production by implementing the immiscible displacement method using carbon dioxide.

The best sources are high-purity, high-pressure natural gas reservoirs that are primarily found in oil and gas exploration. In the USA, the oil basins of Wyoming, Utah, Colorado, and New Mexico have the largest reserves of carbon dioxide.

However, research shows that currently known high-purity natural sources can only supply less than 15% of the total demand. Other natural sources include natural gas deposits containing carbon dioxide released from gas processing plants (Delaware and Val Verde basins in southwest Texas). In addition, all sources where carbon dioxide is a by-product can be used, such as coal, oil and gas power plants, cement and ammonia plants, oil refineries, and ethanol and ethylene oxide plants. This total quantity is more than enough to satisfy existing demand.

The most convenient source must be considered for each specific field to ensure cost-effective implementation of the method.

The method of transportation from the source to the oil field depends on what phase state the carbon dioxide is in—liquid or gaseous. At low injection rates (up to 0.2 million m^3/day), the least expensive method of transport is to transport carbon dioxide as a liquid using existing insulated steel containers at 2 MPa pressure and 0 °F (−17.77 °C). For large long-term projects, it is most economical to transport through a pipeline in gaseous form at a pressure of 10–14 MPa (i.e., above critical), which ensures single-phase flow.

Miscibility pressure: Next, it is necessary to experimentally determine the minimum miscibility pressure.

Analysis of all these parameters allows us to analyze the technological and economic efficiency of the method for a specific field and begin its implementation.

Typically, several **modifications** of the method of enhancing oil recovery by injection of carbon dioxide are used.

Injection of carbonated water: The simplest way to implement the method is by dissolving 3–5% CO_2 in water and then pumping carbonated water into the formation (see Figure 6.41).

At the same time, the good solubility of carbon dioxide in oil ensures its transition into the oil remaining behind the displacement front from the aqueous solution. As a result, the interfacial tension between oil and carbon dioxide is significantly reduced, the volume of oil increases, viscosity decreases, and, accordingly, the phase permeability for oil increases. A significant lag of the carbon dioxide concentration front from the displacement front extends the period of obtaining the effect and, accordingly, the period of field development. At the same time, despite a slight increase in impact coverage, compared to conventional flooding, due to a decrease in the action of capillary forces and improved rock wettability, the oil recovery coefficient changes slightly.

Carbon dioxide slug injection: Another modification of the method is the injection of a carbon dioxide slug in a volume of 10–30% of the pore volume, which is propelled by injected water (see Figure 6.42).

The scheme of displacement of carbon dioxide by a rim is shown in Figure 6.43.

When a carbon dioxide slug is injected into the formation, the following zones are formed:

- single-phase oil flow occurs;
- there is a joint flow of water, oil, and carbon dioxide, accompanied by active mass transfer between them;
- has a joint flow of oil and carbon dioxide, with mass transfer also occurring;
- there is a flow of carbonated water, and oil, depleted in light fractions and therefore inactive, as well as gaseous carbon dioxide;
- there is a flow of injected water and oil, while carbon dioxide can pass from oil to water;
- movement of oil and water takes place.

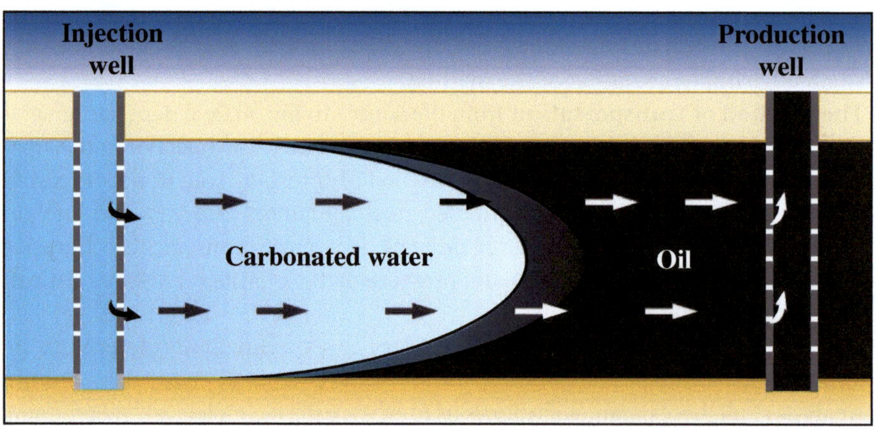

Figure 6.41 Carbonated water injection.

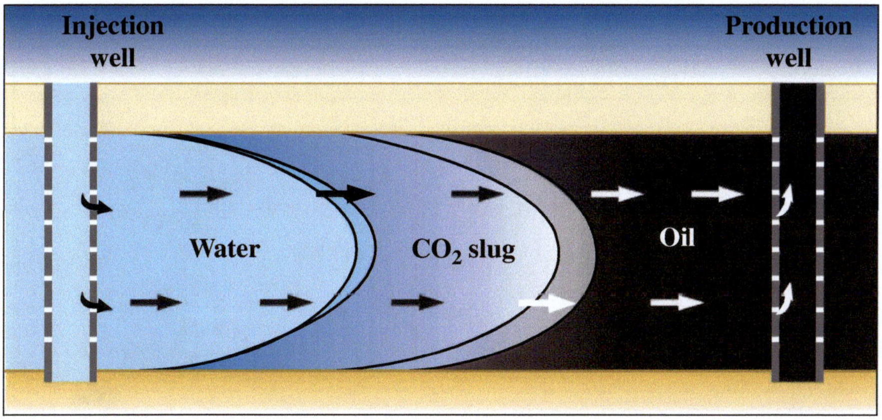

Figure 6.42 Injection of CO_2 slug.

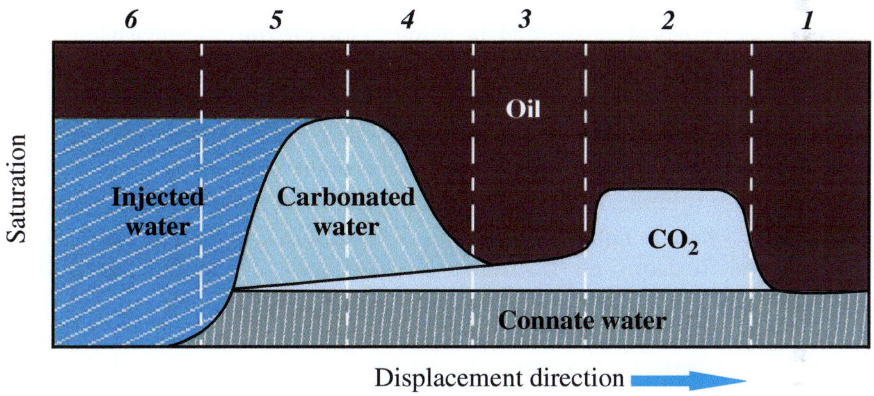

Figure 6.43 Artificial zoning and processes during CO_2/water slug injections.

The best effect when using this modification of the method is achieved when the rim volume is 20–30% of the pore volume. At the same time, the development period is significantly increased, the volume of injected water is reduced and a high oil recovery factor is achieved. The gravitational effect has a significant influence on the process. In this regard, with high vertical permeability, the oil recovery factor is significantly reduced (by 2–3 times).

Alternating injection of carbon dioxide and water slugs: According to numerous studies, the highest efficiency is achieved by a modification of the method in which alternating injection of carbon dioxide and water is carried out (see Figure 6.44).

At the same time, the effectiveness of implementing this modification of the method significantly depends on the correct choice of the size of the injected slugs and the gas-water ratio. Injection of slugs can be carried out either simultaneously or alternately.

The implementation of this modification shows that the optimal ratio of carbon dioxide and water rims lies in the range of 0.25–1. At the same time, a low value

Figure 6.44 Alternating injection of carbon dioxide and water.

reduces the efficiency of the method, bringing it closer to the efficiency when injecting carbonated water, and a high value can lead to an early breakthrough of carbon dioxide into production wells. In a heterogeneous formation, the volumes of fringes should be lower than in a homogeneous formation. In the case of complete miscibility of carbon dioxide and oil, the volumes of the rims can be quite large and reach 20% of the pore volume. With low miscibility, the volumes of the slugs should be small when they are pumped alternately.

Other modifications of the method: There are other modifications to the method in question. To reduce carbon dioxide consumption, it is possible to use alternating injections of carbon dioxide, water, and other more accessible gases. You can add various surfactants to the water, which will ensure the formation of foam in the formation. It is also possible to use polymer additives to the injected water.

The design of the injection well does not differ from the injection of hydrocarbon gases and is shown in Figure 6.15.

Development systems: The field development system when using carbon dioxide injection is practically no different from the development system using conventional waterflooding. The main requirement for the development system is that working agents must be injected via contour or pattern flooding schemes.

The placement of wells also does not differ significantly from conventional waterflooding. However, with high vertical permeability and in the presence of high gravitational segregation of carbon dioxide and water, it is necessary to compact the well pattern.

6.5 Water–Gas Impact on the Formation

The technology of injecting natural or petroleum gas to maintain reservoir pressure and enhance oil recovery in depleted oil fields, particularly in the USA, was used much earlier than waterflooding. At the initial stage, gas was injected into the formations at pressures below the pressure of its miscibility with oil. During the period of time when waterflooding of oil reservoirs was not used, gas injection technology, which contributed to an increase in oil flow rates and, in general, oil recovery from gently sloping deposits from 5 to 10%, and from steeply occurring reservoirs from 15 to 25% in relation to the dissolved gas regime, was considered economically expedient.

Later, when the process of artificial waterflooding in shallow oil reservoirs was widely introduced, it was found that water injection promotes greater oil displacement than injection of gases that do not mix with oil. The low efficiency of gas injection as a displacing agent, due to its breakthrough through highly permeable areas to production wells, a decrease in oil production from wells, and in general the coverage of the formation by the influence, is associated with the low (compared to water) viscosity of the gas.

In order to eliminate the above negative results, combined methods have been developed, including gas and water injection. Since 1951, at the Starogroznenskoye field (Russia), gas injection technology has been used together with edge-to-side flooding to intensify the process of reservoir pressure restoration and increase oil recovery. Regularly prescribed portions of water in the volume of 0.5–1.0 m^3 were injected into gas injection wells to prevent gas breakthrough into production wells. It was found that periodic injection of alternating slugs of water not only eliminates gas breakthroughs but also helps to increase oil recovery from the pilot site. For the first time in the West, WAG technology was implemented in 1957 in Canada at the North Pembina field.

Subsequently, other variants of WAG were developed, such as sequential, simultaneous injection of water and gas, as well as foam WAG, during which a foaming *surfactant is added to the injected water.*

6.5.1 Process Mechanism

The essence of WAG injection is the simultaneous or separate injection of water and gas into an oil reservoir.

Gas injected into the formation, which is a nonwetting phase, moves into large pores and overlying interlayers under the influence of gravitational forces, and injected water, on the contrary, under the influence of capillary forces, occupies small pores of the hydrophilic formation. Taking into account the above characteristics of water and gas and in order to achieve a greater effect from the impact as a result of combining their advantages, it is advisable to perform alternate, cyclic injection. With this technology, it is necessary that the optimal ratio of injected volumes of water and gas be proportional to the ratio of the volumes of small and large pores of the reservoir, taken respectively below and above the average pore size. The use of water-gas mixtures in this sequence will give the maximum positive effect as a result of combining the properties of the injected water and gas.

An increase in reservoir oil recovery with alternate injection of water and gas also occurs due to taking into account the fact that the phase permeability of the wetting phase depends only on the saturation of the reservoir with water. The increase in oil displacement from the reservoir in the presence of free gas occurs by the value of the maximum gas saturation and amounts to 10–15%, at which the gas is in a stationary state.

In practice, various variants of water injection are used, differing in the method of supplying working agents to the reservoir and in the miscibility of the injected gas in the reservoir oil.

Options for water injection, differing in the method of supplying working agents to the formation, are as follows.

- The alternating supply of working agents (see Figure 6.45) includes gas injection accompanied by water injection. At the same time, gas injection increases the efficiency of displacement, and water – the coverage of the formation by the influence.
- Sequential supply of working agents (see Figure 6.46) includes gas injection accompanied by water injection. At the same time, water injection increases both the displacement efficiency and the formation coverage of the impact.
- Hybrid WAG (see Figure 6.47) includes the injection of a large portion of gas, accompanied by the injection of small portions of water and gas.
- Joint injection of working agents (see Figure 6.48). In this case, the working agents are mixed at the wellhead and then the mixture is pumped into the well.
- Separate and joint injection of working agents (see Figure 6.49). In this case, working agents are fed into the well through various pipes, mixing at the bottom.
- Foam-assisted WAG (see Figure 6.50) is carried out by adding a foaming surfactant to the injected water. This ensures the best coverage of the formation by the impact (see Figure 6.51). At the same time, the gravitational separation of working agents and the influence of rock heterogeneity are significantly reduced.

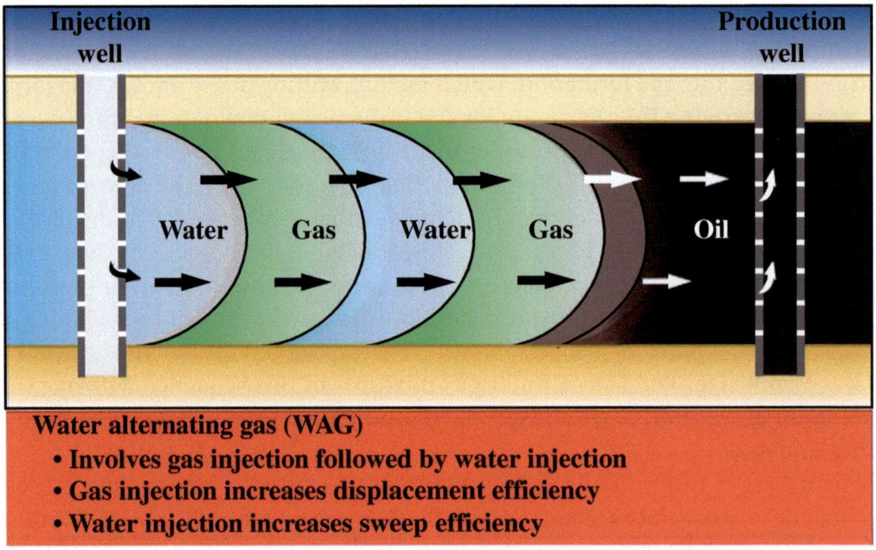

Figure 6.45 WAG process.

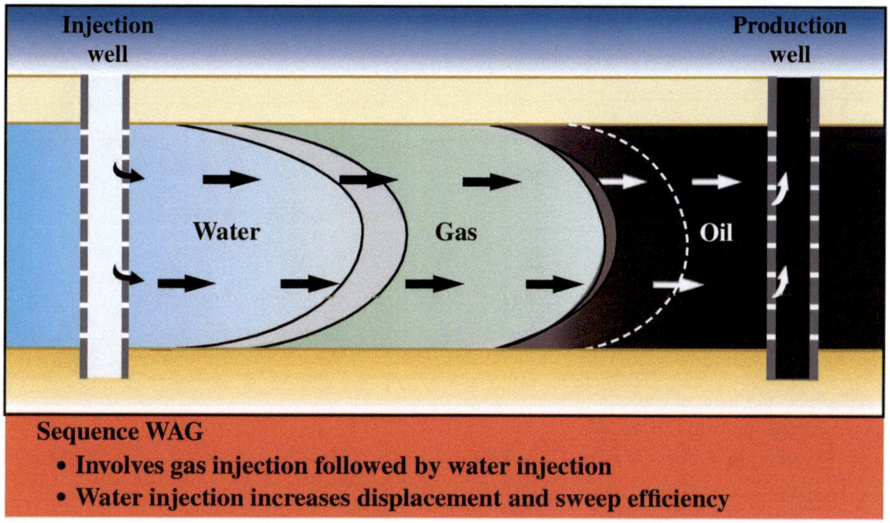

Figure 6.46 Sequential injection of gas and water.

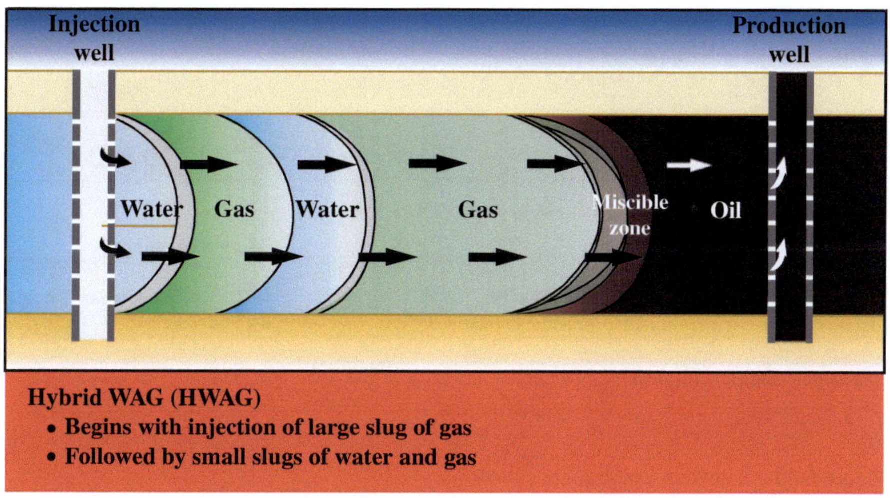

Figure 6.47 Hybrid WAG.

The WAG options for the miscibility of injected gas into reservoir oil are as follows.

- MWAG in the presence of gas and oil miscibility at the minimum miscibility pressure (MPP) (see Figure 6.52). At the same time, the viscosity of oil decreases, and its mobility increases.
- IWAG in the absence of miscibility of gas and oil. In this case, the process is carried out at a pressure below the MMP and is characterized by the presence of a front between the oil and gas phases. The process continues until gas breaks through into the production well.

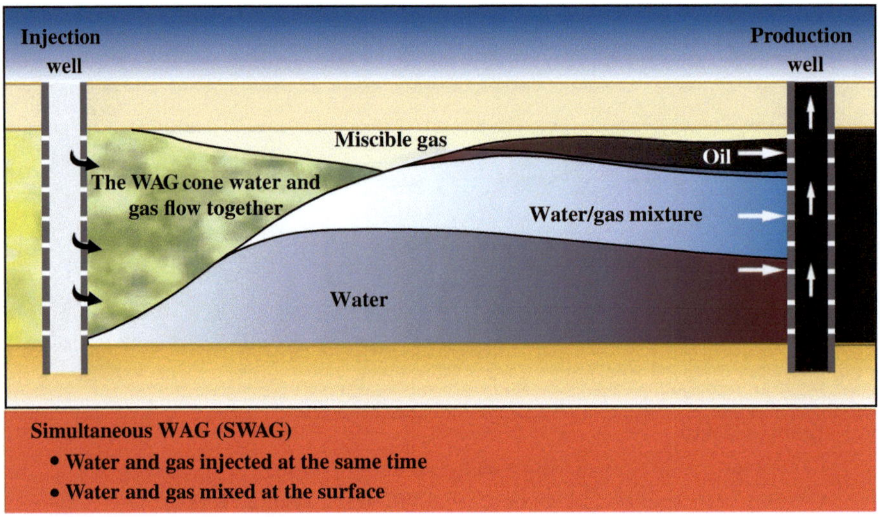

Figure 6.48 Simultaneous WAG.

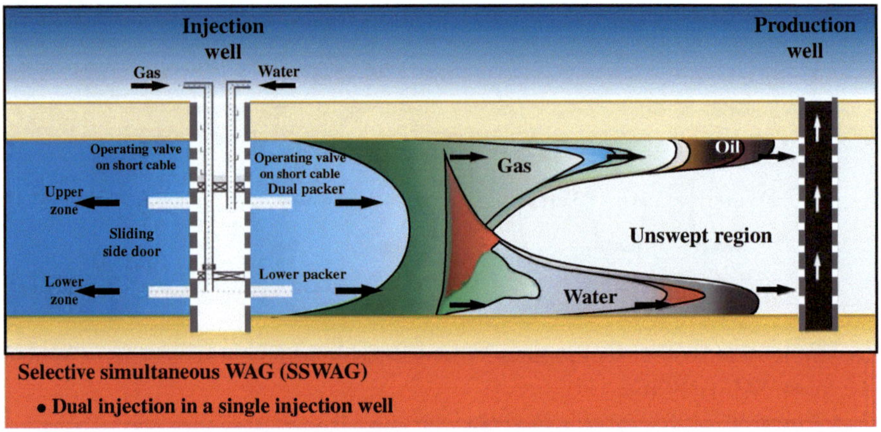

Figure 6.49 Selective simultaneous WAG.

6.5.2 Applicability Criteria

As a result of alternating injection of water and gas, the impact coverage of heterogeneous formations increases due to the blockage of highly permeable areas with a water-gas mixture and the connection of low-conductivity zones to development. It has been established that the final oil recovery of a heterogeneous formation when using any technology of joint injection of water and gas is higher than when oil is displaced only by water or gas. The introduction of water-gas stimulation under optimal options increases reservoir oil recovery by 7–15% compared to conventional waterflooding. The best effect as a result of the use of water-gas stimulation can be achieved

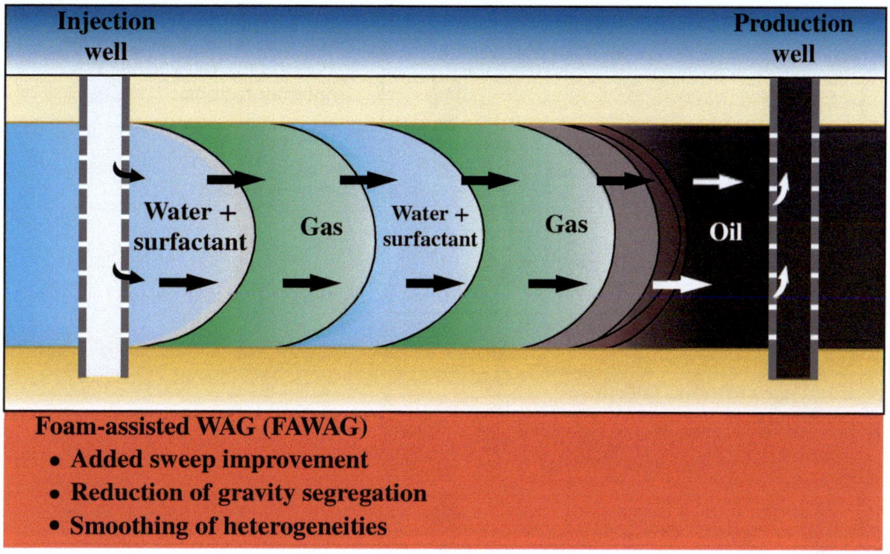

Figure 6.50 Foam-assisted WAG.

in the case of a uniform distribution of the injected gas across the thickness of the flooded reservoir and a simultaneous breakthrough of water and gas to the production wells, which is not always possible to achieve. As a result, in practice, the efficiency of the process is much lower than above, especially in homogeneous formations.

Injecting alternating slugs of water and gas into the formation or simultaneously injecting a water-gas mixture through the same injection well also reduces the effect obtained from the impact. This is due to the fact that after the first cycle of injection of working agents, the phase permeability of the bottom hole zone decreases and, as a result, the injectivity of the injection well decreases sharply: for gas, respectively, by 8–10 times, for water, 4–5 times.

The separation of gas and water in reservoir conditions under the influence of gravitational forces can also negatively affect oil displacement and reservoir coverage. Depending on the heterogeneity of the reservoir and the viscosity ratio of formation fluids, the efficiency of the process may decrease by 10–20%.

It should be noted that for alternating injection of water and gas, the equipment of each injection well becomes significantly more complicated. To eliminate these shortcomings and ensure the process of cyclic water-gas injection with the calculated volumes of working agents introduced into the formation and the required water and gas injection pressures, it is necessary to significantly increase the number of injection wells with corresponding sophisticated wellhead equipment. Reducing or eliminating the negative impact of these factors can be achieved through the use of acceptable technological and technical solutions. For this purpose, it is necessary to ensure the ability to regulate the direction of flow and the beneficial use of gravitational effects through optimal well locations, formation conditions, and well operation parameters (see Figure 6.53).

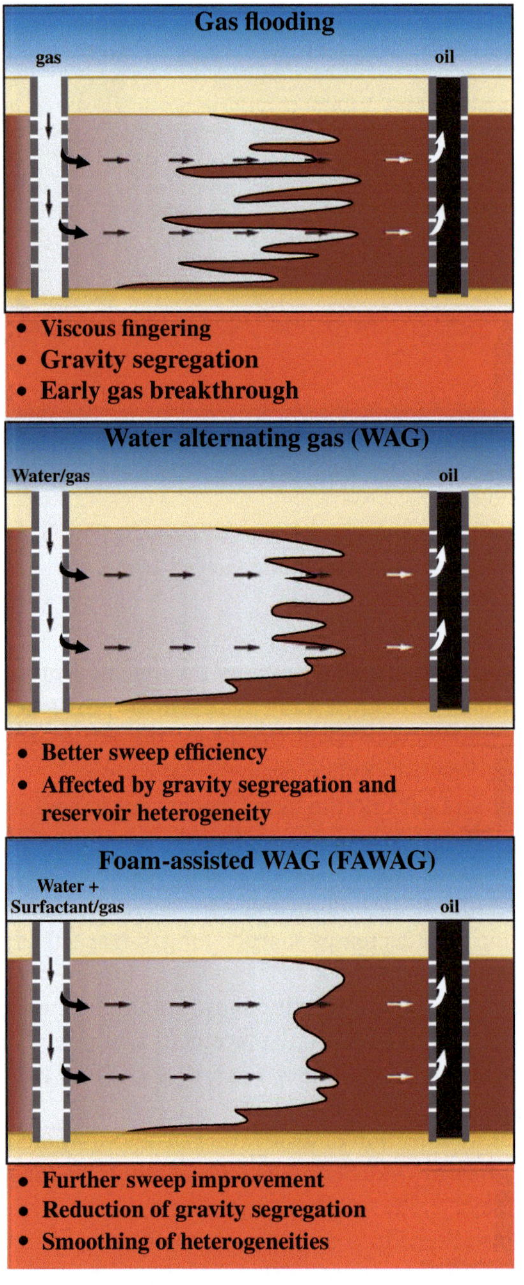

Figure 6.51 Comparison of main processes during simple gas flooding and various WAG implementations.

6.5.3 Field Implementation

In implemented WAG technology projects, air, hydrocarbon gas, nitrogen, acid gas, and carbon dioxide are used as gas.

From the 1950s to the present, about 100 WAG projects have been implemented, the vast majority in the USA; less than 10 were ineffective % of them. Figure 6.54 shows implemented gas injection projects and WGW technology.

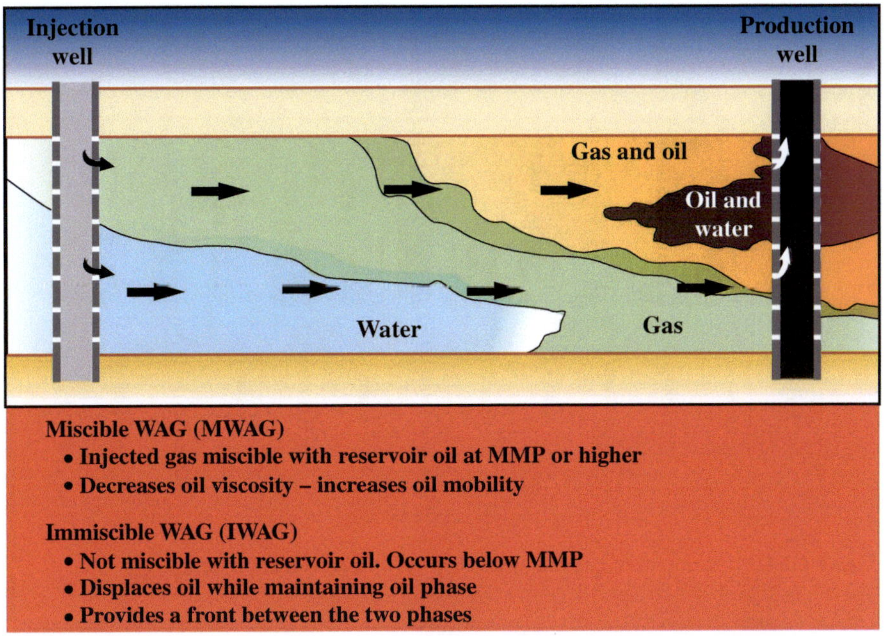

Figure 6.52 Effects of gas/oil miscibility properties during WAG implementation.

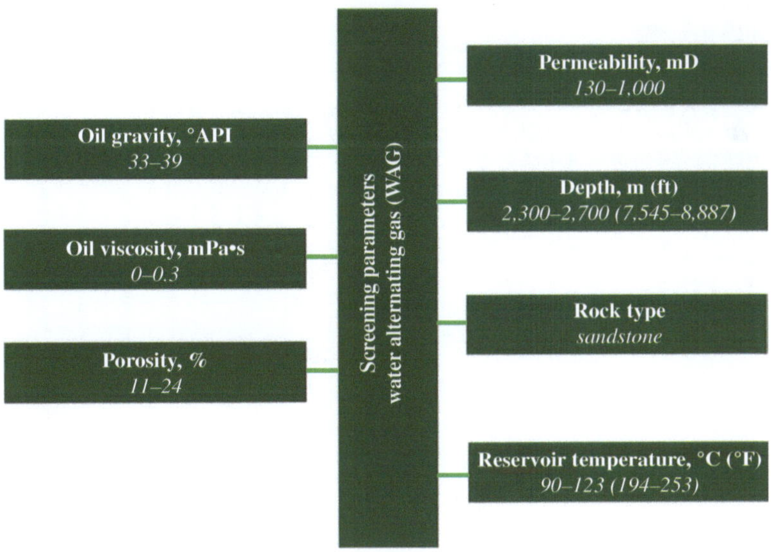

Figure 6.53 Criteria for the applicability of WAG.

As can be seen from the figure, WAG technologies, and miscible gas injection have the greatest efficiency. The maximum additional oil recovery was obtained at the Slaughter Estate and Wasson fields and amounted to 14.9%.

Figure 6.55 shows the results of implemented WAG projects (using hydrocarbon gases) with different methods of supplying working agents.

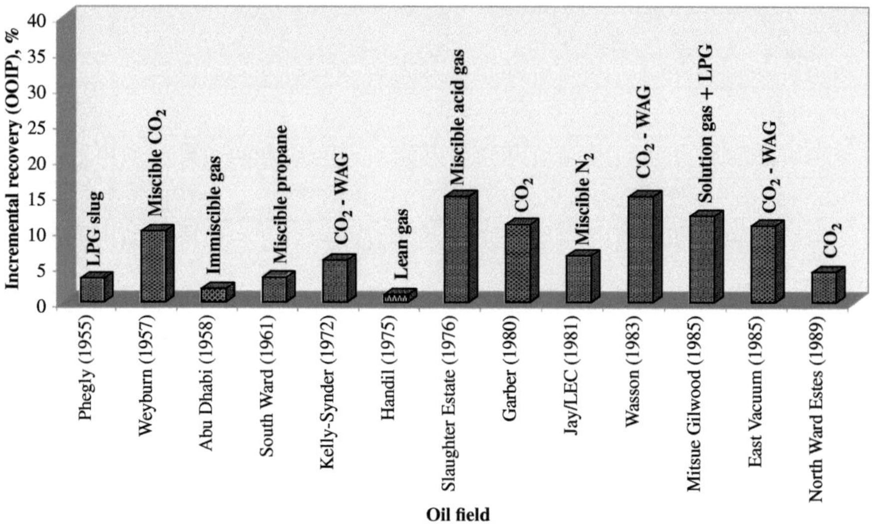

Figure 6.54 Implemented gas injection projects and WAG technologies.

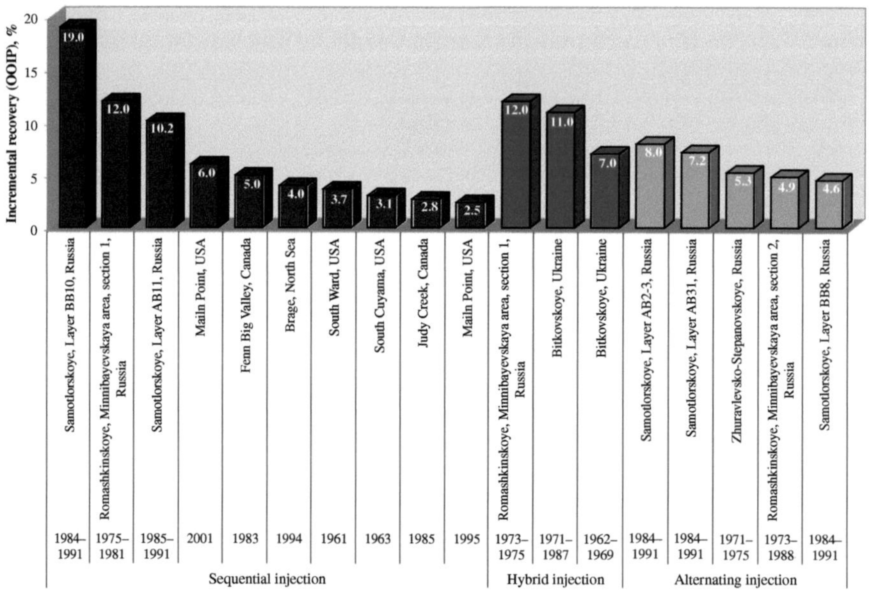

Figure 6.55 Hydrocarbon gas WAG projects with various methods of agent delivery.

As can be seen from the figure, according to the results of 20 implemented WAG projects, regardless of the method of supplying working agents, on average quite high additional oil recovery factors were obtained (2.5–19%).

Figure 6.56 shows the results of implemented WAG projects (using carbon dioxide) at different gas miscibility and reservoir oil.

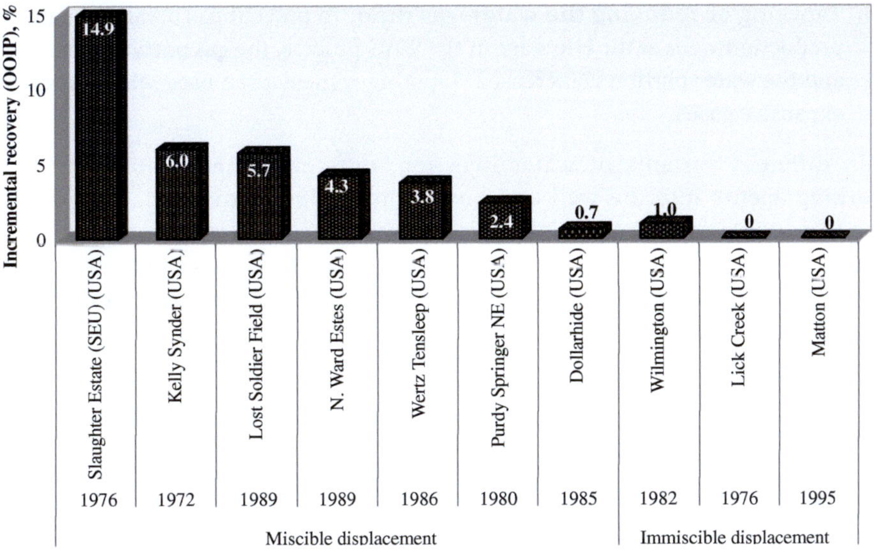

Figure 6.56 Carbon dioxide WAG projects with miscible and immiscible displacements.

As can be seen from the figure, the efficiency of miscible VAG is significantly higher than that of immiscible ones.

Foam-assisted WAG is mainly realized in the North Sea—in the Oseberg, Beryl, Snorre, and Brage fields. The most famous WAG project was carried out at the Snorre field. In this field, over the course of two years—from January 1999 to January 2001—an additional 0.8 million Nm^3 of oil was produced, while the total oil production as a result of the implementation of the event increased by 22%.

6.5.4 Implementation Technology

Typically, WAG technology is carried out on fields that have already been in development for a sufficient time, and both primary and secondary oil recovery have already been implemented.

1. **Gases:** Gases used in WAG projects can be divided into three groups: CO_2, other nonhydrocarbon gases, and hydrocarbon gases. CO_2 is an expensive gas and is typically used in miscible WAG applications. It should be noted that when using CO_2, corrosion has a significant impact on the efficiency of the process. Hydrocarbon gas is produced during the oil production process, and for this reason, most WAG implementations in offshore fields use it. The least used gases in WAG projects are other nonhydrocarbon gases, in particular nitrogen or flue gases.
2. **Injection systems:** Most often, a five-point injection system is used with a fairly close distance between wells.

3. **Tapering or reducing the water–gas ratio:** To prevent gas breakthroughs to production wells at the late stage of the WAG process, the gas portion is reduced and the water portion is increased. Tapering is most often used when injecting expensive gases.

In different variants of water injection, different schemes for supplying working agents into the well and piping the wellhead are used. Figure 6.57 shows possible schemes for supplying hydrocarbon gas and water to the well during their joint injection, as well as piping the wellhead. As can be seen from the figure, both associated and free gas from gas wells can be used in the WAG process.

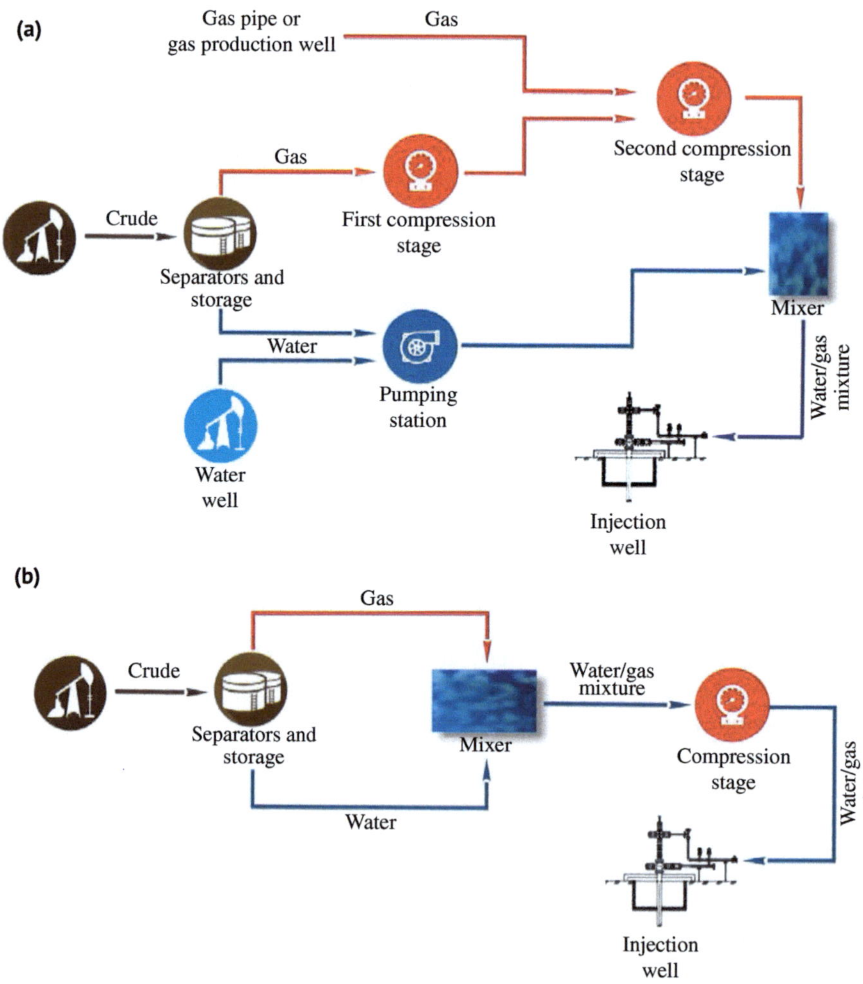

Figure 6.57 WAG surface injection arrangements with predominantly free (a) and associated (b) gas.

Figure 6.58 shows the diagram and piping of the wellhead for other variants of water injection, with separate and jointly separate supplies of working agents using hydrocarbon gases.

Figure 6.59 shows the diagram and piping of wellheads for foamy water injection using hydrocarbon gases.

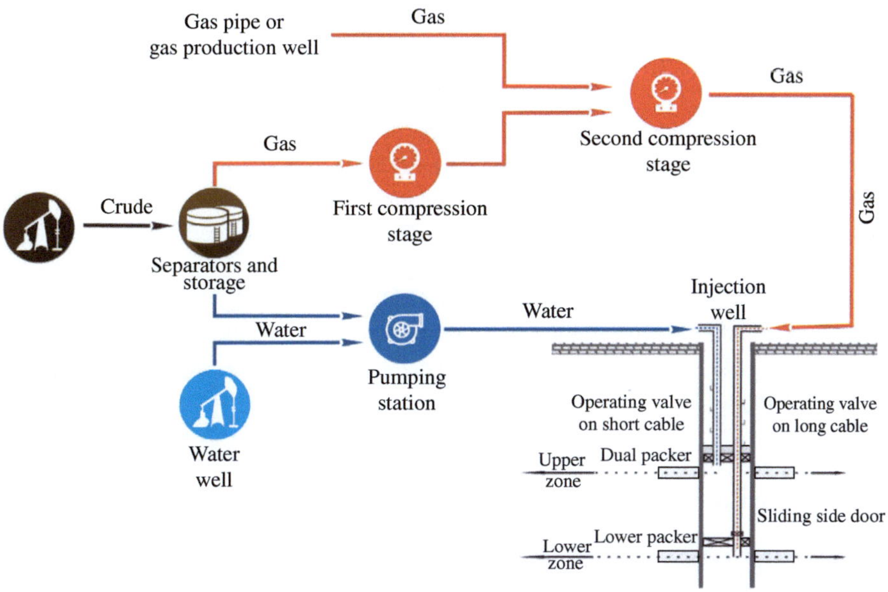

Figure 6.58 WAG hydrocarbons implementation with a separate gas/water injection.

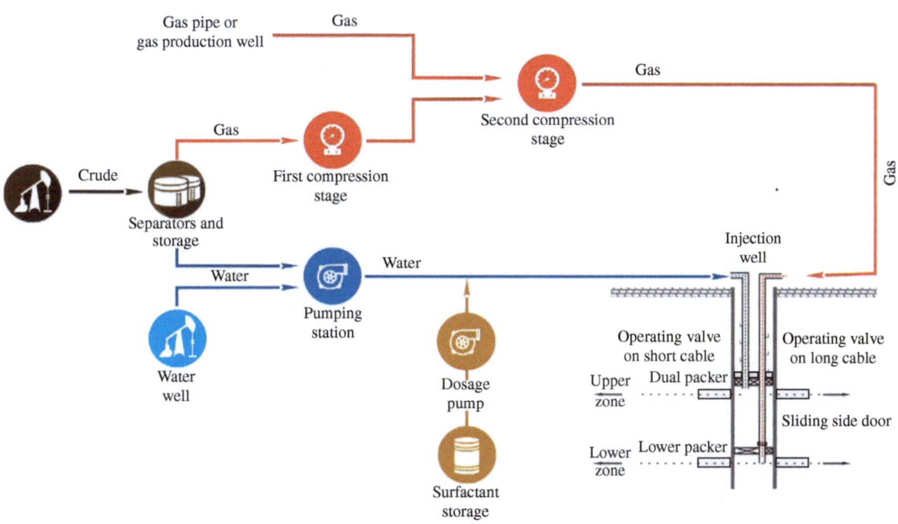

Figure 6.59 Foam WAG hydrocarbons implementation with a separate gas/foam injection.

It should be noted that when using nonhydrocarbon gases, the scheme for supplying working agents and piping the wellhead is identical to the above, with the difference that the gas does not come from the production collection system, but from an independent source. For example, if there is a CO_2 deposit near the field being developed, then the gas can be delivered through a pipeline; nitrogen can be obtained using special generators directly from the air.

Bibliography

Alagorni, A.H., Yaacob, Z.B., & Nour, A.H. (2015). An overview of oil production stages: enhanced oil recovery techniques and nitrogen injection. *International Journal of Environmental Science and Development*, *6*(9), 693–701.

Alvarado, V. & Manrik, E. (2011). *Methods for Enhancing Oil Recovery: Planning and Application Strategies*. Premium Engineering LLC.

Alvarez, C., Manrique, E., Alvarado, V., et al. (2001, October 6–9). WAG pilot at VLE field and IOR opportunities for mature fields at Maracaibo Lake. *Paper SPE-72099-MS Presented at the SPE Asia Pacific Improved Oil Recovery Conference, Kuala Lumpur, Malaysia*.

Anada, H., Sears, J., et al. (1982). Feasibility and economics of by-product CO_2 supply for enhanced oil recovery. *Final Report* (Vol. 1). U.S. Department of Energy.

Balint, V., Ban, A., & Dolesal, S. (1977). *Application of Carbon Dioxide in Oil Production*. Nedra.

Benham, A.L., Dowden, W.E., & Kunxman, W.I. (1960). Miscible fluid displacement: prediction of miscibility. *Journal of Petroleum Technology*, *12*(10), 229–237.

Booth, R. (2008). *Miscible Flow Through Porous Media*. University of Oxford.

Brock, W.R. & Bryan, L.A. (1989, March 6–8). Summary results of CO_2 EOR field tests, 1972–1987. *Paper SPE-18977-MS Presented at the SPE Joint Rocky Mountain Regional/Low Permeability Reservoirs Symposium and Exhibition, Denver, Colorado*.

Carcoana, A. (1992). *Applied Enhanced Oil Recovery*. Prentice Hall, Inc.

Chambers, F.T. (1968). Tertiary oil recovery combination water-gas miscible flood: Hibberd pool. *Producers Monthly*, *32*(1), 17–119.

Christensen, J.R., Stenby, E.H., & Skauge, A. (1998, March 3–5). Review of HW field experience. *Paper SPE-39883-MS Presented at the International Petroleum Conference and Exhibition of Mexico, Villahermosa, Mexico*.

Christensen, J.R., Stenby, E.H., & Skauge, A. (2001). Review of WAG field experience. *SPE Reservoir Evaluation & Engineering*, *4*(2), 97–106.

Clancy, J., et al. (1985). Analysis of nitrogen-injection projects to develop screening guides and offshore design criteria. *Journal of Petroleum Technology*, *37*(6), 1097–1104.

Crump, J.S. (1961). Method of increasing recovery of oil. British Patent No. 868649.

Daltaban, S.T., Lozada, A.M., Pina, A.V., et al. (2008, November 3–6). Managing water and gas production problems in Cantarell; A giant carbonate reservoir in Gulf of Mexico. *Paper SPE-117233-MS Presented at the Abu Dhabi International Petroleum Exhibition and Conference, Abu Dhabi, UAE*.

Farias, M.J. & Watson, R.W. (2007). *Interaction of Nitrogen/CO_2 Mixtures with Crude Oil.* Pennsylvania State University.

Green, D.W. & Willhite, G.P. (1998). *Enhanced Oil Recovery.* Henry L. Doherty Memorial Fund of AIME, Society of Petroleum Engineers.

Groenenboom, J., Kechut, N.I., & Mar-Or, A. (2017, October 17–19). Foam-assisted WAG: Injection strategies to optimize performance. *Paper SPE-186991-MS Presented at the SPE/IATMI Asia Pacific Oil & Gas Conference and Exhibition, Jakarta, Indonesia.*

Holm, L.W. (1976). Status of CO_2 and hydrocarbon miscible oil recovery methods. *Journal of Petroleum Technology, 28*(01), 76–84.

Holtz, M. (2012). *Immiscible Gas Displacement Recovery.* PRAXAIR.

Holtz, M.H. (2016, April 11–13). Immiscible water alternating gas (IWAG) EOR: Current state of the art. *Paper SPE-179604-MS Presented at the SPE Improved Oil Recovery Conference, Tulsa, Oklahoma, USA.*

Jishun, Q., Haishui, H., & Xiaolei, L. (2015). Application and enlightenment of carbon dioxide flooding in the United States of America. *Petroleum Exploration and Development, 42*(2), 232–240.

Juttner, I. (1997). Oil displacement in miscible conditions. *Rudarsko-Geološko-NAFTNI Zbornik, 9*(1), 63–66.

Koch, H.A. & Slobod, R.L. (1957). Miscible slug process. *Transactions of AIME, 210*, 40.

Kokal, S. & Al-Kaabi, A. (2010). *Enhanced Oil Recovery: Challenges & Opportunities.* World Petroleum Council: Official Publication.

Koottungal, L. (2008, April 21). Worldwide EOR survey. *Oil and Gas Journal, 106*(15), 47–59.

Koottungal, L. (2014, April 7). Worldwide EOR survey. *Oil and Gas Journal, 112*(4), 79–91.

Lake, L.W. (1989). *Enhanced Oil Recovery.* Prentice Hall.

Lo, L.L. & McGregor, D.S. (2003, October 5–8). VGV pilot design and observation well data analysis for Hassi Berkine South field. *Paper SPE-84076-MS Presented at the SPE Annual Technical Conference and Exhibition, Denver, Colorado.*

Lyons, W.C. & Plisga, B.J. (Eds.). (2005). *Standard Handbook of Petroleum & Natural Gas Engineering* (2nd ed.). Elsevier Inc.

Manrique, E., Padron, R., Surguchev, L., et al. (2000, October 24–25). VLE WAG injection laboratory field in Maracaibo Lake. *Paper SPE-65128-MS Presented at the SPE European Petroleum Conference, Paris, France.*

Moritis, G. (2008, April 21). More US EOR projects start but EOR production continues to decline. *Oil and Gas Journal, 106*(15), 41–46.

Ning, S.X. & McGuire, P.L. (2004, April 17–21). Improved oil recovery in under-saturated reservoirs using US-WGW process. *Paper SPE-89353-MS Presented at the SPE/DOE Symposium on Improved Oil Recovery, Tulsa, Oklahoma.*

Pautz, J.F., et al. (1990, March). *NIPER-471 Review of EOR Projects Trends and Thermal EOR Technology - Topical Report.* Performed under cooperative agreement No. FC22-83FE60149-DOE. IIT Research Institute, NIPER/DOE.

Ping, G., Zhonglin, W., Guangtian, T., et al. (2003, September 9–11). Case analysis on hydrocarbon alternative gas miscible flooding in Pu Bei oil field. *Paper SPE-80487-MS Presented at the SPE Asia Pacific Oil and Gas Conference and Exhibition, Jakarta, Indonesia.*

Quintella, C.M., de Ondina, C., Dino, R., et al. (2010, April 12–14). CO_2 enhanced oil recovery and geologic storage: An overview with technology assessment based on patents and articles. *Paper SPE-126122-MS Presented at the SPE International Conference on Health, Safety, and Environment in Oil and Gas Exploration and Production, Rio de Janeiro, Brazil.*

Redman, R.S. (2002, May 20–22). Horizontal miscible water alternating gas development of the Alpine field, Alaska. *Paper SPE-76819-MS Presented at the SPE Western Regional/AAPG Pacific Section Joint Meeting, Anchorage, Alaska.*

Sessions, R.E. (1960, July 4). How Atlantic operates the Slaughter flood. *Oil and Gas Journal, 58*(27), 91–98.

Sheng, J.J. (2013). *Enhanced Oil Recovery Field Case Studies*. Elsevier Inc.

Shine, J. & Holtz, M. (2008). Reserve growth & higher recovery using nitrogen gas injection. *Proceedings "2008 Wyoming EOR/IOR Conference."* The Wyoming Enhanced Oil Recovery Institute's: PRAXAIR Inc.

Skauge, A., Aarra, M.G., Surguchev, L., et al. (2002, April 13–17). Foam-assisted HBW: Experience from the Snorre field. *Paper SPE-75157-MS Presented at the SPE/DOE Improved Oil Recovery Symposium, Tulsa, Oklahoma.*

Sohrabi, M., Danesh, A., & Tehrani, D.H. (2005, June 13–16). Oil recovery by near-miscible SWAG injection. *Paper SPE-94073-MS Presented at the SPE Europec/EAGE Annual Conference, Madrid, Spain.*

Spirov, P., Rudyk, S., & Khan, A. (2012, February 20–22). Foam-assisted WAG, Snorre revisited with new foam screening model. *Paper SPE-150829-MS Presented at the North Africa Technical Conference and Exhibition, Cairo, Egypt.*

Stalkup, F.I. Jr. (1984). Miscible displacement. *SPE Monograph Series* (pp. 137–156). Society of Petroleum Engineers.

Surguchev, M.L. (1985). *Secondary and Tertiary Methods for Increasing Oil Recovery*. Nedra.

Surguchev, M.L., Gorbunov, A.T., Zabrodin, D.P., et al. (1991). *Methods for Extracting Residual Oil*. Nedra.

Taber, J.J. (1983, October 5–8). Technical screening guides for the enhanced recovery of oil. *Paper SPE-12069-MS Presented at the SPE Annual Technical Conference and Exhibition, San Francisco, California.*

Taber, J.J., Martin, F.D., & Seright, R.S. (1997). EOR screening criteria revisited. Part 1: Introduction to screening criteria and enhanced recovery field projects. Part 2: Applications and impact of oil prices. *SPE Reservoir Engineering, 12*(3), 189–198, 199–205.

Terry, R.E. (2003). Enhanced oil recovery. In *Encyclopedia of Physical Science and Technology*. Elsevier Science Ltd.

Van Dyke, K. (1997). *Fundamentals of Petroleum*. The University of Texas at Austin.

Van Poollen, H.K. (1980). *Fundamentals of Enhanced Oil Recovery*. PennWell.

Walker, D. (2008). Reserve growth & higher recovery using nitrogen gas injection. *Energy Summit '08*. PRAXAIR Inc.

Yelling, W.R. & Metcalfe, R.S. (1980). Determination and prediction of CO_2 minimum miscibility pressures. *Journal of Petroleum Technology*, *32*(01), 160–168.
Zaks, S.L. (1963). *Enhancing Reservoir Oil Recovery by Gas Injection*. Gostoptekhizdat.
Zhang, N., Wei, M., & Bai, B. (2018). Statistical and analytical review of worldwide CO_2 immiscible field applications. *Fuel*, *220*, 89–100.

7

Chemical Enhanced Oil Recovery (EOR)

7.1 Introduction

Currently, about 90% of oil is produced using waterflooding, and the amount of water injected into the reservoir significantly exceeds the level of oil production. Nevertheless, the average level of oil recovery factor does not exceed 50%, which determines the relevance of the search for new technologies for influencing oil deposits. A technical contradiction when waterflooding an oil field is that maintaining reservoir pressure requires the injection of large volumes of water, at the same time, injection of a large volume of water leads to disruption of the stability of the displacement front and early breakthrough of water to production wells. Overcoming this contradiction is possible by using chemical compounds that change the rheological and surface properties of the fluid injected into the reservoir and ensure both the maintenance of reservoir pressure and the stability of the displacement front.

Chemical flooding is accomplished by adding one or more chemical compounds to the injected fluid. This leads either to a decrease in the interfacial tension between the reservoir oil and the injected liquid or to an increase in the viscosity of the injected liquid, a decrease in its mobility, and, due to this, an increase in the coverage of the formation by the influence.

The main methods of chemical flooding technology are polymer, micellar-polymer, and alkaline flooding. Polymer flooding is used to increase the viscosity of the injected fluid and reduce its mobility, leading to an increase in reservoir coverage by flooding (sweep efficiency). Micellar polymer and alkaline flooding are used to reduce the interfacial tension between the reservoir oil and the injected fluid, which significantly improves the process of displacement of residual oil. Polymer waterflooding is usually used in the early stages of oil field development when there is sufficient oil saturation in the porous medium. Micellar-polymer and alkaline flooding are usually used at the final stage of field development for effective recovery of residual oil.

Methods for Enhanced Oil Recovery: Fundamentals and Practice, First Edition.
Baghir A. Suleimanov and Elchin F. Veliyev.

7.2 Polymer Flooding

As shown above, overcoming the technical contradiction inherent in formation stimulation processes is possible by using non-Newtonian, heterogeneous systems with controlled rheology. Polymer solutions fully possess these properties.

7.2.1 Process Mechanism

One of the main problems when flooding oil reservoirs is the hydrodynamic instability of the front of oil displacement by water. The main factor influencing the hydrodynamic instability of the front is the ratio of oil and water viscosities in reservoir conditions. Indeed, the critical wavelength of the disturbance λ is inversely proportional to the capillary number Ca:

$$\lambda \sim 1/\sqrt{Ca},$$

$$Ca = \frac{\upsilon(\eta_1 - \eta_2)}{\sigma m},$$

υ—filtration rate, η_1, η_2—viscosity of oil and water, respectively, σ—surface tension, m—porosity.

As can be seen from the presented expressions, all other things being equal, the higher the difference in viscosity of oil and water, the higher the capillary number and the lower the critical disturbance wavelength, which leads to early disruption of the stability of the displacement front and rapid breakthrough of water to production wells. In addition, in layered heterogeneous formations, due to the predominant movement of water in a highly permeable interlayer, the coverage of the formation by flooding is greatly reduced.

The solution to these problems is carried out by adding polymers to the injected water, i.e., using polymer flooding. The mechanism of polymer flooding is determined by three main factors.

1. **Thickening of injected water**
   An increase in the viscosity of the injected liquid leads to a significant increase in the hydrodynamic stability of the displacement front.
   When a water-soluble polymer is added to water at a concentration of 0.01–0.1%, the viscosity of the resulting polymer solution increases by 3–4 times compared to the viscosity of water. In addition, during filtration in a porous medium, the apparent viscosity of the polymer solution increases up to 20 times. In this regard, polymer solutions are especially effective in heterogeneous formations and in fields with high oil viscosity.

2. **Surface activity**
   Water-soluble polymers used in polymerization flooding are anionic surfactants. In addition, in the polymer flooding process, surfactant additives are used in the polymer solution. In this case, the resulting complex of polymer and surfactant molecules has a molecular weight higher than the molecular weight of the polymer and therefore a higher viscosity. Reducing the interfacial surface tension allows for the washing of residual oil from the surface of the pore channels behind the displacement front.

3. **Non-Newtonian rheology**

Perhaps the greatest influence on the efficiency of polymer flooding is the non-Newtonian properties of the polymer solution. Water-soluble polymers commonly used in oilfield applications are characterized by shear thinning or pseudoplastic rheology (see Figure 7.1).

However, to increase coverage, solutions characterized by shear thickening or dilatant rheology are preferable (see Figure 7.2).

Indeed, assuming that the flow of a polymer solution in a porous medium obeys a power law, the flow rate can be determined using the following formula:

$$v = -B_0\left|\frac{dP}{dx}\right|^{\frac{1}{n}-1}\frac{dP}{dx},\ B_0 = \frac{n}{3n+1}\eta^{-\frac{1}{n}}2^{\frac{1}{2}(3+\frac{1}{n})}m^{-\frac{1}{2}(1+\frac{1}{n})}k^{\frac{1}{2}(1+\frac{1}{n})}$$

($\frac{dP}{dx}$ is the pressure gradient, k is the permeability of the porous medium, and n is the exponent).

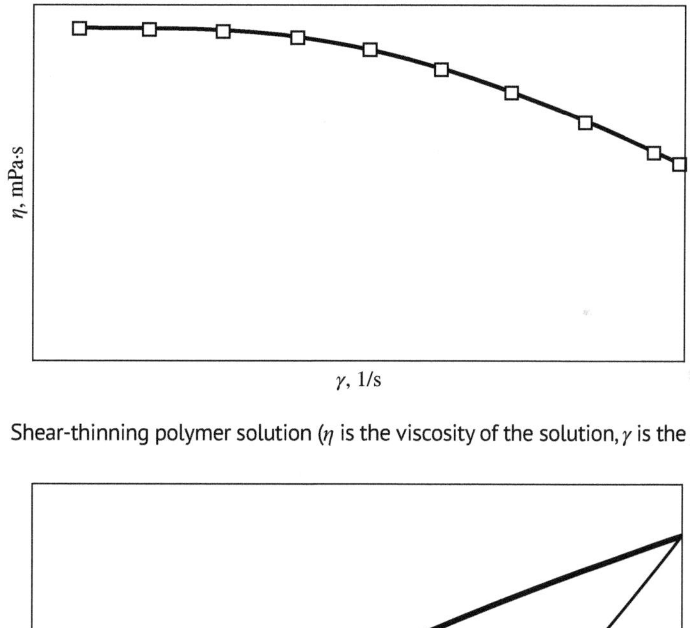

Figure 7.1 Shear-thinning polymer solution (η is the viscosity of the solution, γ is the shear rate).

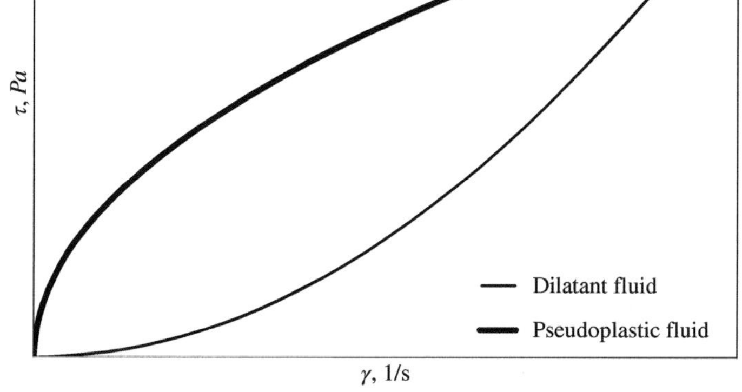

Figure 7.2 Rheology of dilatant and pseudoplastic fluids (τ - shear stress).

Let's consider a simplified flow diagram of a polymer composition in a layered inhomogeneous formation. Let the filtration of the polymer composition occur in two parallel layers with permeability k_1 and k_2 (other things being equal), and $k_1 \gg k_2$ correspond to the power law. Then, according to the above equation, for the expense ratio we obtain $\frac{Q_1}{Q_2} = \frac{v_1}{v_2} = \left(\frac{k_1}{k_2}\right)^{1+\frac{1-n}{2n}}$.

As can be seen from the resulting expression, for a Newtonian fluid ($n = 1$), the flow rate ratio is equal to the permeability ratio; for a dilatant fluid ($n > 1$), the flow rate ratio decreases, i.e., the flow profile levels out, and for a pseudoplastic fluid ($n < 1$), the differences in permeability increase even compared to a Newtonian fluid.

The rheology of a polymer solution in a porous medium differs significantly from the rheology determined in a rheometer. Moreover, dilute polymer solutions, characterized by shear liquefaction, exhibit dilatant rheology, i.e., shear thickening, when measured on a rheometer in a porous medium. This behavior is explained by the retention of polymer molecules by the pore space. In this case, the polymer macromolecules unfold with increasing shear rate, increasing resistance to flow. Retention of the polymer can occur by adsorption or mechanical retention. In layered inhomogeneous and microinhomogeneous porous media, the rheology of a dilute polymer solution has an S-shaped character (see Figure 7.3). In this case, the dilatant nature of the flow is replaced by a pseudoplastic one, but the change in flow rheology occurs at fairly high shear rates, which are not observed in real formations. Thus, in high-permeability zones, there is a higher flow velocity than in low-permeability zones, which leads to an increase in viscosity (decrease in solution mobility), leveling of the filtration profile, and an increase in the formation coverage of the impact.

4. **Polymer adsorption**
As the polymer flows in a porous medium, cohesive water is displaced along with oil. As a result, the polymer solution comes into direct interaction with the reservoir rock, leading to the adsorption of the polymer from the solution. In this case, the concentration of the polymer in the solution decreases, and a wave of water with a low concentration of polymer is formed.

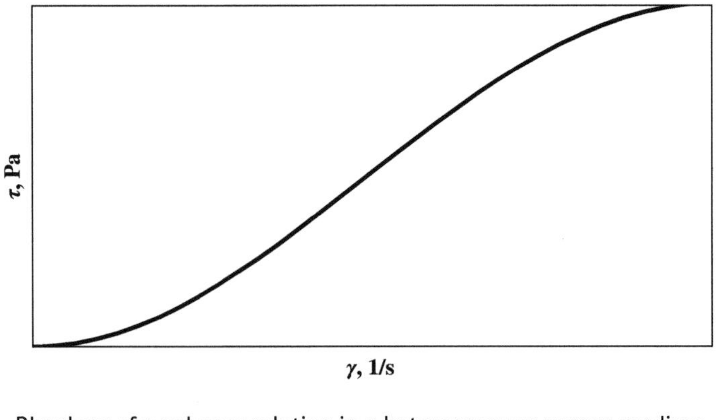

Figure 7.3 Rheology of a polymer solution in a heterogeneous porous medium.

According to various researchers, polymer adsorption can reach 150 g/m^3, which is significantly less than surfactant adsorption. Typically, one of the main requirements for polymers is their minimal adsorption on the surface of a porous medium, as this reduces its losses and consumption. However, this is a simplified idea of the efficiency of oil displacement by a polymer solution. Adsorption plays an important role in the mechanism of polymer flooding, because, like water thickening, it helps to reduce the mobility of the polymer solution (by reducing permeability) and increases the coverage of the formation by the influence. At the same time, adsorption should not go beyond the experimentally determined optimal range, since with high adsorption the front of the polymer movement significantly lags behind the front of oil displacement, which reduces the efficiency of the process due to the fact that oil is displaced mainly by inactive water.

5. **Effect of salts and water hardness**
   Salinity and water hardness have a significant impact on polymer flooding, especially when using hydrolyzed polyacrylamide (PAM, HPAM). The effect of mineralization and water hardness on a polymer solution is expressed in a decrease in the electrostatic repulsion forces in the polymer, which leads to the destruction of the solution structure and a significant decrease in the viscosity of the polymer solution, proportional to the salt concentration. The influence of water hardness has an even more significant effect on the polymer solution, since the effect of divalent ions of calcium and magnesium, other things being equal, is greater than that of monovalent ions of sodium and potassium.

6. **Polymer destruction**
   Macromolecules of polymers are destroyed under the influence of various factors. At the same time, the molecular weight of the polymer decreases and, as a result, its ability to thicken water decreases.

Destruction can be thermal, chemical, microbiological, and mechanical (shear).

Thermal destruction occurs at a fairly high temperature (more than 130 °C).

Chemical destruction is caused by the action of atmospheric oxygen on the polymer. Therefore, the preparation of the polymer solution must be carried out under anaerobic conditions.

Microbiological destruction of a polymer solution can occur in reservoir conditions under the influence of aerobic bacteria that enter the reservoir with injected water and during oil oxidation.

Mechanical destruction occurs when polymer macromolecules rupture at high flow rates in pipes and formations.

Acceptance of polymer solution is an important factor for several reasons. First, the rate of injection of the polymer solution directly affects the economics of the project. Second, work may be required to clean (drainage) injection wells if the polymer solution reduces injectivity. These cleanup efforts can reduce the technological and economic efficiency of polymer flooding. Injectivity decreases with increasing polymer molecular weight, being more favorable when the polymer solution exhibits shear thinning.

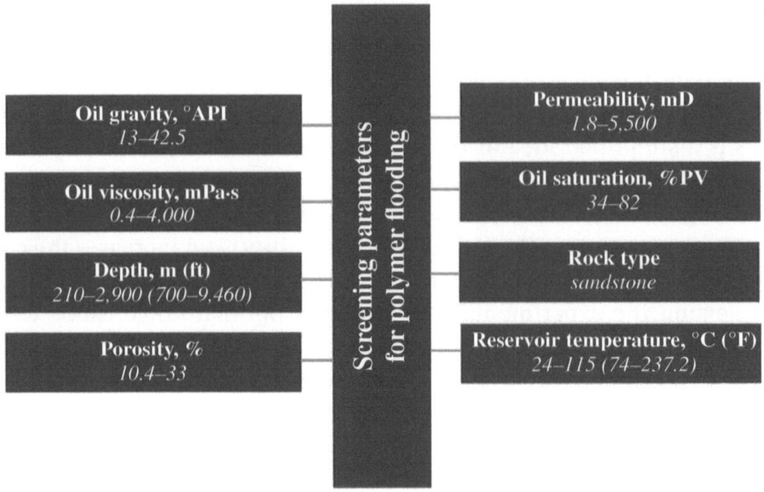

Figure 7.4 Criteria for the applicability of polymer flooding.

7.2.2 Applicability Criteria

The criteria for the applicability of polymer flooding, depending on the properties of the reservoir, are as follows (see Figure 7.4).

- Oil viscosity should preferably not exceed 150 cP, be below 10 cP, and API specific gravity should be more than 13.
- The rock permeability should preferably be higher than 10 mD.
- Reservoir temperature: Low formation temperature is preferred, preferably less than 80 °C (176 °F) and no higher than 100 °C (210 °F).
- Wells with high injectivity and a pressure increase margin should be used.
- The clay content in the formation should be low.
- Low salinity of injected and produced water is preferred.

Polymer flooding is generally successful in sand reservoirs but has been used successfully in carbonate and fractured reservoirs. It is preferable to use polymer waterflooding in layered heterogeneous as well as micro-heterogeneous formations.

It is preferable to use the method in hydrophilic reservoirs, but there are cases of successful application in hydrophobic and mixed wettability reservoirs.

7.2.3 Field Implementation

Since the 1960s of the last century, the active introduction of polymer flooding began. As can be seen from Figure 7.5, 730 projects have been implemented so far in 24 countries. At the same time, the absolute leader here is the USA, where 560 projects have been implemented, i.e., 76% of all implemented projects.

Figure 7.6 shows successfully completed projects. As can be seen from Figure 6.6, the use of polymer flooding can significantly increase the oil recovery factor. The vast majority of polymer flooding uses acrylamide-based polymers (Pusher, PAA-HPAM),

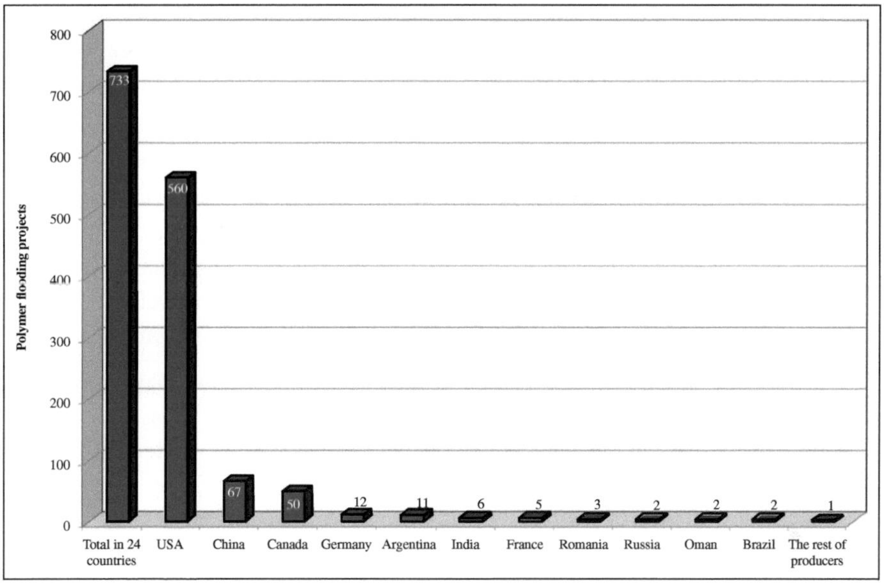

Figure 7.5 Polymer flooding projects in 24 countries.

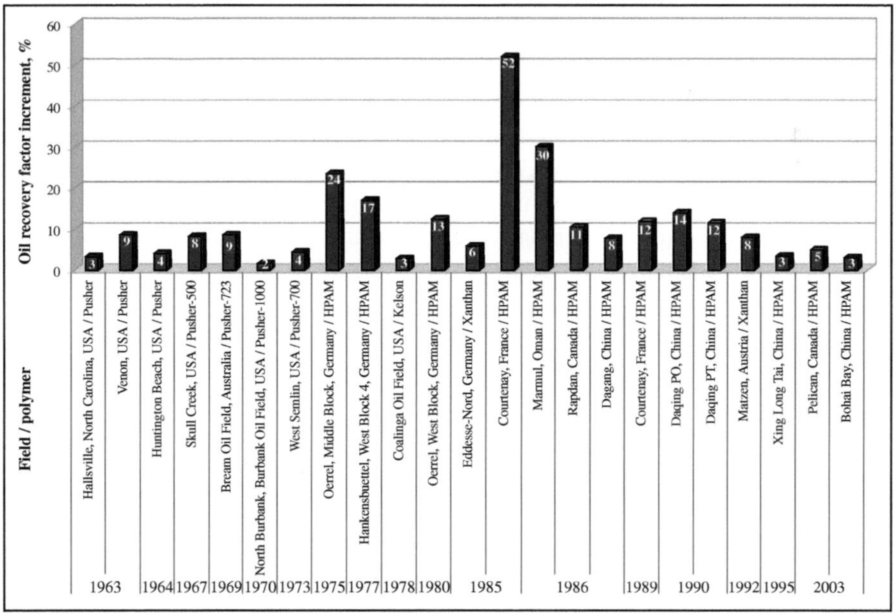

Figure 7.6 Results of polymer flooding projects.

due to their good solubility in water, surfactant, and rheological properties. Some projects used the biopolymer "xanthan." Biopolymers are environmentally safe and resistant to salts and mechanical (shear) destruction.

7.2.4 Implementation Technology

Typically, polymer flooding is carried out using slugs propelled through the formation by water (see Figure 7.7). The size of the rims, depending on the conditions, can be 10–50% of the pore volume. In this case, the dosing of the polymer is carried out using dosing pumps connected to the water supply system of the well.

First of all, one should choose the type of polymer. The polymer must:

- ensure the maximum degree of thickening of the injected water and reduce mobility;
- be easily dissolving in water in the absence or ultra-low content of insoluble sediments;
- have a low degree of retention during the filtration process;
- have shear resistance, chemical, biological, and thermal stability; and
- have sufficient, economically justified pickup capacity.

The type and concentration of the polymer and the volume of the slug are selected based on the salt composition of the formation water, the heterogeneity of the formation, and the pore volume of the field site. It should be taken into account that with high mineralization of water, the concentration of the polymer in the solution should be significantly higher (2–3 times).

Injecting a polymer solution into a reservoir requires pressure significantly higher than that of conventional waterflooding to ensure the necessary economic filtration rate and field development rates. In this regard, the implementation of polymer flooding in formations with low permeability may not be technically feasible.

The placement of wells during polymer flooding does not differ significantly from conventional flooding if the impact is effective and the design parameters of field development are achieved: injection pressure and oil production rates.

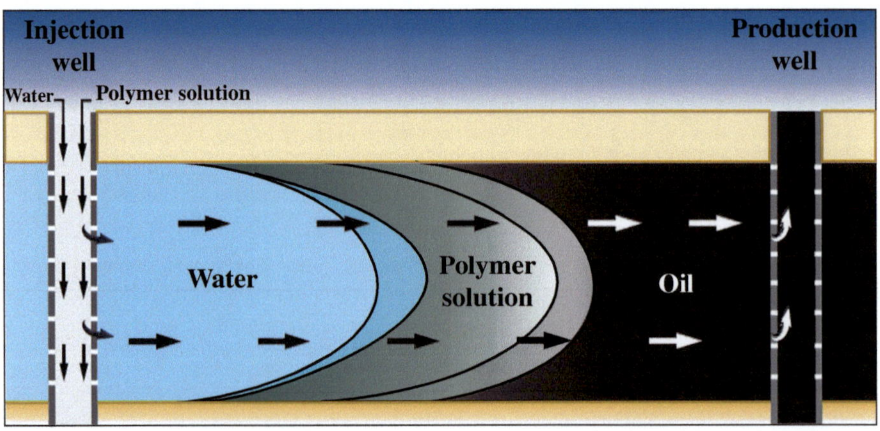

Figure 7.7 Polymer flooding.

The general consensus among experts is that the earlier the development of a field polymer flooding begins, the more effective the process will be. However, two main factors must be taken into account. First, without the characteristics of conventional flooding, it is difficult to give a definite assessment of the oil recovery potential and economic efficiency of polymer flooding. Second, in new fields with a lack of geological information, it is difficult to conduct a full analysis of the applicability of polymer flooding. In this regard, the best solution would be to start using polymer flooding after conventional flooding, when all the necessary characteristics of the object are available.

7.3 Micellar-polymer Flooding

It is well known that water and oil do not mix until a surfactant is added to the solution, which significantly reduces the interfacial tension. For the first time in 1927, Atkinson recommended the use of an aqueous solution of soap to reduce the interfacial tension of an oil-water system in order to displace residual oil. In subsequent years, laboratory and published studies have shown that it is necessary to reduce and maintain interfacial tension at levels between 0.01 and 0.001×10^{-3} N/m to achieve significant increases in oil recovery. Research has shown that the best results are achieved when using micellar solutions, which, along with low interfacial tension, have a higher viscosity, and in some cases, non-Newtonian rheological properties. Micellar solutions are also referred to in the literature as microemulsions. They are translucent, homogeneous, and thermodynamically stable.

Unlike polymer micellar-polymer flooding, as a rule, is carried out in depleted, flooded oil-bearing formations, in which the residual oil is held stationary by capillary forces.

7.3.1 Process Mechanism

Micellar-polymer waterflooding is traditionally used after conventional waterflooding as a tertiary method for enhanced oil recovery (EOR). In this case, the injection of working solutions is carried out in the following sequence (see Figure 7.8).

1. **Water of low salinity is preinjected:** If the formation water has high mineralization, and in offshore fields where seawater is pumped into the formation, this is a common thing; direct contact with the micellar solution is avoided and an aqueous solution of low salinity is first pumped into the formation.
2. **A slug of micellar solution is injected**: The micellar solution is prepared in two versions, namely with a low or high concentration of surfactant. In the first case, a slug of micellar solution is injected into the formation in a volume of 15–60% of the pore volume. In this case, residual oil is gradually displaced from the reservoir. In the second case, a slug of micellar solution is pumped into the formation in a volume of 3–10% of the pore volume. In this case, residual oil is quickly displaced from the formation, and the surfactant concentration in the solution becomes low due to mixing with formation water.

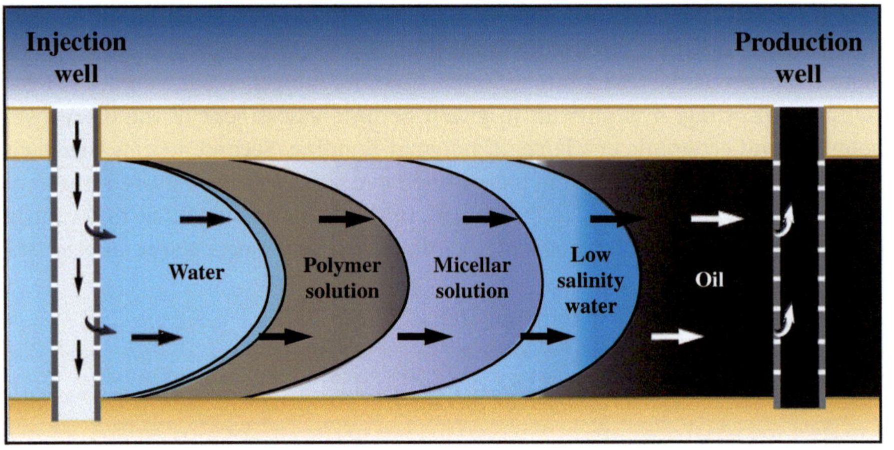

Figure 7.8 Micellar-polymer flooding.

3. **A slug of polymer solution is injected:** This ensures control of the mobility of the micellar solution and prevention of breakthroughs of the water pumped after it.
4. **Waterflood:** This ensures the advancement of the micellar solution in the formation and displacement of oil to the production wells.

The process of micellar flooding at specific fields can vary greatly in detail since the composition of the micellar solution is selected in the laboratory depending on the geological characteristics of the rock and the properties of the fluids saturating it.

As a result of the injection of working solutions in the specified sequence, six zones are formed in the formation (see Figure 7.9) with different water saturation:

- initial water saturation,
- oil bank,
- water bank,
- micellar solution slug,
- polymer solution slug, and
- water zone.

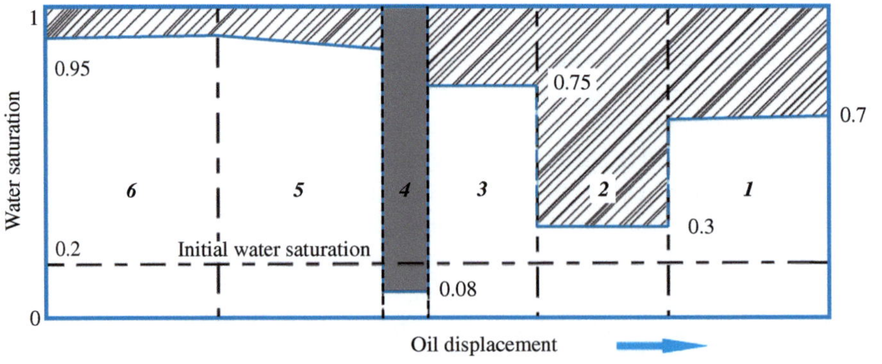

Figure 7.9 Water saturation during the micellar-polymer flooding process.

The indicated distribution of zones is observed in the case of a hydrophilic formation. In the case of a hydrophobic formation, the zones of the oil and water banks are swapped.

The presented mechanism of micellar flooding leads to the following dynamics of oil production. At the first stage, almost only water is extracted (product water cut is up to 99%). At the second stage, the share of oil in production increases sharply to 70%. After passing the oil shaft at the third stage, the water cut of the product increases. Finally, at the fourth and final stage, the micellar solution with residual oil dissolved in it is extracted.

It should be noted that the efficiency of micellar-polymer flooding significantly depends on the completeness of displacement of hard formation water, which greatly changes the properties of micellar and polymer solutions. In this regard, the creation of micellar solutions that are stable and at the same time insensitive to formation salts is of great importance for increasing the efficiency of the process.

7.3.1.1 Structure and Composition of Micellar Solutions

The three main components of micellar solutions are hydrocarbon liquid, surfactant, and water. To ensure the stability of micellar solutions under reservoir conditions (depending on the geological structure, composition of reservoir fluids, and saturation), stabilizers are added to them. To improve viscosity control, water-soluble electrolytes can also be added to micellar solutions. Micellar solutions differ from emulsions due to the microscopic size of the dispersed phase (micelles). In micellar solutions, depending on the energy of interaction of surfactant molecules with hydrocarbon liquid and water, micelles have a spherical or platelet (multilayer) shape with a size of 10^{-6}–10^{-4} mm, while in conventional emulsions they are 10^{-4} mm and higher. The micellar solutions obtained by this method are translucent, homogeneous, and thermodynamically stable.

The micellar solution is mixed with formation fluids, including oil and water, without phase separation, which can significantly increase oil recovery from the formation.

Liquefied gas, kerosene, and crude light oil are usually used as hydrocarbon liquids.

The water used is usually fresh, produced water or water with a certain salt composition.

Substances with high solubilizing ability (the ratio of the volume of oil to the volume of surfactant in solution) are used as surfactants. Typically, petroleum sulfonates and compositions of various nonionic anionic surfactants are used.

Various alcohols (isopropyl, butyl, etc.) are usually used as stabilizers.

Inorganic salts are usually used as an electrolyte.

By changing the content of the micellar solution components, it is possible to obtain a micellar solution with an external oil phase (the solution is miscible only with oil, balanced with water), with an external water phase (the solution is miscible only with water, balanced with oil). Figure 7.10 shows the phase diagram of a micellar solution.

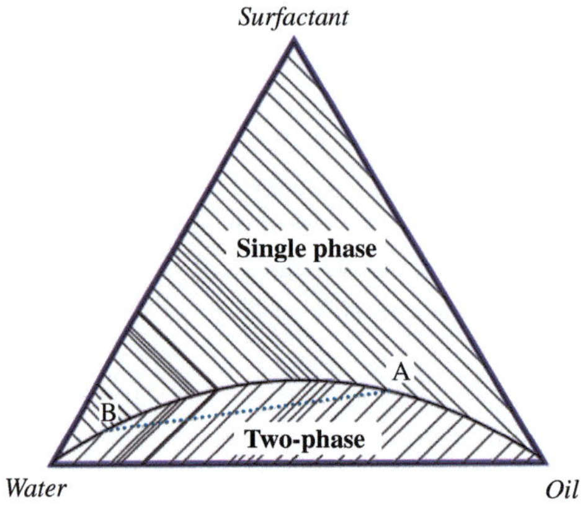

Figure 7.10 Ternary diagram.

As can be seen from the figure, two points (A, B) are shown in the two-phase region: at point A there is a solution with an external aqueous phase, and at point B there is a solution with an external oil phase. One phase is external to oil, and the other is external to water, both in equilibrium and lying at opposite ends of the A-B connecting line. In the single-phase region, all components are mixed. The phase diagram for practical micellar-polymer flooding in a particular reservoir must have a reduced two-phase region as the surfactant slug moving through the reservoir changes its composition to include hydrocarbon liquid and water during miscible displacement.

Laboratory studies have shown that the reduction in interfacial tension required for miscible displacement can only be achieved over a very narrow range of salt concentrations. Indeed, the salinity of the brine affects the phase behavior of micellar solutions, which in turn directly correlates with the interfacial tension.

Stable single-phase micellar solutions exist in a wide range of concentrations of its main components—hydrocarbon liquid, water, surfactant, stabilizer, and electrolyte. There are four types of micellar solutions (see Figure 7.11): (i) single-phase micellar solution, soluble in both water and oil; (ii) micellar solution with an external aqueous

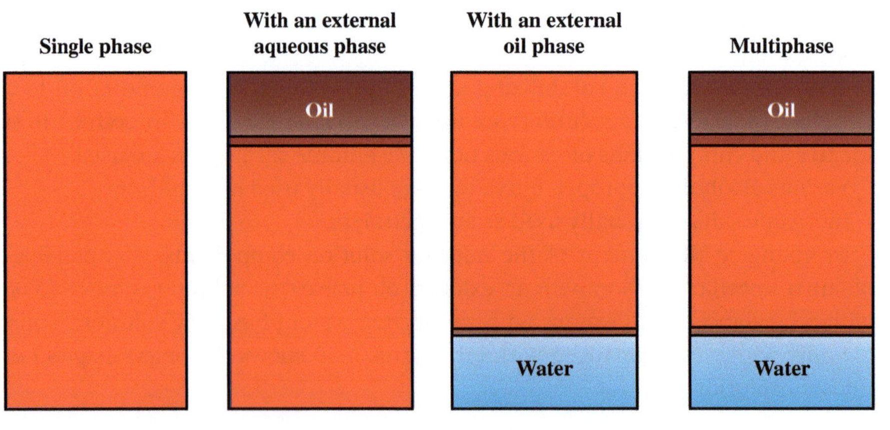

Figure 7.11 Types of micellar solutions.

phase, soluble only in water (water-oil microemulsion); (iii) micellar solution with an external oil phase, dissolves only in oil (oil-water microemulsion); (iv) multiphase micellar solution, insoluble in either water or oil. The phase diagram of micellar solutions (see Figure 7.12) shows the zones of existence of solutions of various types.

The types of micellar solutions are determined by the composition and concentration of surfactants, selected based on the geological structure of the deposits, as well as the properties of formation fluids.

Figure 7.13 shows the volume concentration of the main components of micellar solutions of two types. As can be seen from the figure, stable micellar solutions can be obtained in a wide range of concentrations of the main components.

Figure 7.12 Triple diagram.

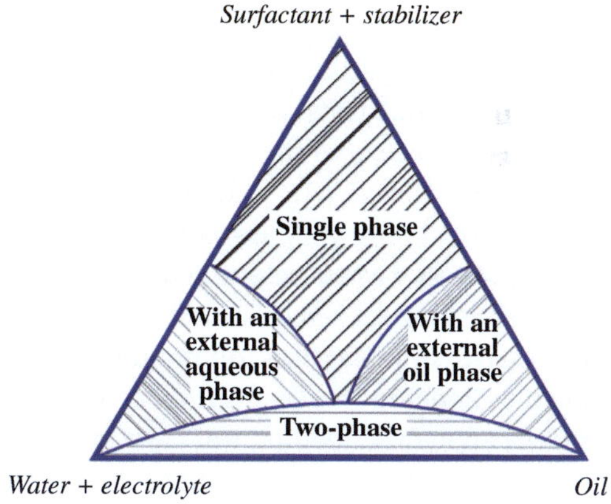

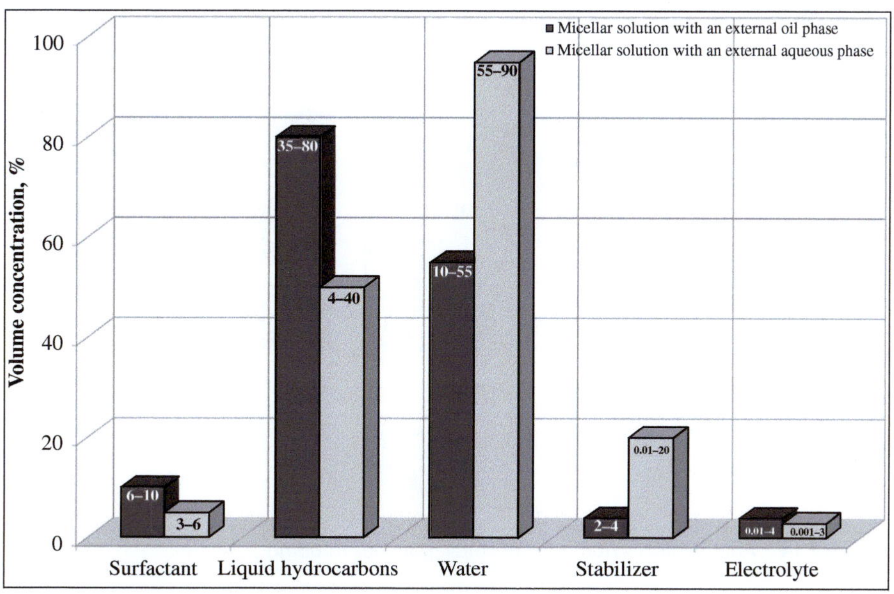

Figure 7.13 Volume concentration of micellar solution components.

7.3.2 Applicability Criteria

The potential application of the micellar-polymer flooding method is quite large. Almost all terrigenous fields with an oil saturation of more than 35% and a reservoir oil viscosity of less than 35 cP can become targets for the use of micellar-polymer flooding.

The main disadvantage hindering the widespread use of the micellar-polymer flooding method is the large need for chemical reagents, which determines its high cost.

The main criteria for the applicability of the method are shown in Figure 7.14.

7.3.3 Field Implementation

Due to the fact that the micellar-polymer flooding method has a relatively high cost, it has not become as widespread as other methods of physical and chemical-EOR. However, there is information about 30 tests of the method in industrial conditions. Industrial tests were carried out in the USA, Canada, China, Romania, Hungary, France, and the former USSR. The results of some of them are presented in Figures 7.15 and 7.16. The first test of micellar-polymer flooding was carried out in the early 1960s by Marathon Oil Corporation at the M-1 site of the Robinson field. The test was successful, which made it possible to significantly increase the scale of testing. As can be seen from Figure 7.16, the increase in oil production was impressive. In most cases, oil production increased by 50% or more when micellar polymer flooding was first introduced. With the exception of two cases, in all fields micellar-polymer flooding was carried out after conventional flooding.

The parameters of the polymer solution, in particular the polymer concentration, have the strongest influence on the efficiency of micellar-polymer flooding. In addition, the quality of the surfactant also has a certain influence on the efficiency of the process.

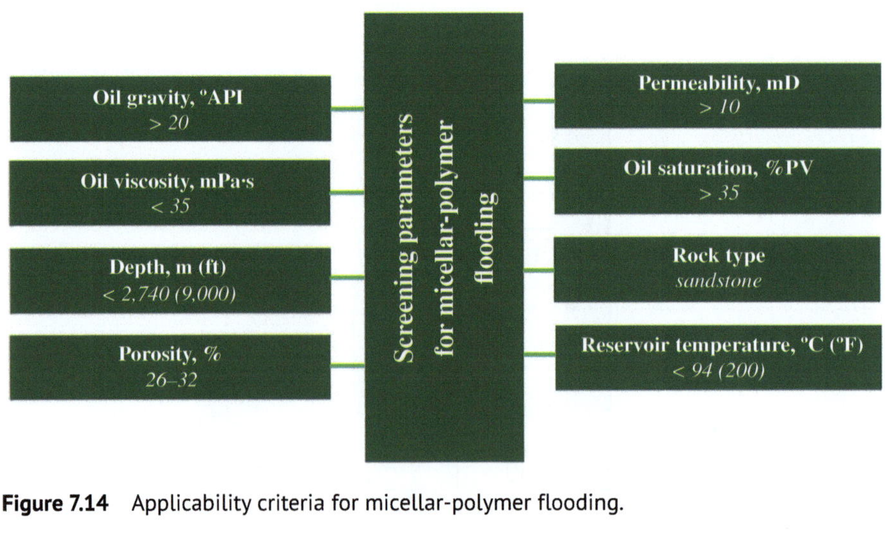

Figure 7.14 Applicability criteria for micellar-polymer flooding.

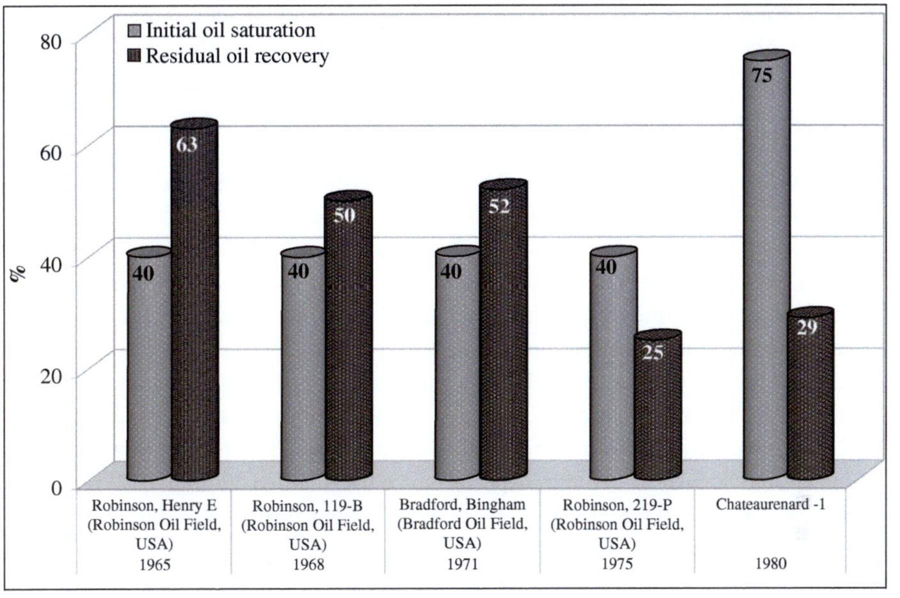

Figure 7.15 Implementation results of micellar-polymer flooding projects.

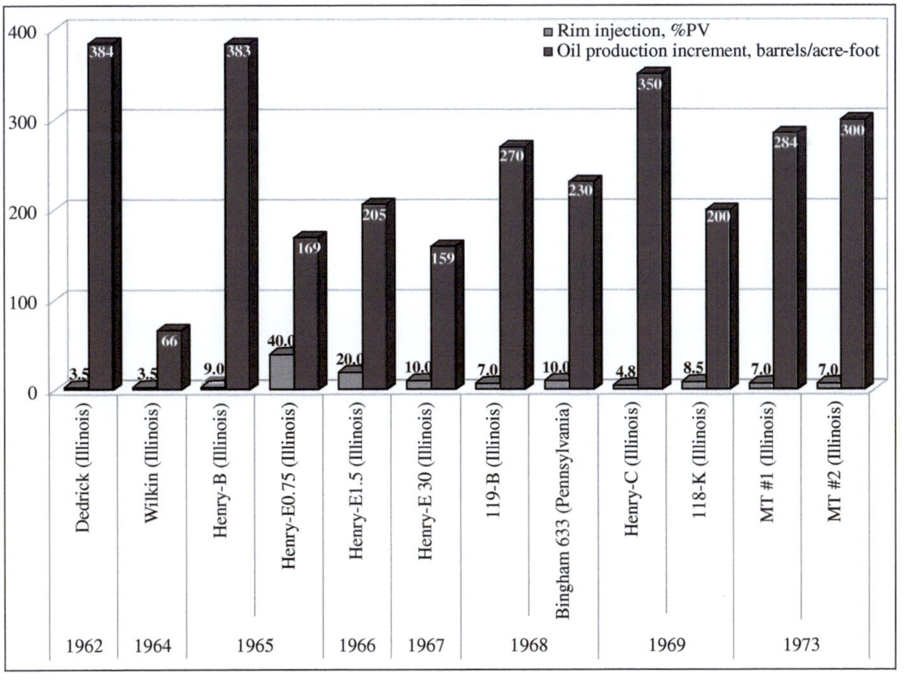

Figure 7.16 Implementation results of micellar-polymer flooding projects.

In the former USSR, trials of micellar-polymer flooding began in the seventies. In 1976–79, preliminary tests of the method were carried out at the Khodyzhenskoye and Romashkinskoye fields. The positive results obtained allowed us to begin industrial testing at the Romashkinskoye field. In 1979, tests began on Yuzhno-Romashkinskaya Square, and in 1983—on Aznakaevskaya Square.

In March 1979, testing began on Yuzhno-Romashkinskaya Square. A rim of fresh water was previously injected into the formation to reduce the salt composition of the formation water. The water cut of the products was 99%. The injection of a slug of micellar solution with an external hydrocarbon phase with a viscosity of 16 cP has begun. The interfacial tension at the oil interface was 0.01 mN/m. The volume of the micellar solution rim was 9.4% of the pore volume. Next, a slug of polymer solution with a viscosity of 5.5 cP was injected into the formation in a volume of 45% of the pore volume. As a result of the use of micellar-polymer flooding, the water cut of the product decreased to 75%, and additional oil production amounted to 2,000 tonnes. The results obtained turned out to be significantly worse than expected since the planned additional oil production was 5,000 tonnes. During the implementation, serious difficulties were discovered: a decrease in the productivity of production and injectivity of injection wells by 8–10 times due to the inversion of the micellar solution and an increase in its viscosity.

One of the micellar-polymer flooding options was implemented in 2004 at the Shengli field (see Figure 7.17). In this case, a polymeric surfactant was used, in which the polymer increases the viscosity of the injected water, while the surfactant acts as an emulsifier.

Since the first injection in June 2004, test results have shown a rapid decrease in water cuts and an increase in oil production rates. Water cut decreased from 98.2% in 2004 to 85.2% in 2007. Oil production increased from 34 tonnes/day in 2004 to 193 tonnes/day in July 2008.

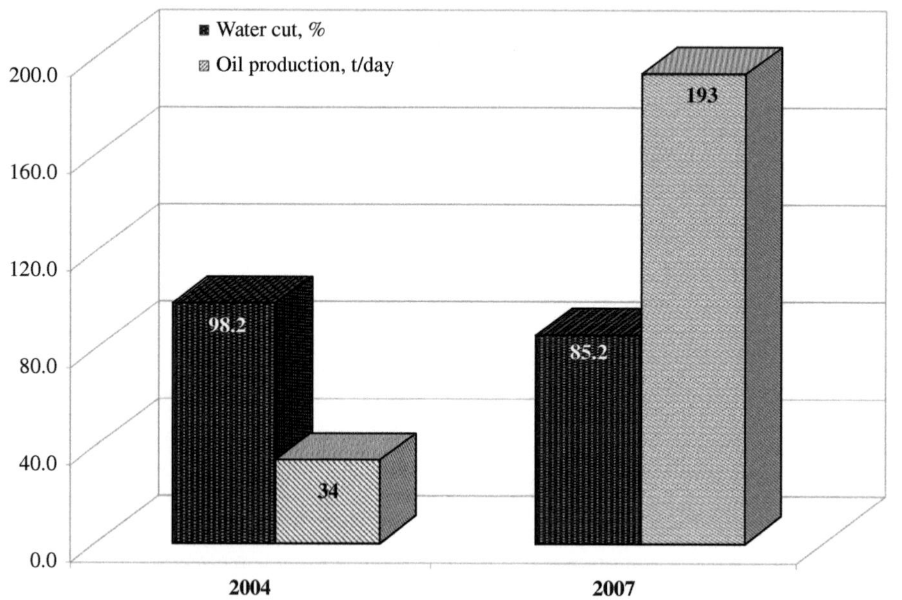

Figure 7.17 Implementation results of micellar-polymer flooding at the Shengli field.

7.3.4 Implementation Technology

The technology for micellar flooding is relatively simple. First of all, laboratory studies are carried out on core material with modeling of reservoir conditions to determine the optimal composition of the micellar solution. In this case, the viscosity of micellar and polymer solutions is assumed to be approximately equal to the viscosity of oil under reservoir conditions. The micellar solution is prepared on-site before use. It is very important to test micellar flooding on a hydrodynamic model of the field, if possible.

7.3.4.1 Injection Sequence, Composition, and Structure of Solutions

Injection of working solutions is carried out in the following sequence.

As shown above, if the formation water has high mineralization, direct contact with the micellar solution is avoided, and an aqueous solution of low salinity is first pumped into the formation. If the mineralization of formation water is low, then a micellar solution is initially injected into the formation. The composition and volume of the solution are determined by the peculiarities of the geological structure of the field, the heterogeneity of the formation, the properties of the fluids saturating the formation, the distribution of reserves over the field area, the size of the selected area of the field, and the location of production and injection wells.

The size of the micellar solution rim, depending on the interfacial tension achieved, is usually 5–50% of the pore volume of the selected area. In this case, the composition of the micellar solution is selected so that the minimum possible interfacial tension is achieved at the front.

After the micellar solution, a polymer solution is pumped in. This ensures control of the mobility of the micellar solution and prevention of breakthroughs of the water pumped after it. The size of the polymer solution rim, depending on the specific geological structure of the formation, is usually 30–60% of the pore volume of the selected area.

Behind the rim of the polymer solution, water is continuously pumped. This ensures the advancement of the micellar solution in the formation and the displacement of oil to the production wells.

7.3.4.2 Well Placement

Due to the low adsorption of surfactants from micellar solutions, it is possible to limit oneself to a sparse network of wells (i.e., the distance between production and injection wells is quite large). This will require higher discharge pressures. In addition, the time to achieve the effect is prolonged, which can reduce the economic efficiency of the process.

In this regard, the most optimal placement of wells is uniform, with a grid density of 12–16 ha/well. In addition, the placement of wells should be single-row and area-based. Multirow systems are unacceptable for micellar flooding since, due to the high viscosity of the solutions, the time to achieve the effect is prolonged.

To achieve the greatest effect, it is necessary to maximize the pace of field development, so that the total period of additional field development using polymer micellar flooding does not exceed 15 years.

7.4 Alkaline Flooding

Alkaline flooding involves the use of aqueous solutions of certain chemicals such as sodium hydroxide, sodium silicate, and sodium carbonate. These solutions will react with components, particularly organic acids, present in some crude oils to form surfactants that significantly reduce interfacial saturation. At the same time, an increase in oil recovery from the formation during alkaline flooding occurs due to a decrease in interfacial tension, oil emulsification, and changes in rock wettability.

Changes in the interfacial properties of oil, water, and rock during alkaline flooding have been known for a long time. The first patents on alkaline flooding appeared in the 1920s of the last century, and practical application was carried out in the Baku oil fields in the 1940s of the last century. Already during the first commercial use, a high displacing ability of alkaline water was discovered. In subsequent years, interest in alkali flooding has increased significantly due to the development of oil fields that contain significant amounts of active polar components.

7.4.1 Process Mechanism

Alkaline flooding is based on the interaction of an alkaline solution with organic acids contained in reservoir oil, with the formation of emulsifying surfactants. The resulting surfactant reduces the interfacial tension by two or three orders of magnitude, increases the wettability of the rock with water, and leads to the emulsification of residual oil and its displacement from the rock surface. This leads to a significant decrease in residual oil saturation and subsequently to an increase in oil recovery from the reservoir.

As experimental studies have shown, with an increase in the content of organic acids in oil, the interfacial tension decreases significantly and can amount to 0.001 mN/m.

7.4.1.1 Oil Activity

The properties of the resulting oil-water emulsion are greatly influenced by the activity of reservoir oil, which is usually divided into three groups based on acidity (see Figure 7.18). In this case, during the interaction of an alkaline solution and active oil, due to low interfacial tension, an "oil in water" emulsion is formed, while low-active oils form emulsions of the opposite type. At the same time, low-active oils do not form stable emulsions; therefore, with increasing water content, their viscosity increases. In the case of active oils, on the contrary, with an increase in water content, the viscosity of the emulsion decreases significantly.

An important factor inherent in alkaline solutions should be taken into account. Regardless of the activity of the oil, when the alkali concentration exceeds 0.04%, the interfacial tension increases.

7.4.1.2 Rock Wettability

An important element of alkaline flooding is the change in rock wettability as a result of the adsorption of organic acids from oil. The use of alkaline solutions

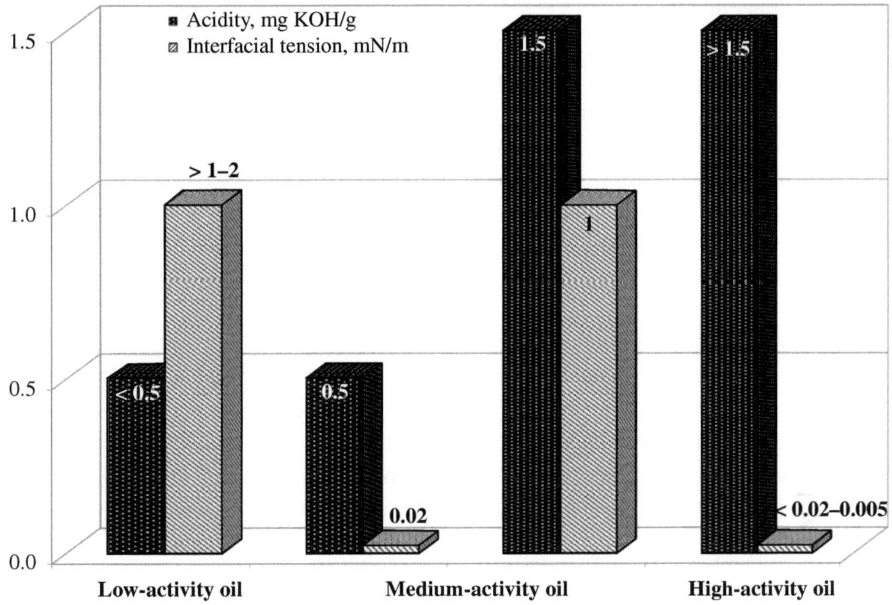

Figure 7.18 Impact of oil activity on interfacial tension.

can significantly reduce the contact angle of formation wetting with water, i.e., hydrophilize the rock. In this case, the advancing contact angle in sandrock can be reduced from 70° to 10°.

When changing the wettability of a porous medium, two cases should be considered.

- **First case:** When displacement occurs in an oil-wet reservoir where the residual oil is in a continuous phase, the alkaline agent changes the pH of the injected water, and the wettability of the rock changes from hydrophobic to hydrophilic. As a consequence, the mobility coefficient of the displacement front decreases, which helps to increase oil recovery from the reservoir.
- **Second case:** Even in hydrophilic reservoirs, given the proper conditions (reservoir temperature, pH, and salinity of the caustic solution), a nonwetting, discontinuous residual oil phase can become a wetting, continuous phase. The presence of water droplets in a continuous phase in a hydrophobic formation increases the pressure gradient, capillary forces are overcome, and the residual oil saturation is reduced.

7.4.1.3 Reservoir Heterogeneity

In heterogeneous formations, emulsification of oil and its capture by small pores contribute to the deflection of the flow of an alkaline solution to areas of the formation not covered by flooding, which significantly increases the coverage of the formation by the influence and leads to a significant increase in oil recovery.

7.4.1.4 Effect of Salts

The presence of salts in the solution significantly changes the process of interaction of oil with an alkaline solution. The presence of calcium chloride increases interfacial tension, reducing alkaline flooding performance. The presence of carbon dioxide also helps to reduce the activity of the alkaline solution due to the formation of soda ash, which does not significantly reduce interfacial tension. Sodium chloride has a positive effect on the activity of alkali, which allows, other things being equal, to significantly reduce the concentration of alkali in an alkaline solution.

7.4.1.5 Influence of Clays

The presence of clays in the rock has a significant influence on alkaline flooding. Clays have a negative impact on the alkaline flooding process. This occurs due to the ion exchange of hydrogen contained in clays with ions (hydroxide) of the alkaline solution, which leads to a decrease in pH. The adsorption of alkali is also different in different clays (see Figure 7.19). As can be seen from the figure, there is no adsorption on quartz sands, but it is high in montmorillonites and especially anhydrites.

In an alkali solution, the swelling properties of clays such as montmorillonite are 2 times greater than in water. Therefore, the presence of clays in the rock significantly affects the efficiency of alkaline flooding. According to experimental studies, when the content of clays of the montmorillonite group is more than 20%, the anhydrous oil displacement coefficient with an alkaline solution is the same as with water displacement, while the anhydrous displacement coefficient increases due to the swelling of the clays, which ensures uniform displacement.

7.4.1.6 Carbonate Reservoirs

Unlike sandy reservoirs, the effectiveness of alkaline flooding in carbonate reservoirs is associated with the presence of nitrogen-containing components in the oil. At the same time, the release of these compounds on the surface of carbonate rock promotes its hydrophilization with all the ensuing consequences.

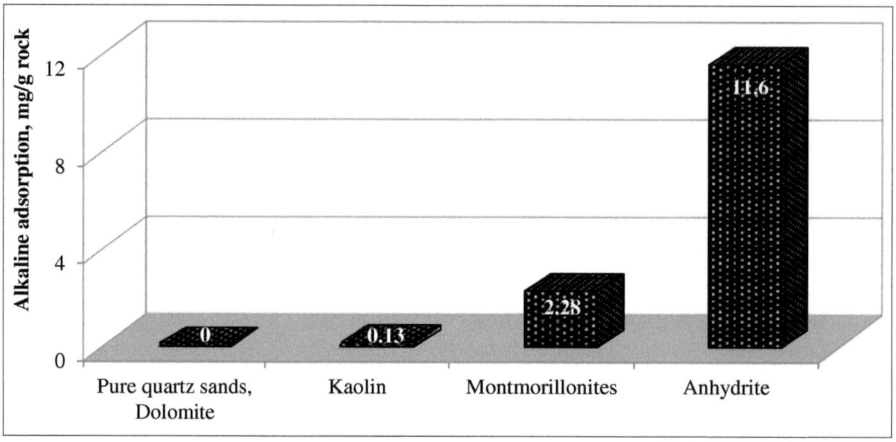

Figure 7.19 Impact of clay type on alkaline adsorption.

7.4.2 Applicability Criteria

General selection criteria, including temperature (less than 94 °C), permeability (from 50 to 250 mD), oil viscosity (less than 200 cP), and low salinity of formation water, coincide with the limits characterizing chemical flooding using surfactants (see Figure 7.20). Reservoirs with large gas cap areas and extensive aquifers should be avoided when selecting candidate fields for alkaline flooding.

There are other special aspects that should be considered when selecting formations for alkaline flooding, such as formation mineralogy, CO_2 content of the oil reservoir, and acid number of the reservoir oil. Below are the main parameters to consider when planning an alkaline flood.

1. **Gypsum:** Formations with anhydrite content of more than 0.1% should be rejected as candidates due to high adsorption from alkaline solution (see Figure 7.19).
2. **Kaolinite:** In fields with high kaolinite content, alkaline flooding can be carried out with alkalis with low pH (8.2–10).
3. **Montmorillonite:** Montmorillonite, due to its high surface area and cation exchange capacity, can absorb most of the introduced alkali due to unfavorable precipitation reactions. Reservoirs with high montmorillonite content are generally not suitable for alkaline flooding.
4. **Rock type:** Sandy formations are most preferable for alkaline flooding. Carbonate rocks are less preferred because high-pH alkalis react with them.
5. **Reservoir oil:** As noted above, the reservoir oil must have a high acid number to achieve low interfacial tension (see Figure 7.18) in alkaline flooding unless the solution contains a synthetic surfactant. However, even in fields containing low-grade oil, alkaline flooding can be successful as a result of oil displacement mechanisms other than reduced interfacial tension.
6. **Carbon dioxide:** The CO_2 content in an oil reservoir is currently considered an important selection criterion. Reservoirs with high CO_2 content and in the presence of formation waters with a pH less than 6.5 are not good candidates for alkaline flooding.

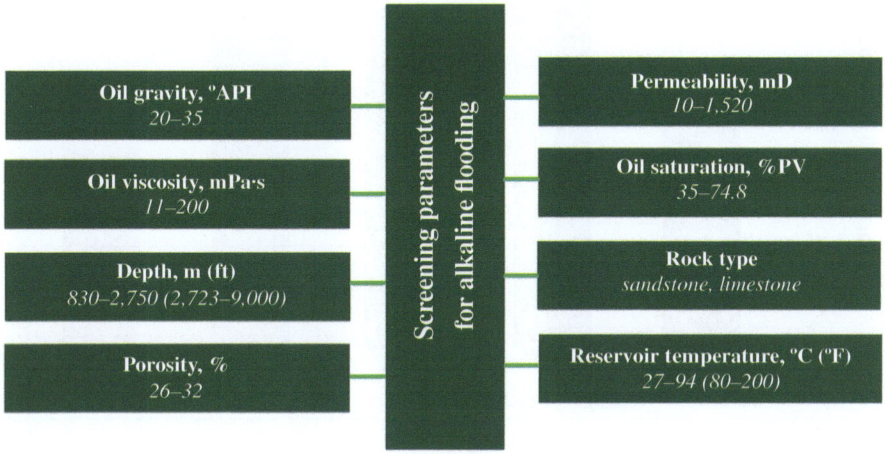

Figure 7.20 Applicability criteria for alkaline flooding.

7.4.3 Field Implementation

Figure 7.21 shows the results of the first implementations of alkaline flooding in the USA in the 1970s. In all these cases, alkaline flooding was carried out after conventional flooding in a variant close to that shown in Figure 7.23. As can be seen from the figure, the residual oil saturation in most cases has changed significantly (up to 31%), which indicates the technological efficiency of the project. Note that in the last century, about 50 alkaline flooding projects were implemented in the USA alone.

The literature reviews production data from 32 implementations of the most common alkaline flooding (ASP) option (see Figure 7.24), both pilot and large-scale, worldwide from 1992 to 2011. Information on other alkaline combinations of waterfloods, such as alkali-surfactant (AS), alkali-polymer (AP), and alkali-surfactant-polymer-foam (ASP-foam), are not given. If we consider these combinations, then in general about 100 pilot and large-scale projects have been implemented using alkalis in the world, many of which were ineffective or the effect of which is difficult to assess. Therefore, we will focus on well-documented and successful projects of the most common alkaline flooding option.

The results of the most successful of the 32 implemented projects are shown in Figure 7.22.

Of the 32 completed projects, 21 projects were implemented in China, seven in the USA, two in India, one in Canada, and one in Venezuela. Only one project in Venezuela was carried out in offshore conditions. As can be seen from the figure, the best result was achieved at the Cambridge field, USA.

All completed projects used a five-point or reverse five-point flooding system.

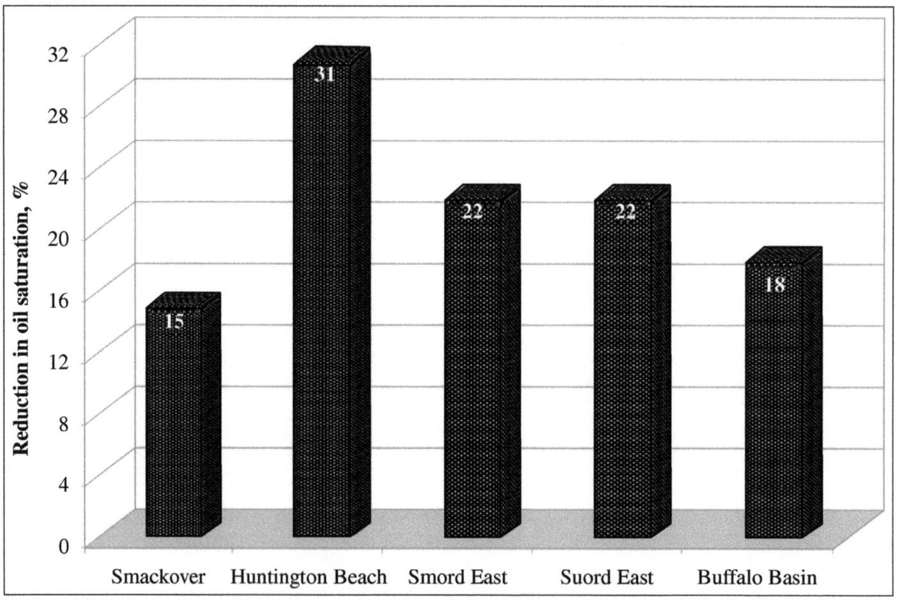

Figure 7.21 Results of implementation of alkaline projects waterfloods.

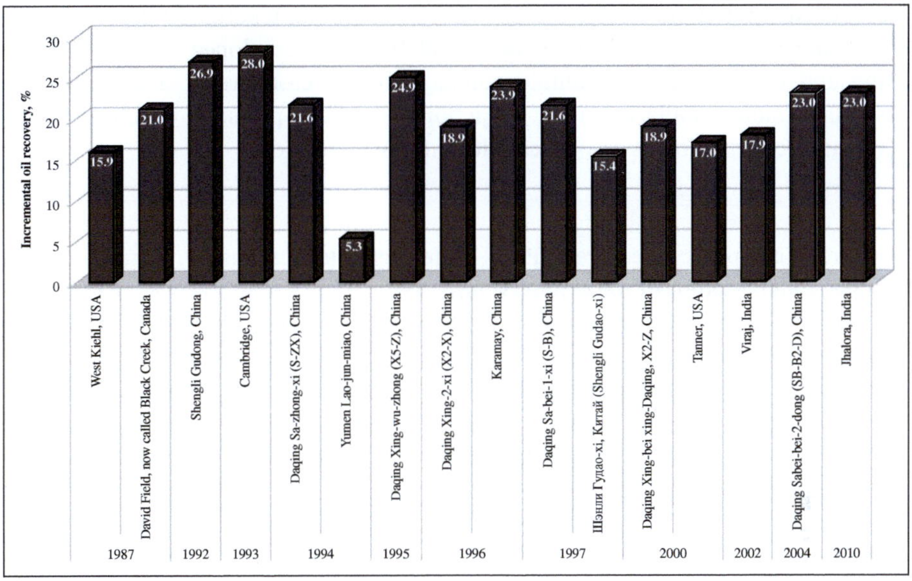

Figure 7.22 Results of alkaline flooding projects.

7.4.4 Implementation Technology

7.4.4.1 Alkaline Flooding Options

Typically, alkaline flooding begins with a preflush of softened water. Followed by the injection of a slug of alkaline solution, in a volume of 10–30%, and moving it through the formation by pumping water. In this case, as a rule, alkaline flooding is carried out after conventional flooding, since it is designed to remove residual oil from the rock surface due to low interfacial tension and oil emulsification. This method is not intended to be implemented in the early stages of field development due to the low potential for increasing impact coverage (see Figure 7.23).

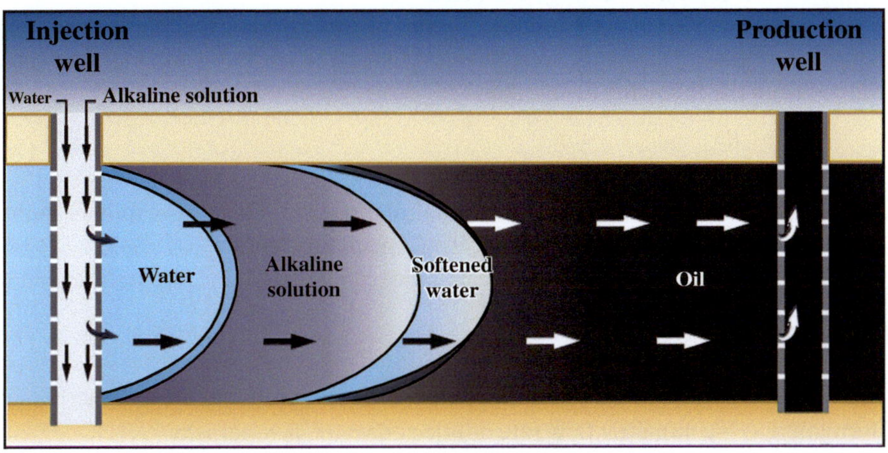

Figure 7.23 Alkaline flooding.

Indeed, let us consider the process of filtration of an alkaline solution with low interfacial tension in a heterogeneous oil reservoir developed with waterflooding and characterized by double porosity. In this case, water is absorbed into small-diameter capillaries under the action of capillary pressure, which, according to Laplace's equation, is associated with interfacial tension:

$$P_k = \frac{2\sigma \cos\theta}{r}, \tag{7.1}$$

where P_k is capillary pressure, MPa; σ—interfacial surface tension, N/m; θ—wetting angle, rad; r—radius of the capillary, m.

As can be seen from Formula (7.1), a decrease *in* σ using alkalis can significantly weaken the absorption capacity of water into small-diameter capillaries under the action of capillary pressure, which can lead to a decrease in oil recovery from low-permeability layers. In highly permeable layers, in which the effect of capillary absorption can be neglected, an alkaline solution, due to low interfacial tension and oil emulsification, will increase oil recovery.

Oil recovery from small-diameter capillaries can be increased by using water characterized by an increased σ value. Fresh water, seawater, formation waters of the chlorine-calcium type, and salt solutions (brine) have high interfacial tension values. However, in highly permeable layers, the washing properties of such water are much worse than those of an alkaline solution.

In this regard, an alternate injection of waters characterized by high and low interfacial tension values, for example, water and an alkaline solution, is used (see Figure 7.23). In this case, water with improved absorbing properties is first pumped in, and then with improved laundering properties. Water characterized by an increased value of interfacial tension is absorbed into capillaries of small diameter and displaces oil from them into large pores. As a result of the subsequent injection of an alkaline solution having improved washing properties, an increase in oil recovery from both low- and high-permeability layers is ensured.

However, this situation is possible only in the case when the speed of water flow in capillaries of small diameter was greater than or at least equal to the speed of water flow in capillaries of large diameter.

Let us estimate the ratio of water flow rates in capillaries of small and large diameters. The speed of water movement in capillaries of small diameter can be determined from the following expression:

$$v_k = \frac{R\sigma}{4\mu\rho L}, \tag{7.2}$$

where v_k is the fluid flow velocity in a small-diameter capillary, R—capillary radius, μ—kinematic viscosity, ρ—liquid density, L—capillary length.

The flow velocity in large-diameter capillaries can be determined from the equation.

$$v_f = \frac{Q}{Fm}, \tag{7.3}$$

where v_f is the flow velocity in large pores, F—cross-sectional area of the flooded formation, Q—volumetric flow rate of injected water, m—porosity.

In order for oil to be displaced from small-diameter capillaries into large-diameter capillaries, the capillary impregnation front must be ahead of or at least keep up with the water flow front in large pores, i.e., the condition must be met

$$v_k \geq v_f. \tag{7.4}$$

The quantities included in formulas (7.2) and (7.3) in real conditions of an oil reservoir can vary within the following limits: $R = 10^{-8} \div 10^{-6}$ m, $\sigma = 10^{-3} \div 10^{-1}$ N/m, $\mu = 2 \cdot 10^{-7} - 2 \cdot 10^{-6}$ m²/s, $\rho = 10^3$ kg/m³, $L = 10 \div 5 \cdot 10^2$ m, $Q/F = 10^{-5}$ m/s, and $m = 0{,}2 - 0{,}35$.

Estimates from Equations (7.2) and (7.3) give the following values:

$$v_k = 2{,}5 \cdot 10^{-12} - 1{,}25 \cdot 10^{-5} \text{ and } v_f = 3 \cdot 10^{-5} - 5 \cdot 10^{-5} \text{м/с}.$$

Thus, the range of changes in flow speed in large pores completely covers the range of changes in water flow speed in capillaries of small diameter, which makes it possible to satisfy the relation (7.4) by adjusting the water injection rate for all conditions.

In addition to this method, there are numerous more expensive options for implementing alkaline flooding. One of them is to inject a slug of polymer solution behind the alkaline solution to control mobility and improve displacement efficiency (see Figure 7.24), of course, if it is economically justified. Due to the complexity of the mineralogy and lithology of petroleum reservoirs, the potential for reaction between alkaline brine, salt water, and oil under reservoir conditions is significant. Therefore, selecting the best injection system for specific field conditions is very important to obtain the best results.

Modern implementations of alkaline flooding use alkaline agents in combination with low concentrations of synthetic surfactant and a polymer solution slug to control mobility.

Another option for alkaline flooding was first implemented in the former USSR, in 1976, at the Trekhozernoye field (see Figure 7.25). This option includes pre-injection of fresh water, injection of hard formation water, and alkaline solution. As a result

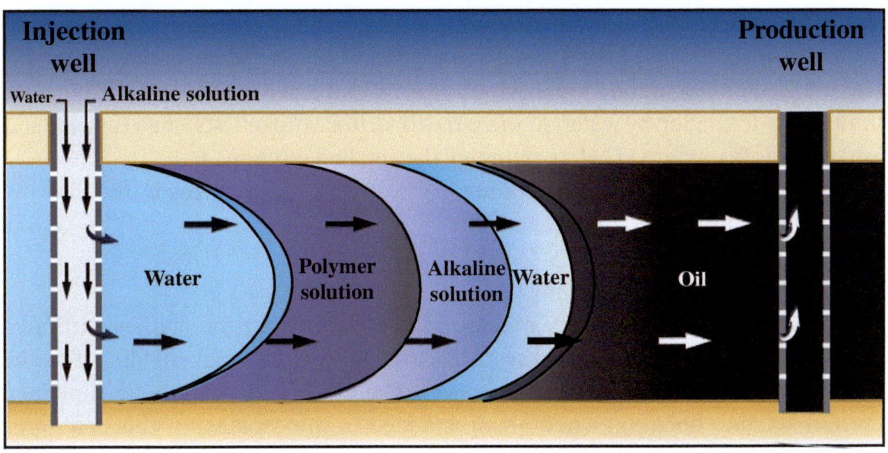

Figure 7.24 Polymer-alkali flooding.

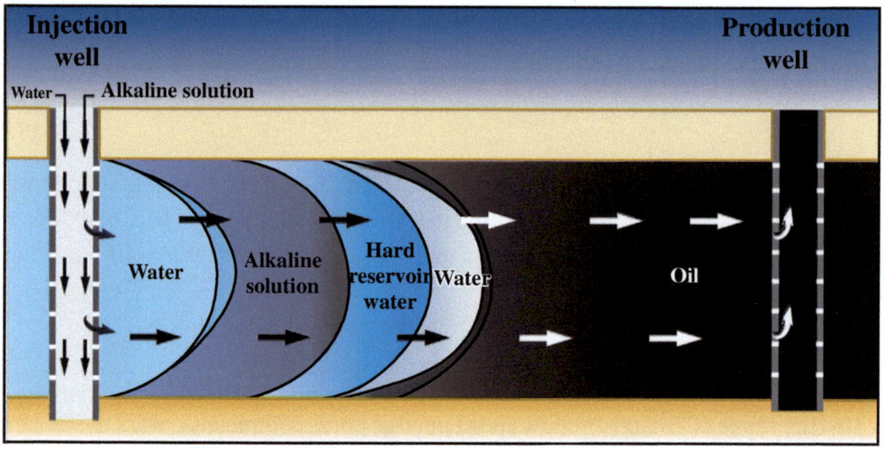

Figure 7.25 Alkaline flooding with precipitate-forming solution.

of the sedimentation reaction between the alkaline solution and hard-formation water, the permeability of highly permeable areas of the formation decreases, and the formation coverage increases. It is also possible for periodic, cyclic injection of liquids in the specified sequence.

7.4.4.2 Preparation of an Alkaline Solution

Caustic soda is used to prepare alkaline solutions. NaOH, soda ash Na_2CO_3, ammonia NH_4OH, water glass Na_2SiO_3.

The concentration of solutions is selected depending on the reservoir and lithological properties of the formation and is determined by laboratory studies on core samples using formation fluids. The main objective of the research is to determine the alkali concentration that provides the minimum possible interfacial tension. Typically, considering adsorption from solution, the concentration of caustic soda is 0.2–0.4%. However, in hydrophobic formations and with high salinity of formation water, higher concentrations of alkali in the solution (up to 5%) may be required to modify the wettability of the rock. As a rule, alkaline solutions are injected in the form of a slug with a volume of 10–30% of the pore volume of the formation (selected depending on the heterogeneity, lithology of the formation, adsorption, and other losses of alkali), propelled through the formation by water. At the same time, to save alkali, softened (usually soda ash) water is first pumped before pumping the alkaline solution.

In some variants of alkaline flooding, cyclic effects and changes in the direction of fluid flows are carried out.

7.4.4.3 Fields with High-viscosity Oils

It is known that high-viscosity oils are more active when interacting with alkalis than low-viscosity oils. However, the use of conventional alkaline flooding does not significantly increase oil recovery. Therefore, it is possible to implement thermo-alkaline flooding, when alkali is added to the coolant. In this case, areal flooding with a dense network of wells is usually used.

7.4.4.4 Well Placement

During alkaline flooding, the system for placing injection and production wells is no different from conventional, polymer, and surfactant in-circuit flooding.

Bibliography

Abbas, S., Donovan, J., & Sanders, A. (2013, July 2–4). Applicability of hydroxyethylcellulose polymers for chemical EOR. *Paper SPE-165311-MS Presented at the SPE Enhanced Oil Recovery Conference. Kuala Lumpur, Malaysia.*

Amyx, J., Bass, D., & Whiting, R. (1962). *Physics of the Oil Reservoir*. Gostoptekhizdat.

Ball, J. and Surkalo, H. (1989). New variations of chemical EOR and their economic potential—the ASP process (alkaline-surfactant-polymer). *Proceedings of the Improved Oil Recovery Conference*. Abilene, *September (DOE/BC/14286-1, March)*, pp. 229–250.

Balzer, D. (1982, November 8–10). Carboxymethylated ethoxylates as EOR surfactants. *Paper Presented at the 2nd European Symposium on EOR, Paris.*

Bird, R., Stewart, W., & Lightfoot, E. (1974). *Transfer Phenomena*. Chemistry.

Burcik, E.J. (1967). Pseudo dilatant flow of polyacrylamide solutions on porous media. *Production Monthly Journal, 31*, 3.

Burcik, E.J. (1968). Use of polymers in the recovery of petroleum. *Earth and Mineral Sciences, 37*, 7.

Carcoana, A. (1992). *Applied Enhanced Oil Recovery*. Prentice Hall, Inc.

Chen, T., Song, Z., Fan, Y., et al. (1998). A pilot test of polymer flooding in an elevated-temperature reservoir. *SPE Reservoir Evaluation & Engineering, 1*(1), 24–29.

De Bons, F.E. & Braun, R.W. (1995, May 15–17). Polymer flooding: Still a viable IOR technique. *Paper Presented at the 8th European IOR Symposium, Vienna, Austria.*

Docher, T. & Wise, F. (1976). Enhanced crude oil recovery potential—an estimate. *Journal of Petroleum Technology, 28*(5), 575–585.

French, T.R. & Burchfield, T.E. (1990, April 22–25). Design and optimization of alkaline flooding formulations. *Paper SPE-20238-MS Presented at the SPE/DOE 7th Symposium on Enhanced Oil Recovery. Tulsa, Oklahoma.*

Han, M., Xiang, W., Zhang, J., et al. (2006). Application of EOR technology by means of polymer flooding in Bohai Oilfields. *Paper SPE-104432-MS Presented at the International Oil & Gas Conference and Exhibition in China.*

Hongyan, W., Xulong, C., Jichao, Z., et al. (2009). Development and application of dilute surfactant-polymer flooding system for Shengli oilfield. *Journal of Petroleum Science and Engineering, 65*, 45–50.

Jay, V., Jim, T., Bob, V., et al. (2000). Alkaline-surfactant-polymer flooding of the Cambridge Minnelusa field. *SPE Reservoir Evaluation & Engineering, 3*(6), 552–558.

Kamal, M.S., Sultan, A.S., Al-Mubaiyedh, U.A., et al. (2015). Review on polymer flooding: Rheology, adsorption, stability, and field applications of various polymer systems. *Polymer Reviews, 55*(3), 1–40.

Lake, L.W. (Ed.). (2007). *Petroleum Engineering Handbook. Vol. V. Reservoir Engineering and Petrophysics*. Society of Petroleum Engineers.

Lake, L.W. & Pope, G.A. (1979). Status of micellar-polymer field tests. *Petroleum Engineering International, 38*(6).

Matheny, Jr., S. (1980). EOR methods help ultimate recovery. *Oil & Gas Journal, 78*(13), 79–124.

Meyers, J., Pitts, M.J., & Wyatt, K. (1992). Alkaline-surfactant-polymer flood of the West Kiehl, Minnelusa Unit. *Paper SPE-24144-MS Presented at the SPE/DOE Enhanced Oil Recovery Symposium.*

Mogollon, J.L. & Lokhandwala, T. (2013, July 2–4). Rejuvenating viscous oil reservoirs by polymer injection: Lessons learned in the field. *Paper SPE-165275-MS Presented at the SPE Enhanced Oil Recovery Conference, Kuala Lumpur, Malaysia.*

Pitts, M., Dowling, P., Wyatt, K., et al. (2006, April 22–26). Alkaline-surfactant-polymer flood of the Tanner field. *Paper SPE-100004-MS Presented at the SPE/DOE Symposium on Improved Oil Recovery.*

Pu, H. & Xu, Q. (2009, March 15–18). An update and perspective on field-scale chemical floods in Daqing oilfield, China. *Paper SPE-118746-MS Presented at the SPE Middle East Oil and Gas Show and Conference.*

Saffman, P.G. & Taylor, F.R.S. (1958). The penetration of a fluid into a porous medium or Hele-Shaw cell containing a more viscous fluid. *Proceedings of the Royal Society of London A, 245*, 312–329.

Savins, I.G. (1969). Non-Newtonian flow through porous media: A review of rheologically complex flow phenomena likely to have a role in mechanisms responsible for behavior observed in non-Newtonian flow through porous media. *Journal of Industrial and Engineering Chemistry 16*(10): R18–R47.

Sheng, J.J., Leonhardt, B., & Azri, N. (2015). Status of polymer-flooding technology. *Journal of Canadian Petroleum Technology, 54*(2), 116–126.

Smith, F.W. (1978). Len-exchange conditioning of sandstones for chemical flooding. *Journal of Petroleum Technology, 30*, 959–968.

Speight, J.G. (2016). *Introduction to Enhanced Recovery Methods for Heavy Oil and Tar Sands* (2nd ed.). Gulf Professional Publishing, Elsevier.

Suleymanov, B.A. (2006). *Features of Filtration of Heterogeneous Systems.* Institute of Computer Research.

Surguchev, M.L. (1985). *Secondary and Tertiary Methods for Increasing Oil Recovery.* Nedra.

Surguchev, M.L., Kuznetsov, O.L., & Simkin, E.M. (1975). *Hydrodynamic, Acoustic, Thermal, Cyclic Effects on Oil Reservoirs.* Nedra.

Surguchev, M.L., Shevtsov, V.A., & Surina, V.V. (1977). *Application of Micellar Solutions to Enhance Oil Recovery.* Nedra.

Thomas, S. and Farouq Ali, S.M. (1992). Micellar-polymer flooding: Status and recent advances. *Journal of Canadian Petroleum Technology, 31*(8), 53–60.

Thornton, S.D. (1986). Reaction of sodium hydroxide with silicate minerals. U.S. Department of Energy Report No. NIPER-129, N-115 Order No. BE86000275.

Zhang, J., Wang, K., He, F., et al. (1999). Ultimate evaluation of the alkali/polymer combination flooding pilot test in Xing Long Tai oil field. *Paper SPE-130113-MS Presented at the SPE Asia Pacific Improved Oil Recovery Conference.*

8

Microbial EOR

8.1 Introduction

One of the most promising methods for enhancing oil recovery is the method of microbiological stimulation of the formation, which is characterized by high manufacturability, low capital intensity, and environmental safety.

There are two main areas of microbiological methods for enhancing oil recovery:

1. Obtaining metabolic products (biosurfactants, biopolymers) on the surface of the earth and injecting them into the formation.
2. Ensuring the activity of microorganisms directly in the formation.

8.2 Introduction to Microorganisms in MEOR

The abundance and diversity of microbial communities in any ecosystem are influenced by the availability of nutrients (food for microorganisms) and various environmental conditions, such as temperature, pressure, pH, salinity, and oxygen levels. These environmental factors can make certain habitats ideal for some microorganisms while being challenging or even hostile for others. Microorganisms interact with their environment by metabolizing available resources to grow and adapt, releasing byproducts that can alter the ecosystem. The key environmental factors that impact microbial growth are temperature, pH, salinity, and oxygen availability. Oil reservoirs, in particular, present extreme conditions with high toxicity, temperature, salinity, and pressure.

8.2.1 Environmental Factors Affecting Microorganisms

8.2.1.1 Temperature

Temperature significantly impacts microorganisms. Geothermal temperature increases with depth, approximately 2–3 °C per 100 m. Deep oil reservoirs often exceed 130 °C, which is beyond the tolerance of most bacteria. However, the temperature drops after water injection during secondary recovery, cooling the

Methods for Enhanced Oil Recovery: Fundamentals and Practice, First Edition.
Baghir A. Suleimanov and Elchin F. Veliyev.

reservoir. Enzymatic activities increase with temperature, leading to faster growth. However, each microorganism has a specific temperature range, known as its cardinal temperature range, within which it can grow:

- *Minimum temperature*: Below this, growth stops as transport processes slow down.
- *Maximum temperature*: Above this, proteins denature and growth ceases.
- *Optimal temperature*: Growth is fastest here due to peak enzymatic activity.

Microorganisms are classified based on their optimal growth temperatures:

- *Psychrophiles*: Thrive below 10 °C.
- *Mesophiles*: Thrive below 50 °C.
- *Thermophiles*: Thrive up to 70 °C.
- *Hyperthermophiles*: Thrive above 70 °C.

Generally, higher temperatures lead to less diverse microbial communities, mainly consisting of thermophilic and thermotolerant species. For instance, microbial communities from oil reservoirs in Texas and Alberta showed high similarity despite different conditions, with common species like *Methanosaetaceae, Methanobacterium, Methanoculleus, Clostridiaceae*, and *Thermotogaceae*. Additionally, abundant archaeal and thermophilic bacteria, such as *Thermus, Thermincola*, and *Thermanaeromonas* spp., have been found in reservoirs with temperatures higher than 90 °C.

8.2.1.2 pH

pH significantly affects microbial communities and their metabolic activities. Most bacteria thrive in neutral pH (6–8), but oil reservoir pH can range from 5 to 8. Surface pH measurements differ from *in situ* pH due to pressure and gas solubility variations. In high-pressure reservoirs, pH can be between 3 and 7. CO_2/bicarbonate systems primarily control the pH, with CO_2 production from microbial fermentation and organic acids influencing pH levels. Minerals in the reservoir rock can dissolve and alter pH, with buffering reactions affecting pH changes. The pH of the reservoir is crucial for microbial enhanced oil recovery (MEOR) practices, influencing the selection of microbial strains and the design of growth media. Different bacteria thrive in different pH ranges, such as *Alphaproteobacteria, Deltaproteobacteria*, and *Actinobacteria* in neutral to alkaline pH and *Pseudomonas* species in low pH. In more acidic environments with a pH of 5.5–6.5, *Gammaproteobacteria, Betaproteobacteria*, and *Epsilonproteobacteria* have been found. The fluid pH of the oil reservoir must be considered when operating MEOR practices, especially when selecting target bacteria and designing biostimulation media. Each microorganism has a preferred pH range, typically spanning about 2–3 pH units:

- *Neutrophiles*: Optimal growth between pH 5.5–7.9.
- *Acidophiles*: Optimal growth below pH 5.5.
- *Alkaliphiles*: Optimal growth above pH 8.

Buffers are often needed in microbial cultures to maintain consistent pH, ensuring stable enzyme activity.

8.2.1.3 Salinity

Salinity in oil reservoirs varies widely, from almost fresh to highly saline (up to 20%). While less studied than temperature, salinity significantly impacts microbial growth, activities, and community composition. For example, Clostridiales are found in wells with 3.8% salinity, while *Petrotoga* and *Desulfotomaculum* species, which degrade hydrocarbons, thrive in 7.2% salinity environments. Secondary oil recovery often involves injecting saline seawater, so MEOR practices must consider whether indigenous or introduced bacteria can adapt to high salinity. Microbial populations tend to decrease significantly in reservoirs with salinity above 10%. Additionally, *Desulfotomaculum* species, known for hydrocarbon degradation, are prevalent even under high-salinity conditions.

Microorganisms in marine environments are often halotolerant or halophilic:

- *Halotolerant*: Can tolerate some salinity (less than 10% NaCl).
- *Halophiles*: Require 3–12% NaCl for growth.
- *Extreme halophiles*: Thrive in 12–30% NaCl concentrations.

Oxygen. Oxygen (O_2) is vital for many microorganisms, but some cannot grow in its absence:

- *Aerobes*: Require O_2 for respiration.
  - *Microaerophiles*: Use O_2 at lower-than-atmospheric levels due to oxygen-sensitive components.
  - *Facultative aerobes*: Can grow without O_2, using aerobic respiration, anaerobic respiration, or fermentation.
  - *Obligate aerobes and microaerophiles*: Only metabolize via aerobic respiration.

Oil fields typically lack oxygen, so facultative aerobes like *Bacillus, Pseudomonas*, and *Escherichia* species are common.

Microorganisms that do not use O_2 are anaerobes:

- *Aerotolerant anaerobes*: Can tolerate and grow in the presence of O_2 without respiring it.
- *Obligate anaerobes*: Cannot survive in the presence of O_2. Common examples include methanogens and *Clostridia*, frequently found in oil reservoirs.

Understanding these environmental factors helps identify which microorganisms can thrive in oil reservoirs, aiding in MEOR.

Biopolymers, biosurfactants, biofilms, biogenic gases, and microbially produced solvents and acids are the main working agents for MEOR. Below is a brief description of each working agent group.

8.2.2 Biosurfactants

Surfactants are surface-active agents that favorably adsorb on surfaces or interfaces within a system. They alter interfacial free energy by attaching at the boundary between immiscible phases. This function is due to their unique structural characteristics, consisting of both lyophilic (liquid-loving) and lyophobic (liquid-hating) parts. For instance, in water, the hydrophobic part of a surfactant associates with water molecules, increasing the free energy of the liquid phase. This makes it thermodynamically favorable for the surfactant to be at the interface rather than surrounded by water molecules. Consequently, surfactants lower surface and interfacial tension (IFT) by concentrating at surfaces and interfaces, reducing the energy state and modifying interfacial properties.

The hydrophobic part of a surfactant typically comprises long-chain hydrocarbon residues, while the hydrophilic part varies based on its electrical charge. Surfactants are classified as anionic, cationic, zwitterionic, or nonionic, depending on the electric nature of the hydrophilic group. Most biosurfactants are amphiphilic compounds naturally produced by microorganisms and modify surface energy similarly to chemical surfactants. They are classified into four types based on their functional groups: glycolipids, phospholipids, polymeric, and lipopeptide surfactants.

Biosurfactants are used in various fields, such as agriculture, food, pharmaceuticals, oil recovery, and CO_2 storage, due to their several advantages over chemical surfactants:

1. *Biodegradability*: Biosurfactants are readily degradable, making them environmentally friendly and suitable for bioremediation, unlike chemical surfactants, which are difficult to degrade and can pose environmental threats.
2. *Low toxicity*: Biosurfactants are generally less toxic than chemical surfactants. For example, rhamnolipids produced by *Pseudomonas aeruginosa* are four times more effective in hexadecane removal, and glycolipids from *Rhodococcus* species 413A are 50% less toxic than the synthetic surfactant Tween 80 in naphthalene solubilization tests.
3. *Producibility from raw materials*: Biosurfactants can be produced using renewable sources such as hydrocarbons and carbohydrates, whereas chemical surfactants are synthesized from chemicals.

These advantages drive extensive research into evaluating and comparing the efficiency of biosurfactants with chemical surfactants. Additionally, the critical micelle concentration (CMC) of biosurfactants is generally lower than that of chemical surfactants, indicating greater efficiency.

Another key advantage of biosurfactants in enhanced oil recovery (EOR) is their emulsification capability. Emulsification involves forming an emulsion, such as suspending small oil droplets in an aqueous phase. High molecular-weight biosurfactants are effective emulsifiers, producing water-in-oil or oil-in-water emulsions in water-oil mixtures. This capability is crucial for altering oil-wet minerals to water-wet status, which enhances oil recovery.

Microbial biosurfactants exhibit a wide variety of chemical structures, including glycolipids, lipopeptides, fatty acids, neutral lipids, and phospholipids. Typically, biosurfactants include hydrophilic parts composed of amino acids, peptides, or polysaccharides, alongside hydrophobic parts consisting of saturated, unsaturated, or hydroxylated fatty acids and amphiphilic or hydrophobic peptides.

Nutrient availability in the growth media significantly affects bacterial growth and biosurfactant production. The types and concentrations of carbon sources, nitrogen substrates, and ions like phosphorus (P), magnesium (Mg), iron (Fe), and manganese (Mn) in the medium are crucial for bacterial surfactant production. Numerous lab tests have optimized biosurfactant production using diverse cultural media. For instance, in biosurfactant production by *Bacillus* spp., glucose and sucrose are preferred carbon sources over sodium acetate, fructose, lactose, and tryptone. Sodium nitrate and ammonium nitrate are effective nitrogen sources, with production being more efficient when these nitrogen sources are limited rather than plentiful in the medium.

Similarly, sophorolipids produced by *Candida* spp. and rhamnolipids produced by *P. aeruginosa* are produced at high concentrations in nitrogen-exhausted media. *Rhodococcus* spp. exhibit maximum growth and biosurfactant production with nitrate as a nitrogen source, consistent with other species. Considering cost and production yields, sucrose and ammonium nitrate show promise as nutrients for stimulating bacterial biosurfactant production, making MEOR applications economically feasible. However, the fundamental mechanisms and optimization of biosurfactant production for MEOR remain not fully understood, warranting further investigation.

Bacillus species. Among various biosurfactant-producing bacteria, *Bacillus* species are considered practical due to their survivability under extreme conditions, such as extreme pH, salinity, high pressure, and high temperature. *Bacillus* spp. are facultative anaerobic bacteria commonly found in soil and aquatic environments. They produce endospores under extreme conditions, extending their viability even for centuries. These endospores are nonreproductive structures that allow bacteria to survive adverse conditions and reactivate when the environment becomes favorable.

Bacillus spp. are widely found in oil reservoirs with temperatures ranging from 19 to 106 °C, pressures from 4 to 27 MPa, and salinity from 2 to 24%. They produce lipopeptide biosurfactants that modify interfacial properties. For example, *Bacillus subtilis* produces surfactin, a powerful lipopeptide biosurfactant. Surfactin is amphiphilic, containing a peptide loop of seven hydrophilic amino acids and a hydrophobic fatty acid chain. Below the CMC, the hydrophobic fatty acid part can extend freely into a solution, making surfactin an effective surfactant. Surfactin can lower the surface tension of water from 72 to 27 mN/m at concentrations as low as 20 μM.

The production of biosurfactants is heavily governed by microbial metabolisms, such as respiration or fermentation processes. Therefore, the survivability and growth of microbes under extreme environmental conditions are prerequisites for the application of biosurfactant-producing bacteria in MEOR practices.

Pseudomonas species. Rhamnolipids, a type of microbial surfactant belonging to glycolipids, are produced by *P. aeruginosa*. They are highly effective in modifying fluid interfacial properties and rock wettability, demonstrating notable surface activity, emulsifying capability, and solubilization potential. Rhamnolipids have four homologs, with one or two rhamnose units linked to one or two fatty acids containing 8–14 carbon atoms. These compounds reduce the surface tension of water from 72 to 31 mN/m, indicating their effectiveness as biosurfactants. The CMC of rhamnolipids ranges from 20 to 225 mg/L in water at 25 °C.

Several factors significantly influence rhamnolipid production:

1. *Environmental factors*: pH, temperature, agitation, and oxygen availability.
2. *Nutritional factors*: Types and concentrations of carbon, nitrogen, and metal ions.

Rhamnolipid production is optimal at a pH range of 6.0–6.4, with a drastic drop when pH exceeds 7.2. Although *P. aeruginosa* is a facultative anaerobic bacterium, the properties of rhamnolipids vary under different oxygen conditions. Aerobically produced rhamnolipids show better surface activity, whereas anaerobically produced rhamnolipids exhibit superior emulsifying activity. Under anaerobic conditions, rhamnolipid production is lower (680 mg/L) compared to aerobic conditions (11,650 mg/L) but still exceeds the CMC. Increased aeration rates also enhance rhamnolipid production. Optimal production conditions include 37 °C, neutral pH, and agitation.

Carbon sources in the culture medium are crucial for rhamnolipid production, with high yields observed using glucose with mineral salt medium. Among nitrogen sources, nitrate is the most effective, and nitrogen-limited conditions can increase production by a factor of 3–4. Iron also promotes rhamnolipid production. *Pseudomonas* spp. are frequently found in oil reservoirs due to their facultative anaerobic nature, allowing them to thrive in microaerobic (microoxic) or anaerobic environments. Numerous *P. aeruginosa* strains isolated from various oil fields adapt well to oil reservoir conditions. Rhamnolipids' stability under varying pH, temperature, and salinity supports their use in MEOR practices. Additionally, *Pseudomonas* spp. can degrade hydrocarbons, using them as a carbon source, making *P. aeruginosa* suitable for bioremediation and MEOR applications.

Several other microorganisms can be utilized for MEOR practices:

1. *Rhodococcus species*: These hydrocarbon-degrading bacteria grow under facultative anaerobic conditions and produce biosurfactants when using hydrocarbons as a carbon source. *Rhodococcus ruber* Z25, isolated from formation brine in the Daqing oil field, China, is a potential model for MEOR applications. This strain produces biosurfactants that reduce the IFT of water and oil, with a CMC of 57 mg/L. *R. ruber* degrades crude oil and produces biosurfactants under both aerobic and anaerobic conditions. Most biosurfactants produced by *Rhodococcus* spp. are glycolipids. *Rhodococcus* sp. strain TA6 exhibits emulsifying activity with various hydrocarbons, including pentane and light motor oil. Glycolipids from Iranian oil-contaminated soils reduce water surface tension from 68 to 30 mN/m and remain chemically stable under high salinity (10% NaCl), elevated temperatures (120 °C for 15 minutes), and a wide pH range (4–10). Core-flooding tests on oil-saturated sand packs have shown that culture broth aids in 70% residual oil recovery.
2. *Acinetobacter species*: These bacteria, frequently found in oil-contaminated soils and reservoirs, can also produce rhamnolipids. For instance, *Acinetobacter junii* BD, isolated from the Xinjiang oil field in northwest China, demonstrates survivability under high fluid pressure conditions (1,088 m deep, 32 °C). *Acinetobacter* spp. produce rhamnolipids that reduce surface tension and emulsify crude oil in oil-water systems.
3. *Arthrobacter species*: Producers of lipopeptide or trehalolipid biosurfactants, *Arthrobacter* spp. are stable at temperatures ranging from 30 to 100 °C and pH levels from 2 to 12. Due to their resilience under extreme environments, they also have potential for MEOR practices.

8.2.3 Biopolymers

Biopolymers produced by microorganisms are widely considered for enhancing oil recovery. A key strategy is selective plugging, which exploits biopolymers to modify the permeability of reservoir formations. This technique involves plugging high-permeability layers, or "thief zones," redirecting water flooding to oil-rich channels to improve oil recovery.

8.2.3.1 Selective Plugging Strategies

Selective plugging can be achieved through two main methods:

1. *Biostimulation*: This method stimulates the indigenous bacterial community within the reservoir, promoting the natural production of biopolymers.
2. *Bioaugmentation*: This involves inoculating the reservoir with specific bacterial cells along with the necessary nutrients to promote bacterial growth and biopolymer production in targeted high-permeability zones. Once biopolymer production begins, additional nutrients can be supplied to sustain and enhance this process.

This section delves into the characteristics of representative biopolymers, highlighting their applications in science, engineering, and their potential in MEOR.

A biopolymer is an assembly of monomers derived from living organisms, including algae, bacteria, and fungi, which consume carbon sources during growth. Biopolymers serve various roles:

- *Energy reserve materials*: Acting as carbon sources.
- *Protective shelters*: Helping microorganisms survive changing environmental conditions.
- *Cell functioning*: Supporting essential cellular activities.

Microbial extracellular polymeric substances (EPSs) are biopolymers synthesized by microorganisms and are categorized into three main types:

1. *Nucleic acids and polynucleotides*: These are genetic materials, such as DNA and RNA.
2. *Polypeptides*: Comprised of amino acids, these catalyze specific biochemical reactions.
3. *Carbohydrates*: Known as polysaccharides, these are the most commonly used biopolymers in industries. Polysaccharides consist of polymeric carbohydrate chains with monosaccharide units, forming viscous hydrogels in water due to abundant surface hydroxyl groups (OH).

For microorganisms, the primary purpose of producing EPS as metabolites is protection. EPS helps shield microorganisms from desiccation and predation by protozoans by surrounding them with a protective barrier. There are two main methods of EPS production:

1. *Intracellular synthesis and export*: Most EPS are synthesized within the bacterial cell and then exported outside the cell wall.
2. *Extracellular synthesis*: Certain biopolymers, like xanthan and dextran, are polymerized and synthesized outside the bacterial cells by secreted enzymes that convert substrates into polymers.

Despite variations in biosynthesis mechanisms among different EPSs, the common steps in bacterial biosynthesis pathways include:

1. *Uptake of resources*: Carbon substrates are assimilated through central metabolic pathways.
2. *Synthesis*: Decomposed substrates are converted into polysaccharides by specific enzymes.

Biopolymers are typically formed through condensation reactions, which dehydrate and link monomers, or via phosphorylated intermediates catalyzed by divalent cation-dependent enzymes. This complex process results in the creation of diverse biopolymers with unique properties beneficial for various applications, including MEOR.

Xanthan gum. Xanthan gum is a well-known biopolymer used in various industries, including oil recovery. Produced by *Xanthomonas campestris*, xanthan gum is notable for its high viscosity and stability under a wide range of temperatures, pH levels, and salinity. This makes it particularly effective for enhancing oil recovery by improving the sweep efficiency of water flooding.

Dextran. Dextran, another polysaccharide biopolymer, is produced by species such as *Leuconostoc mesenteroides*. Dextran is used in various biomedical and industrial applications due to its ability to form viscous solutions. In MEOR, dextran can enhance oil recovery by improving the mobility of injected fluids and altering the permeability of reservoir formations.

Alginate. Alginate is produced by marine brown algae and certain bacteria, such as *Pseudomonas* and *Azotobacter* species. It forms hydrogels that are used in a variety of applications, including drug delivery, wound healing, and as a viscosifier in EOR. Alginate's ability to form gels in the presence of divalent cations, such as calcium, makes it useful for plugging high-permeability zones in oil reservoirs.

While biopolymers hold significant potential for MEOR, several challenges must be addressed to optimize their application:

- *Production costs*: The cost of producing biopolymers can be high, which may limit their widespread use. Research into cost-effective production methods, such as utilizing waste materials as substrates, is ongoing.
- *Environmental conditions*: The effectiveness of biopolymers can be influenced by reservoir conditions, such as temperature, pressure, and salinity. Developing biopolymers that are stable under extreme conditions is a critical area of research.
- *Microbial interactions*: Understanding the interactions between microbial communities and biopolymers is essential for optimizing biopolymer production and application in reservoirs.

Future research aims to address these challenges and enhance the viability of biopolymers for MEOR. This includes exploring genetic engineering techniques to develop microorganisms with enhanced biopolymer production capabilities and investigating novel biopolymers with unique properties suited for specific reservoir conditions.

8.2.4 Biofilms

Biofilms, a common phenomenon in bacterial growth, are mixtures of bacterial cells surrounded by an adhesive matrix secreted by the cells themselves. This matrix consists of polysaccharides, proteins, and nucleic acids. Biofilms serve as a microbial self-defense mechanism, protecting cells from physical forces and toxic molecules while increasing their survival rate. They aid in cell attachment to nutrient-rich surfaces and promote cell viability through interactions between cells.

Biofilm growth is a widespread pattern in water-saturated environments, such as below the water level in reservoirs or below the groundwater table. This growth pattern offers several advantages, including nutrient sharing, electron transfer, and protection from physical and chemical attacks. Naturally occurring biofilms typically comprise multiple bacterial species, forming a complex community that collaborates for survival.

One significant advantage of biofilm growth is its high tolerance to antibiotics, which can be up to 1,000 times higher than planktonic cells. This tolerance is achieved through mechanisms such as blocking antibiotic penetration, inducing genetic variance, and slower cell growth.

Bacteria generally form biofilms rather than growing planktonically for four main reasons:

1. *Self-defense system*: Biofilms provide a physical barrier against external stimuli and enhance the immune system of the cell-biofilm mixture, increasing cell survival compared to planktonic growth.
2. *Microbial nutrient niche*: By attaching to favorable nutrient conditions, biofilms create a nutrient-rich environment where cells can thrive and share nutrients among themselves.
3. *Cell-to-cell communication*: The structure of biofilms facilitates electron transfer and communication between bacterial cells, enhancing the survival chances of the community.
4. *Genetic exchange*: Biofilm formation promotes cell aggregation and physical proximity, increasing the likelihood of genetic exchange among cells and further enhancing survival chances.

Overall, biofilm formation is the default mode of growth for prokaryotic cells, providing numerous benefits for bacterial survival and adaptation in various environments.

Biofilms can modify the wettability of surfaces, including those in oil reservoirs. This modification reduces the capillary pressure of oil in pore throats, making it easier to displace oil. While chemical surfactants and biosurfactants are traditionally used for wettability alteration, biofilms achieve similar outcomes through physical and chemical surface modifications, ultimately increasing oil recovery efficiency.

Biofilms aid in controlling permeability within the reservoir. Biomasses such as bacterial cells, biofilms, biopolymers, and biogas accumulate in pores, causing permeability reduction, also known as bioclogging. Similar to biopolymers, biofilm-induced bioclogging can be utilized for selective plugging in MEOR. Highly permeable formations within a heterogeneous reservoir can be selectively plugged with biomasses, particularly biofilms, enhancing oil sweeping efficiency and facilitating water flooding into unswept, oil-rich zones.

The formation of bacterial biofilms involves several steps:

1. *Initial cell attachment*: Cells randomly move across a surface, and upon collision, adhere to the surface due to attractive forces between the biomass and the interface.
2. *Irreversible attachment*: Attached cells undergo irreversible attachment to the surface, initiating biofilm formation.
3. *Colonization*: Bacterial cells and polysaccharides colonize the surface, forming microcolonies.
4. *Development*: Biofilm growth occurs as cyclic dimeric guanosine monophosphate (c-di-GMP), a second messenger, triggers the transition from planktonic growth to biofilm formation. c-di-GMP regulates the activity of flagella and the production of the biofilm matrix.
5. *Active dispersal*: Biofilms may disperse actively, allowing cells to detach and colonize new surfaces.

Biofilm architecture is highly heterogeneous, consisting of several porous layers with microcolonies of cells, EPS, and flow channels for nutrient and water exchange. Factors influencing biofilm formation and structure include cell-to-cell interactions, microbial diversity, and cell mobility. Mixed cultures with multiple bacterial species often produce thicker and more stable biofilms compared to pure cultures. However, not all mixed cultures result in enhanced biofilm formation, as cell mobility can affect biofilm architecture and stability.

EPS is the main component of biofilms, constituting approximately 50–90% of the organic carbon in these structures. EPS serves as the primary matrix material, providing various physical and chemical properties essential for biofilm integrity and function.

8.2.4.1 Composition and Properties

- *Polysaccharides*: EPS is predominantly composed of polysaccharides, which contribute significantly to its ionic characteristics, generally rendering the biofilm neutral to anionic due to the presence of acidic materials, such as uronic acids and ketal-linked pyruvates. This results in strong ionic bonds with divalent cations, like calcium and magnesium, essential during biofilm formation.
- *Charge variations*: While most biofilms are negatively charged, certain gram-positive bacteria, such as staphylococci, exhibit cationic charge properties.
- *Structural heterogeneity*: The biofilm structure is nonuniform, with the composition and structure of polysaccharides influencing rigidity and solubility. For instance, polysaccharides with 1,3- or 1,4-β-linked hexose form more rigid and less soluble EPS compared to those that are more water-soluble.

8.2.4.2 EPS Production Influences

- *Microbial diversity*: Different microorganisms produce varying amounts of EPS, with older biofilms typically generating more EPS.
- *Growth rate*: Slower bacterial growth rates lead to higher EPS production, aiding in biofilm desiccation prevention.
- *Hydrophobicity and hydrophilicity*: EPS is highly hydratable due to numerous water molecules forming hydrogen bonds, though it can exhibit both hydrophobic and hydrophilic properties depending on specific conditions.

8.2.4.3 Representative Biofilm-bacterial Species

1. ***Pseudomonas aeruginosa***
   - *Characteristics*: Gram-negative, rod-shaped with a monoflagellum, optimal growth temperature 25–37 °C, but can survive up to 42 °C.
   - *Biofilm formation*: Varies by strain, with some producing alginate-rich biofilms that are highly structured and resistant to water flows.
   - *Biofilm morphology*: Changes with conditions such as oxygen availability and flow turbulence, forming mushroom-like or filamentous microcolonies accordingly.
   - *Mechanical properties*: EPS contributes to biofilm's mechanical stability, behaving as elastic solids under shear stress and viscoelastic fluids over longer periods. Calcium ions enhance stiffness through cross-linking.
2. ***Bacillus licheniformis***
   - *Adaptability*: Found in high-temperature oil fields, this facultative anaerobe is halotolerant and thermotolerant, producing various by-products that aid in oil recovery.
   - *EPS Composition*: Includes 26% carbohydrate and 3% proteins, showing pseudoplastic behavior (shear thinning) in solution.
   - *Optimal growth conditions*: Maximum EPS production at 10% NaCl, 50 °C, pH 7, under aerobic conditions. Different strains exhibit high tolerance to salinity and temperature, making them suitable for selective plugging in oil reservoirs.
3. ***Bacillus cereus***
   - *Environmental ubiquity*: A gram-positive, spore-forming bacterium capable of thriving in extreme conditions.
   - *Biofilm formation*: Produces various biofilms and metabolites, including surfactants and enzymes. EPS proteins play a crucial role in cell adhesion and are positively charged at physiological pH.
   - *Biosurfactant production*: Optimal conditions involve specific phosphorus and nitrogen concentrations, with reduced effectiveness at higher salt concentrations.
4. ***Shewanella oneidensis mr-1***
   - *Characteristics:* Gram-negative, rod-shaped, facultative anaerobe commonly found in marine sediments.
   - *Biofilm development:* Forms pink, thin biofilms, using electrically conductive nanowires for electron transport. Capable of reducing metals, which is useful in bioremediation.
   - *Gene influence*: The genes exeM and exeS significantly affect biofilm production, with mutants displaying varying biofilm quantities.

These bacterial species, with their distinct biofilm-forming capabilities and environmental adaptability, are vital for MEOR and other industrial applications. The unique properties of their EPS contribute to the mechanical stability and functionality of biofilms under different environmental conditions.

8.2.5 Biogenic Gases

In EOR, gas injection—whether miscible or immiscible—improves the displacement efficiency of oil by reducing capillary forces within porous media. Even during water flooding, the use of gas can help reestablish pathways for the remaining residual oil, increasing the overall oil recovery rate. Biogenically produced gases, such as CO_2, N_2, H_2, and CH_4, can similarly facilitate the oil recovery process. Initially proposed by Zobell and Johnson, the idea is to inject specific bacteria or stimulate indigenous bacteria to produce gases like CO_2 or CH_4, thereby restoring pressure in the reservoir. A secondary effect of CO_2 injection is the promotion of calcite and siderite dissolution in carbonate reservoirs, further enhancing oil recovery. Various microorganisms can produce these gases, potentially increasing reservoir pore pressure, reducing oil viscosity, swelling the oil, and improving displacement efficiency. However, achieving sufficient gas production in natural anaerobic conditions is challenging, necessitating further optimization studies.

- *CO_2-producing bacteria via fermentation.* The most common microbial mechanism for CO_2 production is fermentation, an energy-yielding anaerobic process where substrates are sequentially transformed through reduction-oxidation reactions. Fermentation involves the enzymatic decomposition of carbohydrates by microorganisms, with no external electron acceptor involved. The redox levels of the substrate and metabolites remain constant, requiring relatively low energy. Typical fermenting bacteria are facultative anaerobes, like *Escherichia* spp.; aerotolerant anaerobes, such as *Lactobacillus* spp.; or strict anaerobes, like *Clostridium* spp. During glucose fermentation to lactate or ethanol and CO_2, pyruvate reoxidizes the NADH produced during glycolysis, restoring redox balance and producing lactate or ethanol and CO_2 as end products. In high sugar concentrations, *Lactobacillus* spp., which is acid-tolerant, often dominates the fermentation process.
- *N_2-producing bacteria via denitrification.* Denitrifying bacteria are capable of denitrification as part of the nitrogen cycle, metabolizing nitrogenous compounds by reducing nitrogen oxides such as nitrite (NO^{2-}) and nitrate (NO^{3-}) to nitrogen gas (N_2) or nitrous oxide (N_2O). Representative denitrifying bacteria include heterotrophic species, like *Paracoccus denitrificans, P. aeruginosa*, and *Pseudomonas stutzeri*, and autotrophic species, such as *Thiobacillus denitrificans* and *Micrococcus denitrificans*. Denitrification typically occurs in anoxic environments where oxygen is limited. These bacteria use nitrate or nitrite as terminal electron acceptors instead of oxygen. Oil reservoirs, generally anoxic, are suitable environments for denitrification processes. Some bacteria, such as *P. denitrificans*, can denitrify under both oxic and anoxic conditions, utilizing enzymes like nitrous oxide reductase. Injecting nitrate to stimulate nitrate-reducing bacteria in reservoirs can enhance *in situ* nitrogen gas production.

- *Representative H_2-producing species.* Biohydrogen production, or the generation of hydrogen gas (H_2) via biological processes, often occurs through fermentation. The process involves the conversion of glucose to acetic acid and H_2, catalyzed by hydrogenases: [FeFe]-hydrogenase and [NiFe]-hydrogenase. Genera known for hydrogen production include *Clostridium, Desulfovibrio, Ralstonia*, and *Helicobacter*. H_2 is crucial for *in situ* CH_4 production by hydrogenotrophic methanogens during CO_2 reduction to methane.
- *Representative CH_4-producing species.* Methanogens produce methane as a metabolic by-product in oxygen-limited environments. Mostly anaerobic, these microorganisms perform methanogenesis, converting substrates like H_2/CO_2, acetate, formate, methanol, and methylamines to methane. Well-known methanogenic genera include *Methanococcus, Methanosarcina, Methanobacterium, Methanobrevibacter, Methanothermobacter*, and *Methanoculleus*. Under anaerobic conditions, hydrocarbons degrade to methane in subsurface oil reservoirs via methanogenic biodegradation.

8.2.6 Solvents and Acids

Microbially produced solvents, such as acetone, butanol, and ethanol, can facilitate MEOR practices by reducing oil viscosity and increasing mobility. These solvents are produced by *Clostridium* spp. and *Zymomonas* spp. through anaerobic fermentation processes, like the acetone-butanol-ethanol (ABE) process. *Clostridium acetobutylicum*, for instance, has been tested for oil recovery improvements, showing stability and functionality under reservoir conditions.

Acid-producing bacteria, which produce organic or inorganic acids, can also assist in MEOR. Acids reduce oil viscosity, improve permeability by dissolving rock minerals, and aid in emulsifying crude oils. These processes are particularly beneficial in carbonate-rich reservoirs. Representatives include *Bacillus, Clostridium*, and *Enterobacter* spp., producing acids, like acetic, lactic, formic, and succinic acids. For instance, *Bacillus subtilis* produces both organic acids and solvents, demonstrating the potential for dual functionality in MEOR. However, economic feasibility is a consideration, especially when the viscosity difference between oil and solvent is minimal, requiring further investigation into the characteristics of microbial solvents and acids in reservoir environments.

8.3 Process Mechanism of MEOR

Even though MEOR mechanisms differ somewhat depending on the technology used, the basic mechanisms remain the same. The essence of the method of microbiological influence on the formation is based on an increase in the oil displacement coefficient as a result of the intensification of the vital activity of microorganisms capable of converting complex organic compounds into simpler ones—liquid (fatty acids, alcohols, solvents) and gaseous (CO_2, CH_4, N_2, H_2) metabolic products that are good oil displacers. Certain groups of microorganisms can form biopolymers and biosurfactants.

The mechanism of microbiological effects is complex and multi-stage. First, in the formation during nutrient flooding, decomposition, and oxidation of hydrocarbons by injected microorganisms occurs in the presence of oxygen, resulting in the formation of fatty, naphthenic, and aromatic acids, alcohols, ethers, and solvents. An increase in the biomass of microorganisms occurs due to a properly selected nutrient medium.

A characteristic feature of oil is its exceptional heterogeneity, due to the presence of organic compounds that differ markedly in chemical composition and structure and are actively exposed to microbial influence. At the same time, individual oil fractions are subject to microbial decomposition to varying degrees. The oxidation products of petroleum hydrocarbons become food substrates for microorganisms deep in the formation in the absence of oxygen. First, the decomposition and oxidation of more easily digestible organic compounds occurs, then those more difficult to utilize, and towards the end, the waste products of previous levels are assimilated into the system.

At the next stage, fatty acids and alcohols are used by other microorganisms, forming acetic acid, carbon dioxide, and hydrogen. In the final phase, methane is formed from intermediate decomposition products, which is the final link in the chain of transformations of organic substances in the formation. The diagram of oil displacement under microbiological influence is shown in Figure 8.1.

The vital activity of microbial cells occurs with the participation of enzymes—organic compounds that catalyze all reactions. Organic substances introduced during nutrient flooding into an oil reservoir undergo a wide variety of transformations under the influence of the functional activity of biofilter microorganisms.

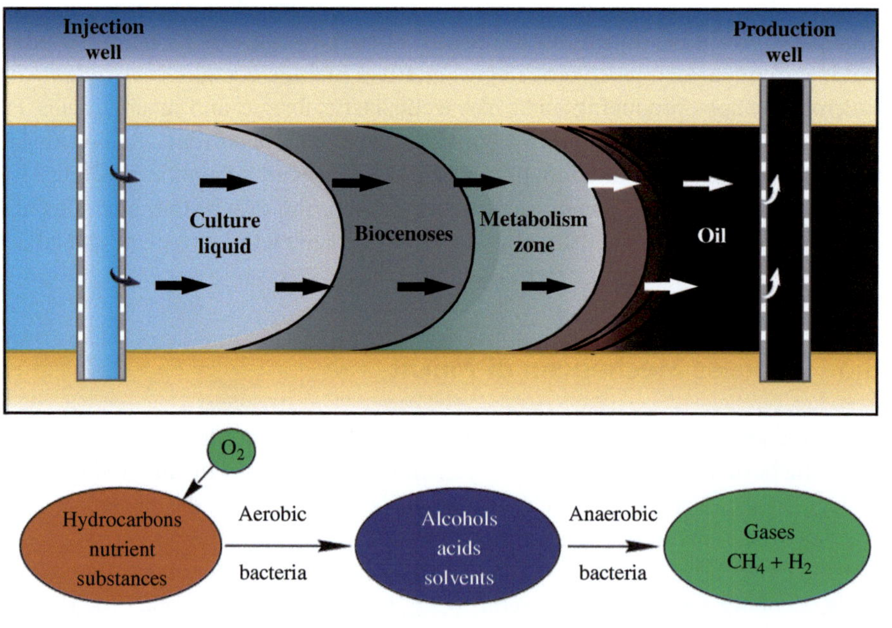

Figure 8.1 Scheme of oil displacement from a reservoir during microbiological enhanced oil recovery.

Thus, during systemic nutrient flooding of an oil reservoir, the following processes occur:

- As a result of the vital activity of aerobic hydrocarbon-oxidizing microorganisms, intermediate oxidation products are formed, including acetate and CO_2;
- Further in the anaerobic zone, because of the activity of other groups of microorganisms (fermenters), hydrogen and methane are released at the last stage.

In microbiological processes, the products of one reaction serve as a substrate for the next. In this regard, compounds formed directly in various zones of the formation have a significant impact on the physicochemical properties of fluids in the "oil-water-porous medium" system. Acids and alcohols dissolve carbonate rocks and prevent salt deposition, which increases the porosity and permeability of rocks. Solvents are directly involved in the extraction of droplet and film oil from rocks, improving the wettability of rocks, etc.

Gases released during the decomposition of organic compounds have a multifactorial effect on formation fluids: pressure, pH, and viscosity of water, oil compressibility coefficient and saturation pressure, density, viscosity and volumetric expansion of oil, the surface tension of oils at the interface with water, phase permeabilities formation fluids, clay swelling, rock permeability, etc.

The resulting microbial surfactants reduce the interfacial surface tension of water at the interface with oil and a solid surface to a level at which globules of residual oil, held in a porous medium by capillary forces, become mobile. Surfactants of microbial origin can have high emulsifying ability and form highly dispersed oil-in-water emulsions, which helps achieve uniform movement of the displacement front. Biopolymers formed because of the vital activity of microorganisms change the rheological properties of formation fluids and contribute to leveling the displacement front.

The processes discussed above in their totality have a significant impact on the final oil recovery of the reservoir. Obviously, the mechanism for enhancing oil recovery is the same as in other EOR methods. However, it is difficult to isolate the dominant mechanism of oil displacement. All the above mechanisms take place simultaneously. Probably, the high efficiency of biotechnologies is not a function of any one factor, as is the case when using physicochemical methods of influence but is determined by the integral synergistic effect of many factors. However, one or another factor or mechanism at each individual site of influence may play a greater or lesser role in the process of displacement. This, in turn, is determined by the specific impact technology, geological conditions of the formation, etc. Figure 8.2 shows the mechanism of the process of EOR using microbiological EOR.

Experimental studies on linear reservoir models revealed that microbiological effects contribute to an increase in displacement efficiency by 15–20% compared to ordinary water.

When using a microbiological method of EOR, based on the functional activity of hydrocarbon-oxidizing microorganisms, in fields producing paraffinic oils, it is possible not only to significantly increase oil recovery, but also to prevent deposits of solid hydrocarbons. In this case, providing the reactive zone with air or oxygen is

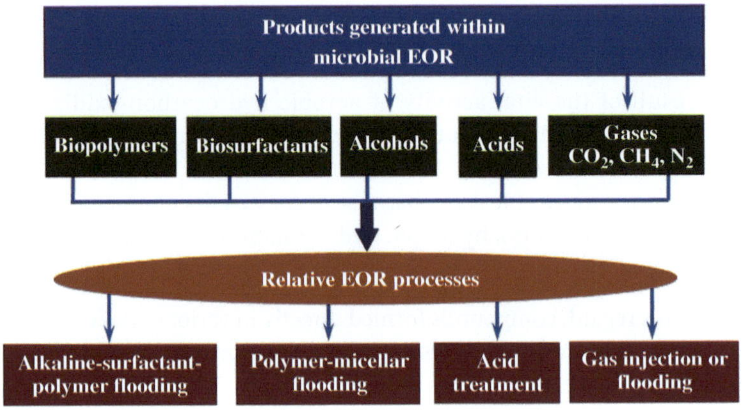

Figure 8.2 Mechanisms of microbiological EOR.

a prerequisite. Microorganisms can use hydrocarbons of different classes of simple and complex structures. Almost all hydrocarbons that make up oil can be subject to microbial action. Thus, microorganisms develop most quickly and efficiently on solid n-paraffins and somewhat more slowly on liquid and gaseous n-paraffins.

It is known that an increase in the concentration of carbon dioxide in the gas mixture will help reduce the amount of precipitation of calcium and magnesium salts. An increase in the carbon dioxide content in an oil reservoir can be achieved by nutrient flooding of the reservoir with a bioreagent containing carbohydrate compounds decomposed by microorganisms during fermentation to CO_2, for example, molasses or whey.

To increase the efficiency of waterflooding of deposits and increase the coverage coefficient under the influence of a formation characterized by sharp geological heterogeneity, the use of biotechnologies associated with the selective action of microorganisms is promising. Microorganisms quickly clog areas of high permeability that receive large amounts of injected fluid. The expected efficiency and selectivity of blocking highly permeable interlayers with microbial biomass are highest when microorganisms and nutrients are simultaneously injected into the formation. At the same time, the biopolymers formed as a result of the vital activity of microorganisms lead to the thickening of formation waters, and, consequently, to the deterioration of their filtration. As a result of these changes, the rapid penetration of water into the reservoir is inhibited, and oil flow rates from wells increase, indicating the effectiveness of the oil displacement process. As a plugging agent, you can use a composition of activated sludge with sediment from primary settling tanks of water treatment systems. In addition to selective plugging, this technology simultaneously involves processes of transformation of organic compounds with the formation of oil-displacing agents.

Flooding of oil reservoirs in some cases may be accompanied by increased activity of sulfate-reducing bacteria (SRB). In this regard, during nutrient flooding, it is necessary to focus on factors limiting the activity of SRB, such as the type of reservoir, salinity and composition of water, and injection of organic substances, which are the substrate for SRB.

The main problems that can be encountered when implementing microbiological methods, in addition to sulfate reduction processes, are associated with the transport of metabolic products deep into the formation, ensuring the necessary concentration of microorganisms for their growth and development, as well as the possibility of optimizing the necessary microbiological activity.

8.4 Applicability Criteria

The geological and physical conditions of objects selected for microbiological influence must meet certain requirements for temperature and pressure of the formation, salinity and composition of formation water. Biotechnologies can be most effectively applied in fields with terrigenous and sandy reservoirs, with sufficiently high porosity (more than 20%) and permeability of the formations (more than 0.04 μm^2), at a formation temperature of less than 100 °C (see Figure 8.3). Highly mineralized formation waters of oil fields prevent the development of microbiological processes. Therefore, the mineralization of water in the formation should not exceed 130 g/L, and the sodium chloride content should not exceed 5–10%. The SO_4 content in formation and injected waters should not exceed 80 mg/L to prevent the development of sulfate reduction. The oil reservoir must be sufficiently flushed to achieve a large oil-water contact area.

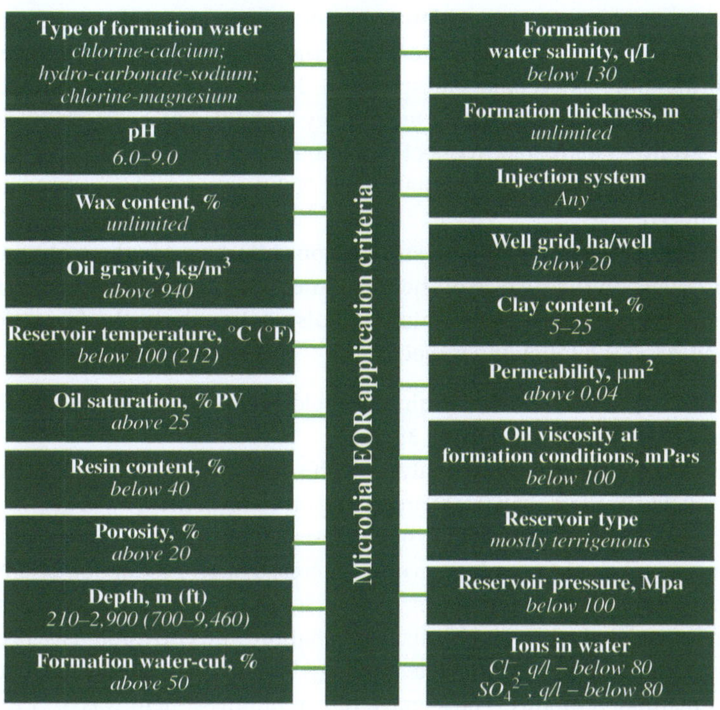

Figure 8.3 Criteria for the applicability of microbiological methods of exposure.

8.5 Field Implementations

The first field experiments on microbiological effects on the formation began to be carried out in the middle of the last century. One of the main factors ensuring the effectiveness of biotechnological methods is correctly selected nutrient substrates and a medium with microorganisms. Micro technologies biological impact on the reservoir can be divided into methods based on the activation of aerobic processes, and methods based on the activation of anaerobic processes. However, aerobic methods simultaneously stimulate the development of anaerobic ones, and vice versa. For this reason, there cannot be purely aerobic or anaerobic processes; there are processes in which one or another type of cultivation predominates.

The following technologies are currently being developed based on the formation of metabolic products directly in the reservoir:

- Microorganism cultures are introduced into the formation along with a nutrient substrate;
- Only a nutrient substrate is introduced into the formation to activate the native formation microflora.

The most widely used technology is "molasses fermentation," which is based on the introduction into the formation of molasses and microorganisms capable of fermenting hydrocarbon substrates and producing biogas containing methane and CO_2. As microorganisms, preference is given to inject mixed cultures of aerobic and anaerobic microorganisms. Valuable substrates for nutrient flooding can be yeast mash, whey, an aerated solution of nitrogen and phosphorus salts, etc. This technology, and in general technologies for ensuring the activity of microorganisms directly in the formation, has become widespread in the fields of Azerbaijan, where it has been introduced since 1980s. The technology was introduced at seven old fields, with the average technological effect (input-output ratio) amounting to 7.2 tonnes per tonne of injected reagent; in total, about 200 thousand additional tonnes of oil were produced.

Technologies of microbiological stimulation (MI), based both on obtaining metabolic products on the surface of the earth and injecting them into the formation, and on ensuring the activity of microorganisms directly in the formation, were implemented in the following projects.

- MEOR application in the Daqing oil field increased oil production by 11.2% while reducing oil viscosity by 38.5%.
- J. T. Portwood presents an analysis of the effectiveness of 322 MEOR projects conducted in more than 2,000 producing oil wells in the United States that used the same MEOR technology. The analysis showed that in 78% of cases, there was a positive effect, with an average increase in oil production of 36%. At the same time, MEOR is environmentally friendly, and general operational problems associated with wax, emulsion formation, and corrosion have been significantly reduced. This MEOR technology has demonstrated cost-effectiveness and manufacturability. The average return on investment was 5 : 1 during the first 24 months, and the average project payback time was six months.

- The successful implementation of MEOR was carried out in the Daqing field, and during three years of implementation of the technology, an increase in oil production was observed in 77% of production wells located in the impact zone with a technological effect (input-output ratio) of 1 : 5.2.
- In Canada, in Southern Alberta, a MEOR was carried out at the X field, the effect of which continued throughout the year, while the average monthly production from one well increased from 21 barrels per day in July to 51 barrels per day in September 2014. During September 2014, approximately 14,000 additional barrels of oil were produced.

8.6 Implementation Technology

Most of the developed microbiological methods for enhancing oil recovery are based on the introduction of production waste into the formation as bioreagents. In this regard, the capital costs of the process consist of the costs of delivery and injection of bioreagents. To implement the microbiological influence process, no special equipment is required. The volume of slug pumped into the injection well must be at least 0.1% of the pore volume. After injecting the bioreagent, the wellhead is sealed for an incubation period (1–2 weeks), then it is put into operation.

The process is monitored by field research methods of the quantity and composition of production from producing wells, hydrodynamic studies, and geophysical methods of development control. For the microbiological processes occurring in the formation not to fade, it is necessary to periodically introduce microorganisms and nutrient substrates into the formation. Nutrient flooding cycles are determined based on laboratory and field studies.

To improve microbiological methods for increasing oil recovery and obtaining an integrated effect, it is promising to combine them with other technologies. It is known that hydrocarbons are extremely reduced organic compounds and can be oxidized by microorganisms mainly in the presence of free oxygen in the air, i.e., under aerobic conditions. In the anaerobic zone, in the absence of oxygen, certain groups of microorganisms (fermenters) consume carbon dioxide, forming hydrogen and methane. With an increase in the concentration of oxygen and carbon dioxide, the rate of their consumption increases to a certain critical value. With a further increase in gas concentration, the rate of their consumption does not change. In this regard, when combining microbiological and gas methods of influencing the formation (air injection, CO_2) or alternating them, a synergistic effect can be obtained. To study the processes in detail and identify the dominant mechanisms of integrated methods of influence, additional research is necessary. The greatest efficiency is achieved in cases where, when combining methods, their mutual influence on the mechanism of each of them is manifested.

The source of energy for microbial cells, as is known, is the components of organic polymers and surfactants, which significantly accelerate metabolic processes. At the same time, for the successful occurrence of biochemical decomposition reactions, the presence in formation waters of trace element compounds K, Na, Ca, Mg, etc., which are part of alkaline solutions, is necessary. Therefore, with physicochemical methods of influencing the formation, namely when injecting solutions of polymers, surfactants, alkalis, micellar solutions in combination with bioimpact or when alternating these technologies, a synergistic effect will be observed with a certain pattern.

It should be noted that microbiological methods for enhancing oil recovery compared to traditional physical and chemical methods are low-cost, technologically advanced, and can be used in deposits with hard-to-recover oil reserves. Biotechnologies not only contribute to a significant increase in oil recovery, but also reduce the impact of water inflow on operational performance. Along with this, microbiological methods make it possible to solve issues of creating waste-free technologies and improving the environmental situation of developed deposits. The above provides grounds for further improvement of the developed biotechnologies in order to create new modifications and expand the scope of their use.

Bibliography

Akintunji, F., Marcotte, B., Sheehy, A., & Govreau, B. (2012, April 14–18). A Texas MEOR application shows outstanding production improvement due to oil release effects on relative permeability. *Presented at Improved Oil Recovery Symposium, Tulsa, OK.*

Cusack, F., Brown, D.R., & Costerton, J.W. (1985). Field and laboratory studies of microbial/fines plugging of water injection wells: Mechanism, diagnosis and removal. University of Calgary.

Guo, H., Li, Y., Zh, Y., et al. (2015, August 11–13). Progress of microbial enhanced oil recovery in China. *Paper SPE-174697-MS Presented at the SPE Asia Pacific Enhanced Oil Recovery Conference, Kuala Lumpur, Malaysia.*

Havemann, G.D., Clement, B.G., et al. (2015). New microbial method promises in EOR. *Journal of Petroleum Technology*, *67*(03), 32–35.

Ibragimov, K.M., Abdullayeva, F., & Guseynova, N.I. (2015, November 4–6). Experience of microbial enhanced oil recovery methods at Azerbaijan fields. *Paper SPE-177377-MS Presented at the SPE Annual Caspian Technical Conference & Exhibition, Baku, Azerbaijan.*

Li, Q., Kang, C., Wang, H., Liu, C., & Zhang, C. (2002). Application of microbial enhanced oil recovery technique to Daqing Oilfield. *Biochemical Engineering Journal*, *11*, 197–199.

Portwood, J.T. (1995, April 2–4). A commercial microbial enhanced oil recovery technology: Evaluation of 322 projects. *Paper SPE-29518-MS Presented at the SPE Production Operations Symposium, Oklahoma City, Oklahoma.*

Sheng, J.J. (2013). *Enhanced Oil Recovery Field Case Studies* (1st ed.). Gulf Professional Publishing.

Town, K., Sheehy, A.J., & Govreau, B.R. (2010). MEOR success in south Saskatchewan. *SPE Reservoir Evaluation & Engineering Journal*, *13*(5), 773–781.

Zahner, B., Sheehy, A., & Govreau, B. (2010, April 24–28). MEOR success in Southern California. *Presented at IOR Symposium, Tulsa, OK.*

Zahner, R.L., Tapper, S.J., Marcotte, B.W.G., et al. (2011, July 19–21). What has been learned from a hundred MEOR applications. *Presented at the SPE Enhanced Oil Recovery Conference, Kuala Lumpur, Malaysia.*

9

Forefront EOR

9.1 Introduction

This chapter examines the latest technologies for enhancing oil recovery, testing of which began already in the 21st century, and active use in the last decade. The exception is high-pressure air injection (HPAI), which began testing in the eighties of the last century, but the active use of the method dates back to the new century.

9.2 In Depth Fluid Diversion

Traditional flow diversion methods for the purpose of leveling the filtration profile and formation coverage are based either on increasing the viscosity of the working agent or on sedimentation and gelation near the injection well, which leads to a significant increase in discharge pressure. In addition, the formation of an insulating zone near the well significantly reduces the amount of additionally produced oil due to the rapid flow around the installed zone by the injected water. These disadvantages are overcome in methods for leveling the injectivity profile, based on the deep diversion of the injected liquid using new generation polymers. They are called delayed activation polymers action polymers (DAP).

9.2.1 Injection of Thermoactive Polymers

The Bright Water (BW) project is the most developed technology for the application of thermoactive polymers at the moment. The BW project was created in 1997 by an industry consortium (BP, ChevronTexaco, and Ondeo Nalco Energy Services) as a potential research program to develop a swellable polymer. The objective of the project was to create a polymer composition of delayed activation with a high degree of swelling to improve the formation sweep factor.

The first laboratory samples of the products were received and tested by BP in early 1998.

BW is a long-chain, temperature-sensitive polymer that produces submicron-sized particles. Essentially, BW is compressed into a ball that is a polymer chain

Methods for Enhanced Oil Recovery: Fundamentals and Practice, First Edition.
Baghir A. Suleimanov and Elchin F. Veliyev.

with a core (kernel) in the central part. These balls freely propagate in the rock matrix until the trigger starts the mechanism of increasing particle size. The trigger is the reservoir temperature. The swollen particles block the pore channels, reducing the permeability of absorption zones, and in some cases, completely blocking them. Because of this mechanism of action, BW is often compared to popcorn. In this case, the filtration profile is leveled, the reservoir coverage is increased by the influence, and, as a result, oil recovery is increased. The main advantage of the technology is the depth deflection of the injected liquid, which can significantly increase the coverage of low-permeability oil-saturated interlayers.

A distinctive feature of BW is its one-component composition, which prevents separation. When using multicomponent formulations, different densities of the constituent components are a common cause of separation, leading to several undesirable consequences. To prevent this during exposure, the density of the injected liquid should be close to the density of water. To maintain the injectivity profile, the viscosity of the injected system must match the viscosity of the injected water.

The thermal properties of rocks, the temperature of the injected fluid, and the reservoir temperature determine the conditions for swelling and filtration. The diameter and rheological properties of the particles before swelling must correspond to the size of the pore channels in the formation. Research has shown that the average particle diameter must be less than one-tenth of the average pore throat size, and physical properties such as density must be sufficient to ensure efficient transport with water flow. After swelling, the BW particles should have an average diameter of the order of or greater than one-quarter of the average diameter of the pore channels. Both conditions depend on the BW concentration. At a lower concentration of introduced particles, their sizes are small and close to the diameter of the pore channels, which in turn provides better filtration compared to highly concentrated BW solutions.

9.2.1.1 Process Mechanism

The mechanism of BW action can be divided into three distinct stages:

Stage 1. Injection of nonactivated solution (kernel solution) BW occurs at relatively high injection rates and low temperatures of 10–50 °C. At this stage, the size of the injected particles is small (particle diameter does not exceed 1 μm), which helps prevent a decrease in the well's injectivity.

Stage 2. Particles move through the pore channels in the injected water flow at a gradually decreasing speed, simultaneously increasing the temperature gradient between the injected water and the formation temperature. At this stage, there is no change in particle size.

Stage 3. Activation of the solution upon reaching a given temperature (activation temperature BW depends on the brand of product used). Due to the absorption of water, the polymer particles swell, blocking highly permeable water-flooded intervals of the formation. There is a deviation of the injected water in the direction of low-permeability oil-saturated intervals. As a result, due to an increase in reservoir sweep by displacement, oil recovery from the reservoir increases.

9.2.1.2 Applicability Criteria

The following criteria are considered by most researchers to be the main ones when selecting candidate wells for the successful implementation of BW technology:

- early water breakthrough, leading to increased water cut in products;
- high heterogeneity of the formation (a desirable condition is that the permeability of highly permeable zones is 5 times higher than the permeability of zones not covered by displacement);
- porosity of the highly permeable zone is more than 17%;
- permeability of the highly permeable zone is more than 100 mD;
- low formation fracturing;
- reservoir temperature in the range of 50–150 °C;
- the passage time of the injected fluid flow from the injection to the production well is more than 30 days; and
- mineralization of the injected water is below 70 g/L (70,000 ppm).

9.2.1.3 Field Implementations

Between 1999 and 2000, several candidate wells were considered for pilot testing of the BW technology. High-formation water cut with large volumes of injected water and low oil recovery factor was considered as the main criterion for the presence of high-permeability zones. As a result, the Minas field, located on the island of Sumatra in Indonesia, was chosen for pilot testing. Testing at the Minas field began in November 2001. The initial geological oil reserves of the Minas field were estimated at 8.7 billion barrels, of which 50% were recoverable, and the water cut was 97%.

The purpose of the pilot test was to test the effectiveness of *in situ* blocking of high-permeability zones, as well as the feasibility of efficient logistics for reagent supply and injection into the formation.

According to the published results, a blocking plug was obtained in the formation at a distance of 38 m from the injection well, which led to a small additional increase in oil production (the percentage of increase in oil production was not specified).

A second pilot test was carried out in the North Sea in mid-December 2002. The possibility of using this method in offshore fields under conditions of limited production area of small oil platforms was confirmed. It is noteworthy that it was possible to maintain the well's injectivity when injecting BW, despite the relatively low permeability of the rock of 395 mD (0.395 μm^2). Unfortunately, the field changed hands before the test results were received, and the question of the effectiveness of the measure remained open.

The first full and complete pilot test was carried out at the Milne Point field in Alaska.

Since August 2008, BP has implemented this technology with varying degrees of success in seven fields operated by the company alone or jointly.

To date, 21 successful implementations of the technology out of 25 are known. The volume of additional oil production amounted to more than 9 million barrels at a processing cost of less than $6 per barrel.

9.2.1.4 Implementation Technology

Typically, waterflooding thermoactivated polymers are carried out using slugs propelled through the formation by water.

The size of the rims, depending on the conditions, can be 10–50% of the pore volume. In this case, the dosing of the polymer is carried out using dosing pumps connected to the water supply system to the well.

The polymer concentration and slug volume are selected based on the formation temperature, formation water composition, formation heterogeneity, and pore volume of the field site.

In contrast to traditional polymer flooding, when injecting thermoactivated polymers, the injection pressure and injectivity of the well practically do not change compared to conventional flooding. In this regard, the method can be used in formations with low permeability, in which traditional polymer flooding is technically infeasible.

The placement of wells does not differ from conventional waterflooding if the impact is effective and the design parameters of field development are achieved: injection pressure and oil production rates.

The consensus among experts is that the earlier the method is used in field development, the more effective the process will be. However, it must be taken into account that, without the characteristics of conventional waterflooding, it is difficult to give a definite assessment of the oil recovery potential and economic efficiency of the method. In new fields, with a lack of geological information, it is difficult to conduct a full analysis of the applicability of thermoactivated polymers.

In this regard, the best solution would be to start using the injection of thermoactivated polymers after the usual flooding.

9.2.2 Injection of Colloidal Dispersed Gels

Colloidal dispersed gels (CDG) were developed by TIORCO, Inc. The chemical composition of these gels is represented by low concentrations of partially hydrolyzed polyacrylamide (HiVis 350) and a solution of chelated aluminum citrate (Tiorco 677) as a cross-linking agent. CDGs have a long gelation time, which allows them to be used for deep deflection of injected liquid. The mechanism of action is based on reducing the permeability of highly permeable water-flooded layers by swollen gel particles.

A low concentration of polymer and cross-linker prevents the formation of a cross-linked gel system. The key characteristic of CDGs that distinguishes them from cross-linked gel systems is the absence of a continuous intermolecular network (Figure 9.1). CDGs are a suspension of gel particles that arise mainly under the influence of intramolecular forces. As a rule, hydrolyzed polyacrylamide is used as a polymer to prepare this system.

9.2.2.1 Process Mechanism

Gel formation of CDG occurs at pressures below the threshold. At higher pressures, CDG has fluidity and filtration characteristics similar to a polymer solution. Below threshold pressure, gelation occurs, a concomitant decrease in the permeability of high-permeability water-flooded layers, and deviation of the

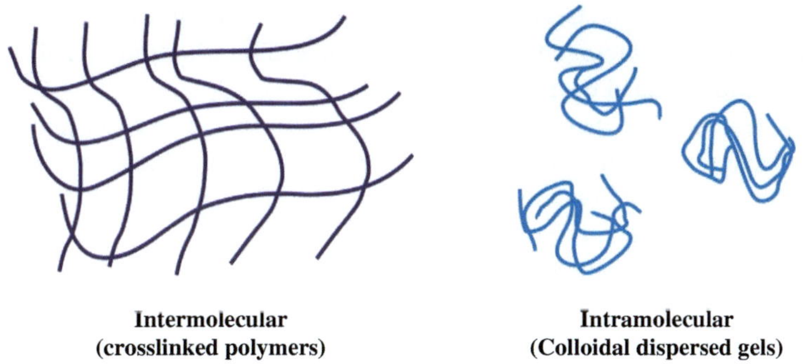

Figure 9.1 Difference between cross-linked and colloidal dispersed gels.

injected fluid to low-permeability oil-saturated intervals of the formation. The pressure in the near-wellbore zone is usually higher than the threshold pressure, which facilitates the penetration of the CDG solution beyond this zone into the depths of the formation. Other factors affecting the rate of gelation are: polymer concentration, polymer- cross-linker ratio, water salinity, and temperature. The gelation time is controlled and can range from several hours to weeks.

CDG is recognized by experts as the most effective enhanced oil recovery (EOR) method for displacement of viscous oils.

9.2.2.2 Applicability Criteria

In field conditions, this technology has a number of undeniable advantages. First of all, it should be noted the economic efficiency in the production of large volumes of CDG, due to the low concentrations of the reagents used. Slower gelation time ensures deep penetration into the formation. However, there are some limitations associated with the use of this technology.

- Most researchers consider it most appropriate to implement this technology in fields at a late stage of development, although there are examples of successful implementation in fields at an early stage of development.
- High-precision injection is required exclusively into the target high-permeability formation since when entering low-permeability interlayers, complete blockage and, accordingly, a drop in permeability are possible.
- The presence of a large positive skin effect in the well or very low permeability of the near-wellbore zone.
- High mineralization of formation water: the upper limit of salinity is 5,000 ppm.
- High Reservoir Temperature: The upper-temperature limit for this technology is 94 °C.

9.2.2.3 Completed Projects

El Tordillo (El Tordillo Field) is the field with the largest number of CDG injections (11 wells). According to published data, the increase in the oil recovery factor was 3.5%.

The LomaAltaSur field is located in the province of Mendoza in the Neuquén River basin in Argentina. The productive reservoir is the Grupo Neuquén formation

(Grupo Neuquen), which is a heterogeneous, multilayered sandstone. The high coefficient of formation heterogeneity and the presence of interlayer flows led to the need to introduce methods of deep fluid deflection.

A pilot test of EOR using CDG was carried out on the LAS-58 injection well. Water injection began in 2002, followed by rapid water breakthrough. The LAS-58 well section includes 10 reacting production wells (offset producers). The characteristics of reservoir rock and oil are given in Table 9.1.

The field trial was carried out in two stages. In Phase I, 186,000 barrels of CDG were injected between July 14, 2005 and February 2, 2006. During Phase II, from April 11 to October 31, 2007, another 192,200 barrels of CDG were injected. The total volume of CDG injected in two stages was 2.96% of the pore volume, which is considered insignificant by industry standards. The data presented below refers to the first stage of the pilot survey. Figure 9.2 shows the change in the injectivity profile before and after injection of CDG. Injectivity was reduced in the main absorption zone (zone 3) and increased in zones 1 and 2. In several layers (A170a, B120-B120a, and B150-B150a) there was no injectivity at all before the injection of CDG.

Figure 9.3 shows the change in production of 10 production wells until October 2007. Production dynamics will stabilize in November 2005, approximately three months after the start of the first stage. As of October 2007, the increase in oil production was 21,194 m^3 (133,292 barrels), and water production was reduced by 64,710 m^3 (406,949 barrels; see Figure 9.4). During both stages of CDG implementation, wells from the idle stock and drilling of new wells were not carried out, so the increase in oil production was ensured only by the use of CDG.

Cretasos (DC) oil field, located in Colombia, was discovered in 1969 and is operated by ECOPETROL. The DK field is classified as heterogeneous reservoirs with a Dykstra–Parsons coefficient above 0.7. Reservoir permeability varies

Table 9.1 Characteristics of reservoir rock and oil of the LomaAltasur field (LomaAltaSur).

Reservoir characteristics	
Pollecturer type	Sandstone
Area, acre	57.3
Reservoir temperature, °C	45
Average depth, m	450
Effective thickness of the formation, m	35
Permeability, mD	20–1,000
Dykstra–Parsons coefficient	0.89
Porosity, %	18.8
Geological oil reserves, m^3	2,169,000
Oil characteristics	
Density, API	21
Viscosity, cP	30

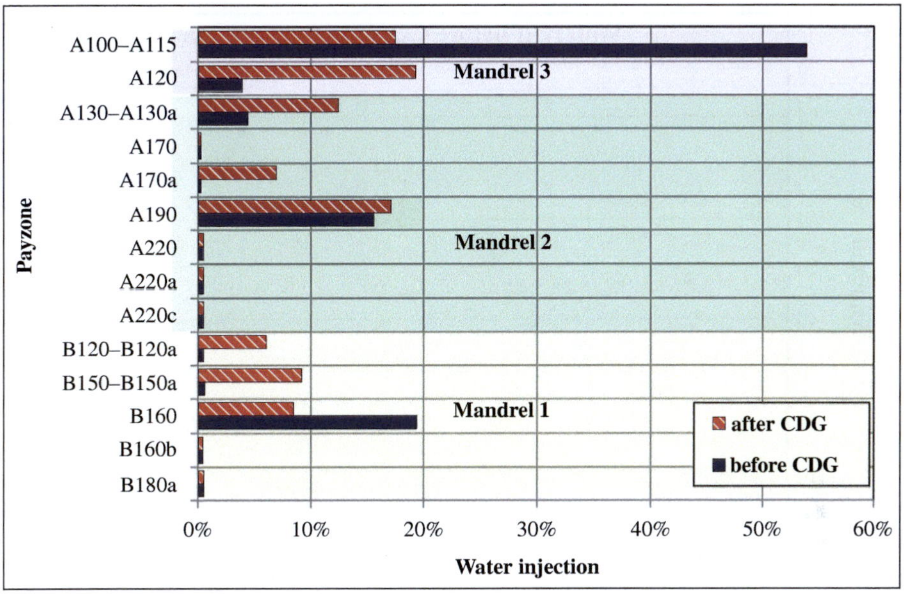

Figure 9.2 Change in the injectivity profile of the LAS-58 well after injection of CDG.

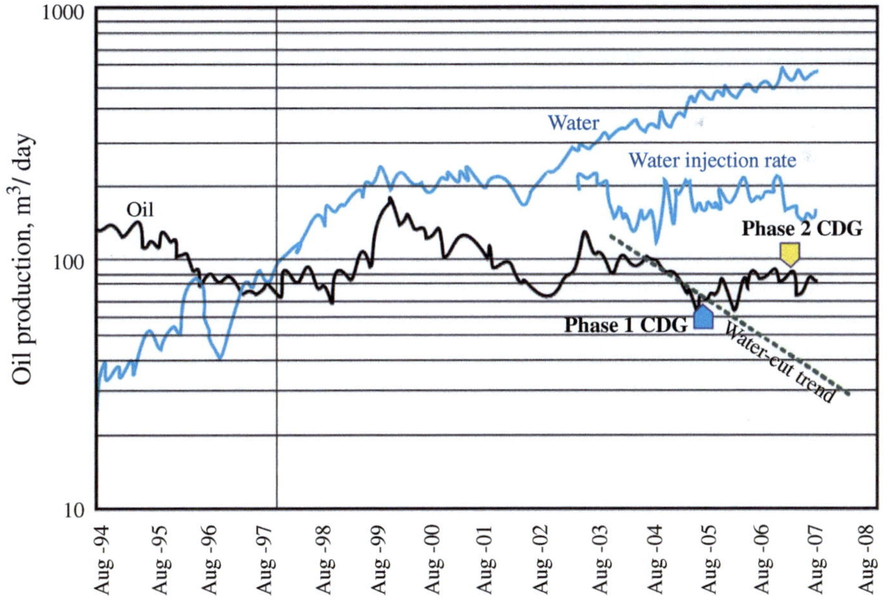

Figure 9.3 Dynamics of changes in oil production for 10 responding production wells.

from 50 to 200 mD, and porosity—from 14 to 18%. The wettability of the rock varies from mixed type to weakly oleophilic.

Oil production began in 1969, and water injection began in 1985. Despite the positive results obtained with water injection, the water cut of extracted products increased from 30 to 96%. Similar results were explained by a decrease in permeability

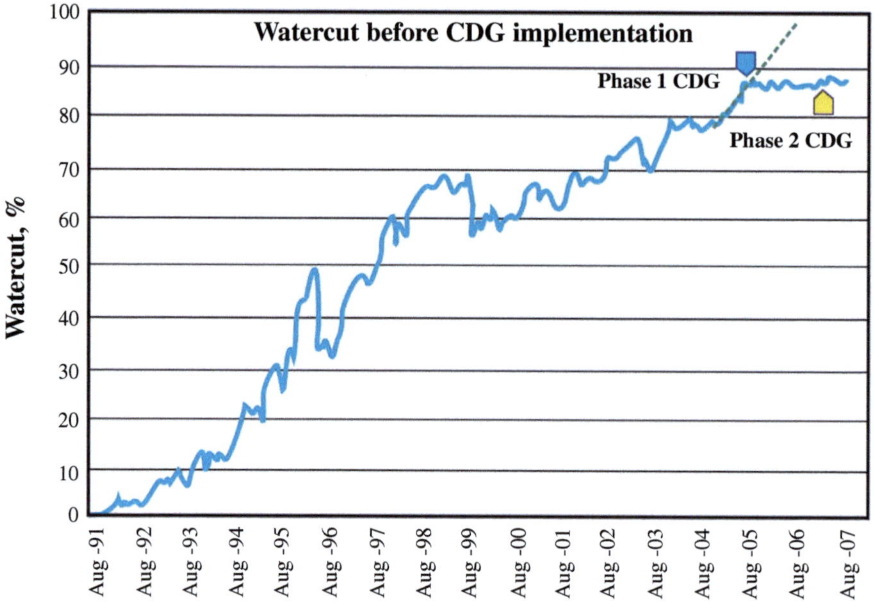

Figure 9.4 Dynamics of changes in water cut of produced fluids for 10 reacting production wells.

and a high mobility coefficient. Oil density is 22–24°API, and viscosity in reservoir conditions varies from 10 to 12 cP. The current oil recovery factor was estimated at 32%, which prompted a comprehensive assessment by the operator of potential EOR options applicable to the field. EOR using CDG was chosen.

As of September 2012, well DK-03 had injected a total of 437,000 barrels of CDG, resulting in 50,000 barrels of additional oil production and a 10% water cut reduction. The expected additional oil production over 10 years is 350,000 barrels, which corresponds to a 5% increase in oil recovery factor.

9.2.2.4 Implementation Technology

The polymer included in the CDG is supplied in the form of a powder, which is then added to the injection water flow in the concentration required to obtain the target viscosity. A typical equipment wiring diagram for this process is shown in Figure 9.5. The polymer powder is stored in a container and distributed by a metering screw in the prehydration unit. Then the solution enters the container for complete hydration and dissolution. Finally, this solution is pumped and diluted to the desired concentration.

Polymer dissolution and hydration systems, as well as pumps, are housed in 6- or 12-m containers adapted to local field conditions. As a rule, these containers are filled with nitrogen to prevent the penetration of oxygen, which negatively affects the viscosity of the polymer solution. Depending on the water supply pressure, there are two types of equipment piping.

At low water injection pressure, the piping shown in Figure 9.6 is used, and at high pressure, the piping shown in Figure 9.7 is used.

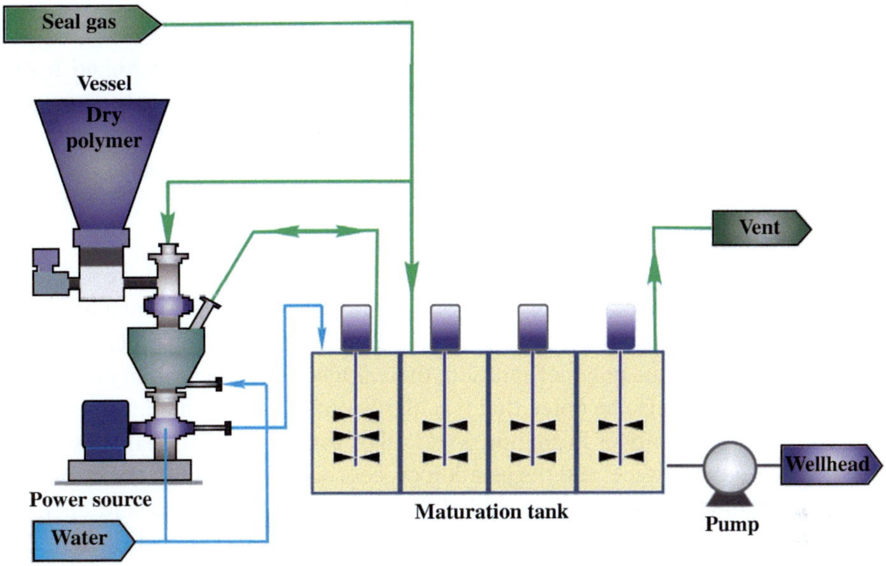

Figure 9.5 Scheme of piping of ground-based equipment when pumping CDG.

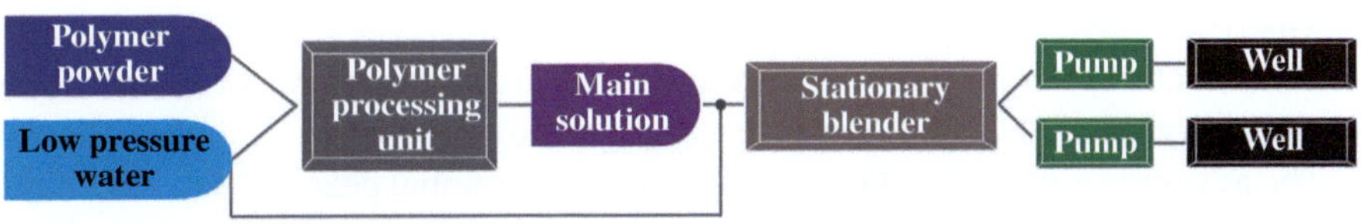

Figure 9.6 Piping diagram for ground-based equipment at low water injection pressure.

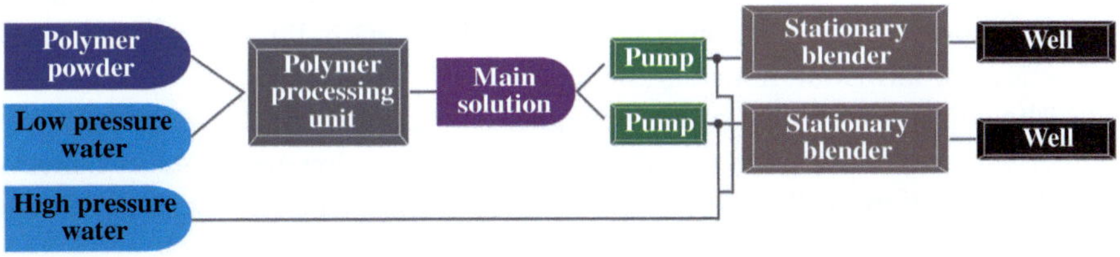

Figure 9.7 Piping diagram for ground-based equipment at high water injection pressure.

This difference in equipment piping is caused by the need to ensure complete hydration of the polymer solution before injection into the formation. In general, the equipment piping scheme may undergo some changes depending on field conditions and the availability of the necessary equipment. However, the key elements of the circuit remain virtually unchanged.

The placement of wells when implementing EOR using CDG does not differ significantly from conventional waterflooding, if the impact is effective and the design parameters of field development are achieved: injection pressure and oil withdrawal rates. It should be noted that, as with traditional polymer flooding, in-line flooding is preferred.

9.2.3 Injection of Preformed Particle Gels

Preformed particle gels (PPGs) are whitish-yellow or brown cross-linked hydrophilic polymer systems that absorb water up to several hundred times their original volume. Since PPGs are relatively inexpensive, easy to operate, and environmentally friendly, they are successfully used in operations to level the displacement front.

9.2.3.1 Process Mechanism

This technology allows us to overcome some of the disadvantages inherent in gel systems formed in reservoir conditions. For example, gel systems formed in the formation have poorly controlled gelation time. In addition, when separately pumping the cross-linker and the polymer solution, there is also a high probability of no gelation at all. It should also be noted that it is highly sensitive to formation conditions, especially to the salinity of formation water. PPG, unlike conventional gel systems, are formed on the surface, and the gelation process does not occur in the formation. The one-component composition and slight sensitivity to physical and chemical conditions in the formation, such as pH, salinity, multivalent ions, hydrogen sulfide, and temperature, distinguish them from similar compositions.

With the correct selection of the chemical composition, PPG forms a tampon that successfully diverts the flow of a displacing agent (water, surfactant, or CO_2) into low-permeability zones of the formation. Today, various types of PPG are commercially available: micromillimeter; microgels; and pH-sensitive and swelling, submicron size. Types of PPG differ in particle size, gelation time, and swelling coefficient.

The advantages of using PPG, compared to conventional gels, include the following aspects.

- C are synthesized on the surface, which makes it possible to quite accurately regulate the gelation time. Due to the fact that gelation does not occur in the formation, but only swelling of gel particles, PPG is little susceptible to the effects of formation minerals and liquids.
- The adjustable particle size makes it possible to pump PPG into fractured reservoirs, minimizing the penetration of the gel into low-permeability zones of the formation. At the same time, the relatively large particle sizes allow them to easily penetrate into cracks and large pore channels.
- As a rule, PPG is a one-component system, which, accordingly, significantly simplifies the process of injection into the reservoir. Injection of PPG does not require additional equipment for mixing and dissolution, as when injecting cross-linkable polymer systems.
- PPG, unlike cross-linked polymer systems, are insensitive to the mineralogical composition of the mixing water, which makes it possible to successfully use both seawater and formation water for these purposes.

9.2.3.2 Applicability Criteria

The following generally accepted criteria for the applicability of PPG are distinguished:

- reservoir temperature below 120 °C;
- the presence of high-permeability channels or cracks in the formation;

- high injectivity of injection wells;
- high oil flow rates and water cuts in production wells;
- low recovery factor.

Despite a number of advantages of using PPG, there are also some limitations. For example, due to the relatively large particle sizes, PPG practically does not penetrate into porous blocks, that is, the presence of cracks and channels in the formation is a prerequisite. It should be noted that the presence of channels and cracks with ultra-high permeability is also undesirable due to the high probability of removal of PPG particles.

9.2.3.3 Field Implementation

The use of PPG for displacement front leveling was first initiated in 1997 by Petro's Research Institute for Petroleum Exploration and Development (RIPED), China. Since this period, PPG, alone or in combination with other methods, has been applied to approximately 4,000 wells in most of China's oil fields.

PPGs were successfully used in the Zhongyuan field (China), operated by SINOPEC under conditions of increased salinity of formation water and high temperatures (104 °C). The use of cross-linked polymer systems under these conditions was impossible. The reservoir was sandstone with large vertical heterogeneity and was also subject to waterflooding for a long time. As a result of the successful implementation of PPG technology, it quickly became the dominant technique for leveling the displacement front in this field.

This technology has also been widely used in the Dajing oil field, which is one of the largest oil fields in China. Low temperatures (45 °C) and salinity of the formation water (4,500 mg/L) contributed to the extreme efficiency of polymer flooding over a certain period. However, as in the previous example, serious problems were observed with the vertical heterogeneity of the formation, which was the reason for the application of the PSG technology from the beginning of 2004. However, unfortunately, to date, there is no information about how successful this implementation has been.

In many fields, such as Dangang, Zhongyuan, and Shengli, a combined injection of PPG in combination with weak polymer gels has been carried out. In the United States, EOR using PPG has been used since the early 2000s. EOR using PPG was implemented at the Anton Irish field in West Texas, operated by Occidental Petroleum Corporation. CO_2 injection was started as an EOR in 1997, but soon there was a breakthrough of both CO_2 and water. After several unsuccessful attempts to introduce various technologies for conformance improvement, the choice fell on PPG. The implementation results were quite successful. At the same time, it was possible to significantly reduce the production of CO_2 and water.

It should be noted that EOR using PPG has also been quite successfully used in the fields of Texas and Alaska.

9.2.3.4 Implementation Technology

Injection of PPG, as noted earlier, does not require complex surface equipment; the scheme shown in Figure 9.8 is the most commonly used arrangement in the application of PPG.

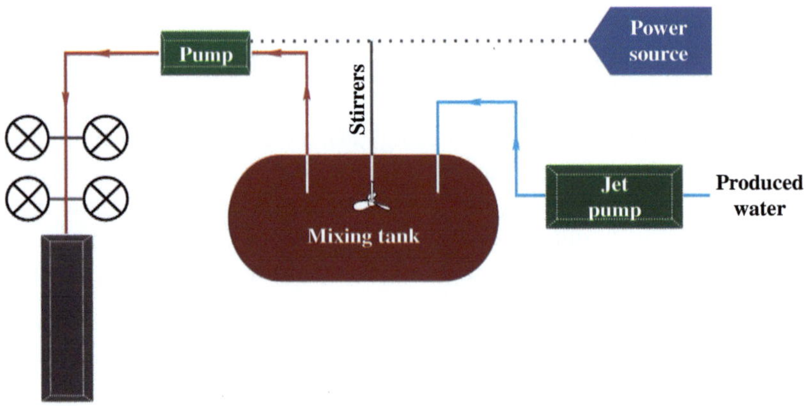

Figure 9.8 Strapping scheme used when using PPG.

At the first stage, a small slug of highly concentrated PPG solution containing large particles is pumped into the formation to form an impermeable crust in areas of low permeability and, accordingly, prevent the penetration of subsequent PPG flow into them.

At the second stage, PPG is usually injected at low concentration and low flow rate to provide the necessary time to penetrate deep into the formation. With a high solution concentration and high injection speed, a sharp increase in injection pressure is possible.

Pressure is a key factor in PPG injection and must be monitored in real-time in order to make timely adjustments to the originally designed injection process (primarily adjusting particle size and strength). Such an adjustment is necessary and is a very convenient tool and also another competitive advantage compared to the injection of cross-linked polymer systems. As a rule, with the successful implementation of this technology, along with an increase in oil recovery factor and a decrease in water cut, a slight increase in injection pressure is also observed. Although it should be noted that when the PPG solution penetrates deep into the formation, the increase in pressure may be insignificant.

The placement of wells when implementing technology using PPG is practically no different from conventional waterflooding.

9.3 Injection of Low-salinity Water

BP and Shell almost simultaneously filed an application for an international patent and registration of the LoSal and Designer trademarks Water accordingly. It should be noted that the Designer brand Water, announced by Shell, was a technological scheme for pumping water with low salinity and hardness, consisting of connecting water injection lines with a nanofiltration unit and then with a reverse osmosis unit. Around the same period, the term "smart water" has also been used in a similar context when injecting low-salinity water (LSW) into chalk oil formations in the North Sea.

Despite the well-known dependence of the permeability of a porous medium on the ionic strength of the filtered solution, according to which when filtering solutions with high ionic strength (i.e., high mineralization), the permeability of the porous medium is higher than for solutions with lower ionic strength, the mineral composition of the injected water as a means of increasing oil recovery, as a rule, has not been the subject of attention of specialists. Traditionally, waterflooding used the nearest available water source, which in offshore conditions meant injecting seawater. Laboratory studies related to the modeling of offshore reservoir conditions have also typically been carried out exclusively using (synthetic) reservoir or seawater. Thus, the salinity of injected water was taken a priori as something unchangeable, depending on external factors, to which it was ultimately necessary to adapt both in field and laboratory conditions. With this approach, the mineralogical composition of water was not considered as a factor influencing oil recovery. However, in the 1990s, Norman Morrow and his colleagues at the University of Wyoming changed the situation by publishing the results of laboratory experiments showing an increase in oil recovery factor when injecting LSW compared to injecting seawater. The salinity of injected water thus became a subject of interest to researchers and began to be considered as a factor in increasing oil recovery factor. However, despite conducting a large number of studies in this direction, the scatter of the results obtained was quite large: from an increase in oil recovery factor to the complete absence of additional production. This fact did not allow us to make an unambiguous conclusion about the effectiveness of the injection of low-mineralized water. The question of whether it is most effective to use low-mineralized water as a secondary or tertiary EOR also remains open due to the contradictory data of various researchers. At that time, the main evidence of the effectiveness of LSW injection was data from laboratory experiments on displacement on the core material, but, unfortunately, the lack of a standardized approach to research and the repeatability of the results obtained left a fairly large number of open questions.

To date, there are already results of the field implementation of this technology, and various theories have been proposed to explain the mechanism of the process, but there is still no consensus among specialists regarding its effectiveness. Of course, the presence of theories that explain the mechanism of action of this technology is important when searching for selection criteria that could help predict reservoirs in which this technology would be most effective.

9.3.1 Process Mechanism

To date, several hypotheses have been put forward to explain the effectiveness of the injection of LSW. All hypotheses, with the exception of those based on osmotic pressure, are based on various chemical reactions occurring at the interface of the "reservoir-fluid" system, leading to a change in the type of wettability of the rock.

The first hypothesis explained the mechanism of action of low-mineralized water by the formation of hydrophilic areas on the surface of the rock as a result of the release of clay particles from its surface. Released clay particles as a result of transport by the injected fluid flow block narrow pore channels of the rock, redirecting filtration flows into larger pore channels filled with oil. Such a mechanism is often classified as a physical method due to the lack of a chemical explanation of the ongoing processes in this theory. Essentially, it comes down to changes in rock permeability and pressure distribution in the formation. Otherwise, all proposed hypotheses are based on changes in rock wettability.

To date, two theories have gained the most popularity to explain the reasons for changes in wettability:

- theory of multicomponent ion exchange (MIE),
- theory based on changes in pH.

The theory of MIE, first put forward by Lager, is essentially based on the expansion of the double electronic layer of clay when low-mineralized water is injected, which leads to a change in wettability to hydrophilic. It should be noted that the proposed theory is not without some unclear points. The concentrations of Ca^{2+} and Mg^{2+} ions in water with low salinity play an important role. However, the required concentrations of these ions to achieve additional oil production are not indicated. The recommended concentration of these divalent ions should be lower than in formation water (but not equal to zero). However, this condition was violated in many field implementations of the method, and yet an increase in oil recovery was still observed. It is also unclear how the influence of pH is taken into account since the exchange of H^+ ions is of great importance in addition to the influence of other cations (Na^+, K^+, Ca^{2+}, and Mg^{2+}).

Lager identifies the following necessary conditions for the operation of the MIE mechanism:

- the presence of clay in the rock composition is mandatory, especially kaolinite; g line with positive zeta potential is unfavorable for improved oil recovery;
- the presence of reservoir water containing a certain concentration of Ca^{2+} ions;
- the presence of oil containing polar components.

pH theory, originally proposed by Austad et al. (2010), is based on the better solubility of the polar components of oil in LSW. Currently, this idea has been reworked, and in its current version, it is based on the mechanism of changing wettability. In it, as in theory MIE, it is assumed that clay particles are separated from the rock, leading to the formation of more hydrophilic areas, while near areas covered with clay, there is a local increase in pH, since Ca^{2+} ions adsorbed on the clay are replaced by H^+.

Austad identifies the following necessary conditions for the implementation of his proposed mechanism:

- the presence of clays with a high cation exchange capacity (for example, montmorillonite is better than illite; mica is better than kaolinite, etc.);
- the presence of oil containing polar components.

9.3.2 Applicability Criteria

The following generally accepted criteria for the applicability of LSW injection are distinguished:

- presence of clay in the formation;
- the presence of highly mineralized water in the formation (formation or seawater from previous injection);

- the salinity of injected low-mineralized water must be below 5,000 mg/L;
- presence of polar components in oil;
- The wettability of the reservoir should be oleophilic or intermediate (mixed).

In addition to these five points, the MIE theory requires that the Ca^{2+} concentration in the injected water be lower than that in the produced water, although this requirement is not as widely accepted as the above points.

Temperature as a factor when injecting water of low salinity is, as a rule, not considered, but a number of researchers indicate that the effectiveness of the method depends on both the temperature of the injected water and the formation temperature.

9.3.3 Field Implementation

The first pilot test was carried out in 2008–2009 at the Endicott field (Endicott, Alaska, USA) by BP. Endicott is the region's third largest field, with oil reserves estimated at approximately 1 billion barrels. It was put into operation in 1987.

Initial test results from four wells at Prudhoe Bay and Endicott fields showed high efficiency of LSW injection. Furthermore, the oil recovery factor increased by 6–12% in the areas of the treatment after application.

Further testing was carried out in the K3A-2 formation of the Endicott field. The site included one production and one injection well with 317 m (1,040 ft) spacing, 9–14 m (30–45 ft) formation thickness, 20% porosity, and 100 μm permeability. In this case, the formation contains 12% clay, mainly kaolinite. Initially, seawater was pumped up to 95% of the water cut of the extracted product, and then water with reduced salinity was pumped in a volume of 1.6 times the pore volume for 10 months (the composition of water with low salinity is shown in Table 9.2), then seawater was pumped.

Table 9.2 Composition of formation, produced, sea, and injected water at the test site of the Endicott field, concentration of components is given in mg/L.

Component	Produced water	Produced water	Sea water	Injected water
Barium	7	0	0	0
Bicarbonate	2,000	1,868	147	6
Calcium	320	194	402	17
Chloride	17,275	14,946	18,964	821
Iron	10	2	0	0
Magnesium	48	360	1,265	55
Potassium	110	177	386	17
Sodium	11,850	9,190	10,812	468
Strontium	24	7	7	0
Sulfate	63	570	2,645	115
pH	6.5	7.0	7.7	
Total dissolved solids	32,000	28,000	34,644	1,500

After two months, an increase in oil production and a decrease in water cut were observed, and the increase in oil production was accompanied by the entry of injected water of reduced salinity into the well. In general, the actual increase in oil recovery factor was in good agreement with the predicted values, but the drop in product water cut (from 95 to 93%) was less than predicted.

Despite the absence of iron in the formation and injected water, a sharp increase in its concentration to 3–4 mg/L was observed. BP specialists regarded the decrease in water cut and increase in iron concentration as confirmation of the MIE theory. According to their assumptions, the iron covering kaolinite binds the polar molecules of oil to the clay. When LSW was injected, this relationship was disrupted and led to the release of iron and oil molecules.

Overall, the results of the pilot survey were successful. Oil recovery factor increased by 10% with continued injection of highly mineralized water.

Injection of LSW as secondary EOR was carried out at the Omar field in Syria by Al-Furat (a Shell subsidiary). This is one of the few documented field-wide LSW floods.

Omar is a field containing light oil, put into operation in 1989. A sharp decrease in reservoir pressure was observed at the field, indicating the absence of abutment water. Injection of river water with a salinity of 500 mg/L began in 1991. The salinity of the formation water was 90,000 mg/L, with a high content of divalent ions (5,000 mg/L), and the clay content (mainly kaolinite) was 0.5–4%.

Analysis of the core material showed the oleophiol type of rock wettability. An increase in oil recovery factor was observed in experiments on spontaneous impregnation of the core material with LSW, accompanied by further injection of highly mineralized water. At the same time, the increase in oil production (compared to the injection of highly mineralized water) amounted to 24% of the pore volume.

According to Al Furat (and Shell), measurements and observations from 21 wells in Oman provide comprehensive evidence of wettability changes occurring at the reservoir scale. The analysis shows that a change in wettability will give an expected increase in oil recovery of 17% compared to the injection of highly mineralized water. However, a forecast of 5–15% seems more realistic.

9.3.4 Implementation Technology

Generally, seawater is used for waterflooding. The treatment water is filtered, deaerated, biocide is added, and in some cases desulfurized to prevent the formation of sulfate deposits. The essence of all such treatments is to reduce incompatibilities between injected and produced water.

Sometimes water sources with the optimal salinity can be found in the aquifer layers at the field itself. To apply LSW injection in marine conditions, desalinated seawater is most often used. On a small scale in offshore fields, the use of seawater desalination plants is a fairly common practice. There are two main methods of seawater desalination: thermal and membrane. Each of these methods has its own disadvantages and advantages. Thus, some experts consider it inappropriate to use thermal methods in marine conditions, since the equipment for implementing the

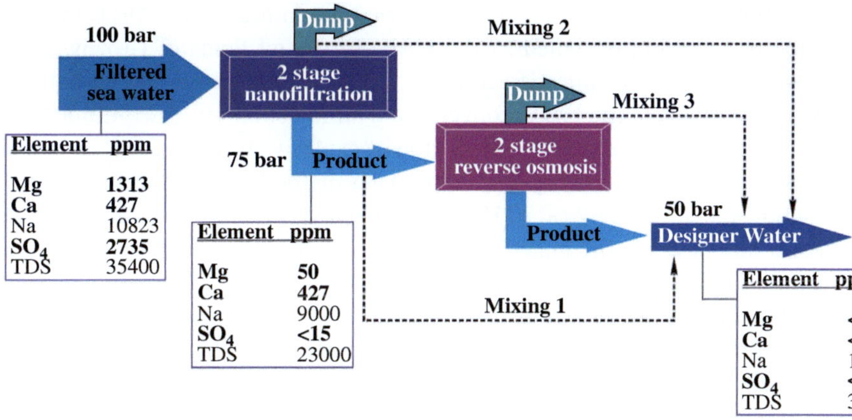

Figure 9.9 Technological scheme for obtaining the required composition of Designer Water.

method requires large production areas that are not available in marine conditions. The membrane method of desalination of seawater (i.e., the reverse osmosis method) does not have such disadvantages and does not require a large production area and energy consumption. The disadvantage of the membrane method is that reverse osmosis produces almost fresh water, leading to swelling of clays. Partial desalination, or mixing fresh water with seawater, is ineffective due to the increase in sulfates, which is also extremely undesirable. The solution was proposed by Shell specialists and consisted of a new technological scheme for seawater desalination, registered under the name Designer Water (Figure 9.9). The process is based on a 2-stage treatment of seawater: nanofiltration and reverse osmosis. In this case, the first reduces water hardness, including the removal of SO_4^{2-}, and the second reduces salinity. The end result is water with a salinity of 500 mg/L, a value below that required for LSW injection (typically 1,000–5,000 mg/L), so small amounts of nanofiltered water or waste from each stage are back-mixed to obtain the required formulation of Designer Water (see Figure 9.9). To prevent swelling and flocculation of the clay, divalent ions are added (but without SO_4^{2-}).

Such a process plant can be installed on an offshore platform or even on a ship or floating facility (allowing its reuse).

The cost of desalinating seawater on a large scale is about US$0.80–2.20 per m^3 of water injected for onshore and offshore fields, respectively.

The placement of wells when implementing the LSW technology is practically no different from the usual water flooding.

9.4 High Pressure Air Injection

Gas EORs are widely used in oil fields. However, despite their high efficiency, gas EOR is significantly inferior to the commonly used waterflooding. The main limiting factor is the lack of available gas agents. In this regard, the use of air, which is ubiquitous and requires minimal costs for its injection into the reservoir, is of interest.

Air injection into the reservoir was first carried out in the USA in 1911. In 1928, the technology was implemented in the USSR in the oil fields of Grozny. From the 1950s to the 1970s, air injection into the reservoir was used in the oil fields of Azerbaijan. However, in the 1980s, conventional air injection was discontinued due to low efficiency and the risk of the formation of explosive mixtures.

In the 1970s, the USSR developed a thermal gas method for enhancing oil recovery, which is known in the West as HPAI. In recent years, HPAI has been increasingly used. Compared to the injection of carbon dioxide (CO_2), nitrogen (N_2), and natural gas, the injection of high-pressure air has important advantages due to its lower cost and easy availability.

Currently, results from HPAI studies indicate that crude oil can react with oxygen in the air to form carbon dioxide and carbon monoxide (CO). However, at low reservoir temperatures, the oxygen concentration remains high, and due to the low rate of the oxidation reaction, it enters production wells. This may lead to accidents due to the possibility of explosive mixtures being formed. In this regard, oxidation acceleration catalysts are used at low reservoir temperatures, so the technology is not limited to use in high-temperature reservoirs.

9.4.1 Process Mechanism

The HPAI mechanism is based on the *in situ* transformation of air into an effective displacer due to oxidative processes leading to an increase in reservoir temperature and the formation of carbon dioxide and CO.

Oil oxidation. Oxidation is the process of interaction of any substance with oxygen. Hydrocarbon oxidation reaction (in particular, alkanes) has the following form:

$$C_nH_{2n+2} + O_2 = nCO_2 + (n+1)H_2O + Q.$$

When hydrocarbons are completely oxidized, carbon dioxide and water are formed, but in practice, when there is a lack of oxygen, CO is also formed.

The oxidation process can be accompanied by combustion. The necessary combustion conditions are the following: the presence of a combustible substance, oxygen, and a temperature not less than the self-ignition temperature of the combustible substance.

The auto-ignition temperature of oil is at least 300 °C. Considering that during HPAI oxidation occurs at a lower temperature, oil combustion does not occur. In this regard, HPAI differs significantly from *in situ* combustion. In HPAI, the displacing agent is the gases released during the oxidation of oil (carbon dioxide and monoxide), so it is a gas method rather than a thermal one. During high-pressure combustion, liquid-phase oxidation of oil occurs at 100–250 °C, while oil combustion when implementing the *in situ* combustion method occurs at temperatures up to 1,000 °C.

The characteristics of the displacement front also differ significantly. The main difference is that with HPAI, the area where the oxidation reaction occurs can reach several tens of meters, which is significantly larger than with *in situ* combustion. In this regard, the probability of gas breakthrough to production wells during HPAI is significantly lower than with *in situ* combustion, which indicates a higher safety of the HPAI method.

In addition to the above, the HPAI method is not limited, like *in situ* combustion, to heavy and viscous oil, while the HPAI method can be used in fields containing light oil. Another advantage of the HPAI method is a significantly lower injection rate than with *in situ* combustion. This allows the use of less efficient compressors (it is possible to use mobile compressors), which significantly reduces the economic costs of implementing the method.

The most important feature of the HPAI method is the process of oil oxidation upon contact with air. In this case, the oxidation process must occur as quickly as possible so that atmospheric oxygen does not reach the production wells. An essential factor ensuring the fulfillment of this condition is temperature. Table 9.3 shows the time period over which oil oxidation occurs. As can be seen from Table 9.3, with increasing temperature the induction period decreases significantly. If at a temperature of 60 °C the induction period is about six days, then at a temperature of 150 °C the induction period decreases by 65 times and is only about two hours.

Figure 9.10 shows the dependence of oxygen content on temperature in the formation, obtained based on numerical modeling.

As can be seen from Figure 9.10, the size of the zone of complete absorption of atmospheric oxygen is significantly less than the distance between the producing and injection wells.

Thus, the high rate of oil oxidation reaction ensures the safety of the HPAI method.

In fields with low reservoir temperatures, implementation of the method is also possible. In this case, it is necessary to use a catalyst to accelerate oil oxidation, which allows reducing the concentration of remaining oxygen. One of the effective catalysts for oil oxidation is chlorine dioxide (ClO_2). In addition, there are projects in which nitrogen is injected together with air, allowing the oxygen concentration to be reduced to a safe value.

Table 9.3 Results of assessing the induction period of the autoxidation reaction oil from Jurassic formations.

Temperature, °C	Induction period, days
150	0.09
140	0.13
130	0.19
120	0.29
110	0.45
100	0.71
90	1.15
80	1.92
70	3.30
60	5.84

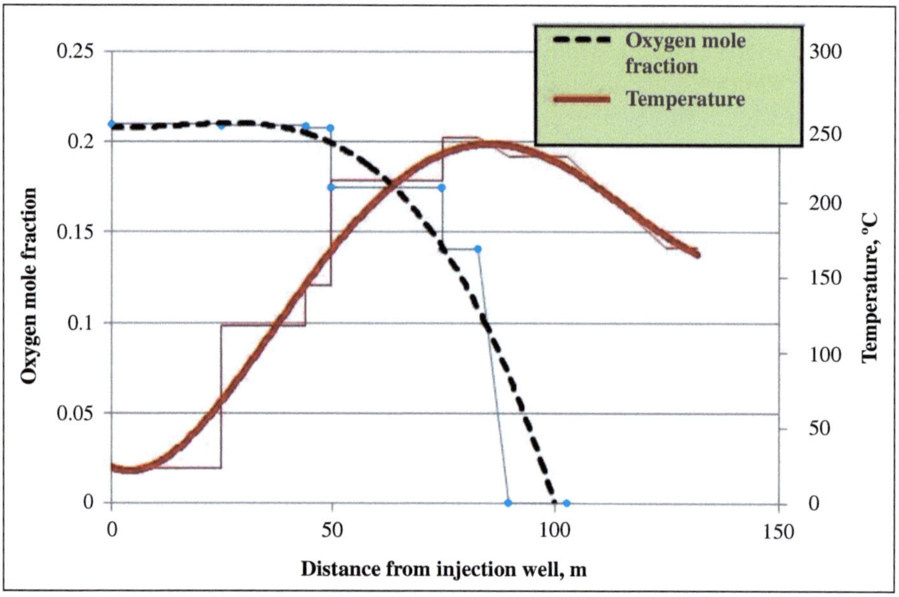

Figure 9.10 Dependence of oxygen content on temperature in the formation.

9.4.2 Applicability Criteria

The results of the field application of HPAI show that the method is most effective with the following values of the main development indicators:

- permeability 10^{-2}–10^{-3} μ m^2;
- porosity—less than 15%;
- after completion of natural displacement and depletion regimes, as well as after waterflooding;
- well grid density (100 ha/well);
- viscosity 1–3.5 mPa·s;
- reservoir thickness 1–10 m;
- oil density 830–860 kg/m^3;
- depth 2,000–3,000 m and above;
- reservoir temperatures 85–110 °C;
- the average increase in oil recovery factor is 0.15–0.17.

9.4.3 Field Implementation

Experience in air injection at low reservoir temperatures (less than 60 °C). Injecting air into the reservoir was first implemented in the USA in 1911. In 1928, the application of the method began at the Grozny oil field, and within 12 months, 6,495 tons of additional oil were produced. In the same year, the use of air injection into the reservoir began at the Azerbaijani Surakhani field. Application of the method led to uncontrollable gryphon formation in production wells. In this regard, the use of the method was discontinued. The method was revived in the 1950s, but due to a lack of high-pressure compressors, the process was abandoned in the 1960s. The depth of the affected horizons was 300–950 m. As a result of the implementation of the method, daily production increased by an average of 35–40%, and in some wells—by 150–200%.

Despite the fact that the potential for increasing oil recovery by injecting compressed air into low-temperature formations was considered low, nevertheless, in 1945–1973, the method was applied in 34 areas of 10 fields in Azerbaijan, including Balakhani-Sabunchi-Ramana, Bibiheybat, Binagady, Lokhatan-Puta-Kushkhana, Kala, Chakhnaglar, Shabandag, Ateshgah, and Sulutepe. As a result of the application, 1,278,684 thousand m^3 of air were injected, and an additional 1,363 tonnes of oil were extracted. The average air consumption per tonne of oil was 938 m^3. An analysis of the application of the method in the fields of Azerbaijan showed that the implementation of the method was effective only in five fields. These deposits account for 92.4% of total technological efficiency, with the best result obtained at the Bibiheybat field (53.7% of total technological efficiency). The method was applied here at seven sites in 1947–1952.

For seven years, compressed air injection was carried out at the Buzovna-Mashtaga field. However, technological efficiency from the use of the method was not obtained.

Research results have shown that increased oil recovery due to compressed air injection lasts no more than five years. If heavy components are present in the reservoir oil in significant quantities, this period is reduced to 1.5–2 years. The main reason for the low efficiency of this method is the low reservoir temperature (below 60 °C). At the same time, the oil oxidation time increases significantly, and the likelihood of air breakthrough to production wells increases.

Experience in air injection at high reservoir temperatures (more than 60 °C). The method was successfully tested in the 1980s and 1990s in fields in Russia, Azerbaijan, Ukraine, Belarus, and the USA. As a result of HPAI at these fields, oil production increased by 1.5 times or more, and the oil recovery factor increased by 64 % with complete utilization of oxygen.

To carry out the oxidation process in low-temperature formations, for the first time in August 1987 at the Kala field (formation temperature –30.9 °C) at well No. 797, a process of increasing the temperature by chemical means was carried out (a composition based on chromic anhydride and organic solvents was used). After the temperature increased, air injection began. 7.98 million m^3 of compressed air was pumped into the well, resulting in additional oil production of 1,296 tons. In 1989, the method was implemented at well No. 1240. 1.06 million m^3 of air was pumped into the well, resulting in an additional 146 tonnes of oil produced. Since 1993, air injection into well No. 545 began. A total of 110 thousand m^3 of compressed air was pumped, and additional oil production amounted to 1,300 tons.

Based on the research carried out, Tables 9.4 and 9.5 show the main indicators of the implementation of high-pressure injection in light oil fields and in carbonate reservoirs, respectively. The results of the use of high-pressure injection in light oil fields and carbonate reservoirs indicate a high potential for increasing oil recovery when using the method.

Table 9.4 Results of HPAI implementation at light oil fields.

Field	Depth of the field, m	Productive layer thickness, m	Porosity, %	Permeability, μm^2	Oil viscosity, cP	Oil density, kg/m^3	Reservoir temperature, °C	Results
Skhodnitsa (Ukraine)	340–470	15	20	0.1	3	820	18	Oil production increased by 5–8 times for some wells, and by 3 times for the area.
Kala (Azerbaijan)	820–900	12	24	0.13	10	790	36	Increase in annual oil production – 24%, reduction of water cut – up to 34%.
Gnedintsy (Ukraine)	1,750–1,800	4–15	19.7	0.19	1.6	784	48	Oil recovery increased, production increased 2–4 times. Oxygen was fully utilized.
Vyshanskoe (Belarus)	2,500	20	9	0.04	8.5	850	56	Due to oxidation reaction additional oil production amounted to 270 thousand tonnes and the oil recovery factor was 0.221.
Sredny Nadym (Russia)	2,750	2	8	3	—	—	107	Additional oil production – 30 thousand tonnes.
Deli (USA)	1,037	2.4	31	1.07	3	825	57	During the test period 26 thousand tonnes of oil was produced, i.e., 50% of the remaining reserves. Oil production increased by 4 times.
Sloss (USA)	1,900	304	20	0.19	0.8	830	93	Up to 43% of residual reserves were produced, including additional production of light oil fractions and gas phase – over 30%. Oxygen was fully utilized.

Table 9.5 Results of implementation of HPAI projects in fields with carbonate reservoirs.

Field	Project start date	Porosity, %	Permeability, mD	Depth, m	Oil viscosity, cP	Temperature, °C	Oil recovery, %	Annual production, total/due to EOR, thousand tons/year
Medical-Pol (USA)	1985	17	15	2,900	2	110	42	14/14
Western Medicine-Pol (USA)	2001	17	10	2,900	2	102	34	51/51
Northern Chedar Hills (USA)	2002	18	10	2,740	2	102	53	572/572
Buffalo	1979	20	10	2,580	2	102	64	20/20
West Buffalo	1987	20	10	2,580	2	102	64	12/12
South Buffalo	1983	20	10	2,580	2	102	64	48/48
West Chedar Hills (USA)	2003	17	10	2,740	2	102	53	43/43
Southern Medicine-Pol (USA)	2003	17	10	2,800	2	106	40	22/22
Pennel 1 (USA)	2002	17	10	2,680	1.44	93	48	21/8
Pennel 2 (USA)	2002	17	10	2,680	1.44	93	46	78/5
Little Beaver (USA)	2002	17	10	2,530	1.44	93	28	83/38
Total								964/833

As a result of the use of high-pressure water pumps at various fields around the world, 148 thousand tonnes of oil were extracted in 2003, 645 thousand in 2005, 774 thousand in 2006, 971 thousand in 2008, and 964 thousand tonnes of oil in 2010.

Figure 9.11 shows the dynamics of implemented projects in the USA. As can be seen from the figure, the number of implemented projects is constantly increasing, which indicates the great potential for the use of HPAI.

The experience of using HPAI shows that the method allows increasing oil recovery by up to 15%.

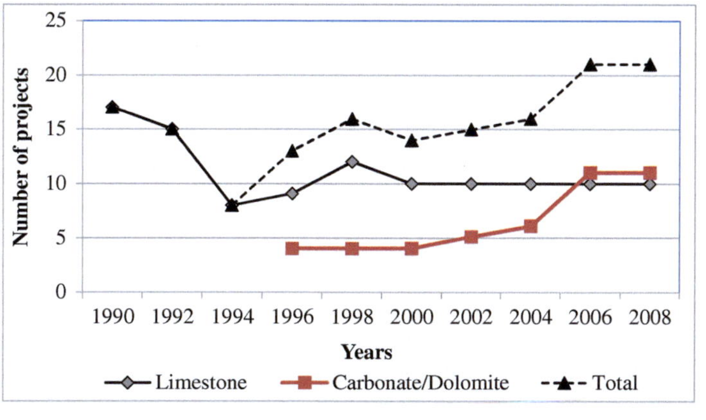

Figure 9.11 Dynamics of implemented HPAI projects in the USA.

9.4.4 Implementation Technology

Calculation of the volume of injected air. When implementing a project for injection of nitrogen and flue gases, as well as for other gases, first of all, it is necessary to calculate the need for the production of a unit of additional oil. In this case, the volume of injected air is calculated based on reservoir pressure and temperature. The calculation can be done using the formula shown in Section 6.2.4.

Compressor equipment. To compress to the design pressure and pump air into the design volume, special **compressor equipment is required**. Field implementation experience shows that to implement the process, it is necessary to pump in at least 20 thousand m^3/day of air at a pressure of at least 20 atm. The choice of compressor is based on the calculation of the required air volume and injection pressure.

Borehole design. The main requirement for the design of injection wells is to prevent rapid wear of the casing material as a result of corrosion and high temperature. Therefore, casing pipe shanks must be made of special heat-resistant materials.

Injection is carried out through tubing, which is usually lowered to the upper holes of the filter. The annulus is covered with the drilled packer, which is lowered onto the tubing. This protects the casing from high injection pressure. The design of the injection well is shown in Figure 6.15.

Control of the composition of produced gas. A serious problem in implementing the technology is caused by air oxygen breakthroughs into production wells. To prevent a breakthrough, it is necessary to control the gas injection rate, gas factor, and composition of the produced gas. If an increased oxygen content is detected in the produced gas, to prevent a breakthrough, air injection is limited by injecting nitrogen, as well as carrying out insulation work.

9.5 Overview of Organic Oil Recovery Methods

Organic oil recovery (OOR), a subset of microbial enhanced oil recovery (MEOR), has seen notable adoption with 183 applications in North American waterfloods between July 2007 and 2011. The OOR process typically includes the following steps:

1. Field screening initially
2. Sampling wells and conducting lab analyses
3. Applying nutrients in a single well to check microbial activity *in situ*
4. Conducting pilot tests in representative waterflood sectors if needed
5. Implementing across the entire field

A total of 44 treatments were executed in 41 production wells, and 123 in 41 injection wells. As of January 1, 2012, these applications reported an 88% success rate in boosting oil production in treated and adjacent wells, with an average increase of 102% from baseline to peak post-treatment production rates.

Unlike previous methods, the OOR process does not introduce new microbes but instead utilizes existing ones in the reservoir, analyzing them through advanced fluid studies. These microbes are capable of altering their interactions with trapped or hard-to-extract oil, positioning themselves at the oil-water interface to affect the oil's flow properties. This results in the formation of microemulsions at the interface,

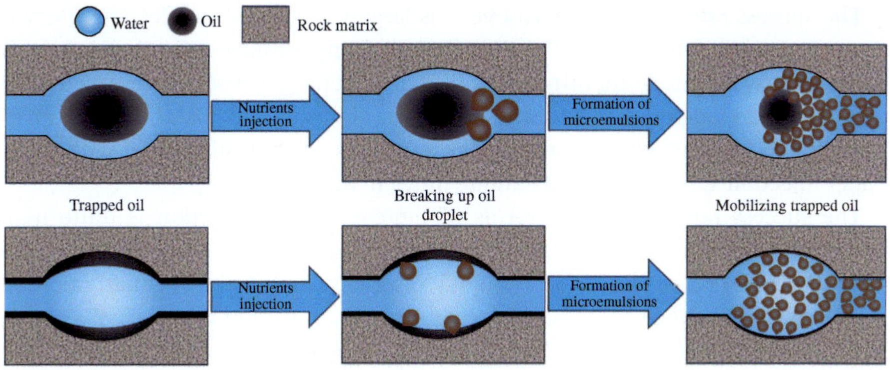

Figure 9.12 Schematic illustration of the enhanced organic oil recovery process through nutrient-assisted mobilization.

reducing interfacial tension and facilitating the release of oil into the reservoir's flow pathways, as illustrated in Figure 9.12.

Customization to each reservoir's unique microbial ecosystem is crucial, aiming to disperse nutrients throughout via the water injection system. In more permeable areas, the newly liberated oil, water, and microbes can form transient microemulsions, enhancing the efficiency of waterfloods. Laboratory studies suggest that this OOR process can retrieve an additional up to 10% of the original oil in place in waterflood projects.

The first application took place in the Alton field in Australia, targeting oil fields employing conventional water injection for secondary recovery. The process is suitable for crude oil extraction but not for natural gas or condensate fields, nor for immobile oil types. In mature waterfloods where oil exists as immobile droplets, this OOR process helps release trapped oil by resaturating production channels in the reservoir rock, altering relative permeability to enhance oil flow and decrease water movement.

An extensive review of over 180 treatments over four years shows that OOR can be more widely applied than initially anticipated, with low risk. The review highlighted that successful applications are feasible with thorough scientific evaluation and prior lab testing. Key observations include:

1. Identifying suitable reservoirs with the right type of microbes and movable oil is essential.
2. OOR is effective across a wide range of oil gravities (16–41 API).
3. Controlling microbial growth is crucial to avoid reservoir plugging or damage.
4. Microbes can thrive under extreme conditions, such as high temperatures and salinities.
5. OOR has been successful even in dual-porosity reservoirs.
6. It can also help reduce reservoir souring.
7. It's effective even in reservoirs with low permeability down to 7.5 mD.
8. Sometimes, very mature waterflood projects do not show an oil response due to low residual oil saturation.

The success rate of OOR, around 90%, is largely attributed to effective reservoir screening. Identifying reservoirs already containing suitable microbes is critical since no new microbes are introduced. Initial screening focuses on reservoir characteristics like temperature, salinity, and pH levels, as well as historical data on oil mobility and waterflood response. Ideal reservoirs are typically those with active water injection systems, which facilitate nutrient distribution.

The success rate of OOR in terms of increased oil production is quite high, approximately 90%. This success is largely due to an effective screening process. Critical success factors include identifying reservoirs that already contain microbes and movable oil. Since the process does not involve injecting microbes, it's crucial that the reservoir naturally harbors them. The typical initial screening criteria are as follows:

1. Reservoir temperature below 80 °C (180 °F)
2. Water salinity less than 10% (100,000 ppm TDS)
3. Neutral pH ranging from 6 to 8

Additionally, historical data should demonstrate that heavier oils are capable of movement and that there is enough pressure differential in the reservoir to allow flow. Evidence of a waterflood response strongly suggests the presence of movable oil.

Ideal reservoirs for OOR are those with active water injection or water drive, as water aids in distributing nutrients throughout the reservoir. While a preferred reservoir permeability is greater than 50 mD, the main factors to consider are injectivity and the ability of the released oil to flow to the producers.

The second stage of screening involves sampling the fluids produced from wells and conducting rigorous laboratory analyses. These fluid samples are examined to assess whether resident microbes can be manipulated. If the appropriate microbes are present, lab work develops a nutrient formulation specific to that reservoir.

For all applications, the reservoir oil saturation before treatment is unknown. Although most sandstone reservoirs are likely water-wet, specific wetting conditions and the interfacial tension between oil and water are often not known. All documented applications, except one, involve active waterfloods. In one instance of an inactive waterflood, low reservoir energy may have contributed to the lack of response.

OOR is adaptable to a broad spectrum of oil gravities, effective generally in reservoirs with oil gravity starting from 20 API and above. Nonetheless, it has also been successfully applied in reservoirs with oil gravity as low as 16 API. The definitive lower limit for oil gravity that supports successful OOR application remains undetermined. The highest recorded gravity treated was 41 API in the Devonian sandstone in Alberta, Canada, while the lowest was 16 API observed both in the Sparky sandstone in Alberta, Canada, and the Upper Topanga Sandstone (Miocene) offshore California. These reservoirs all showed positive reactions to waterflooding.

Impressive outcomes in extreme gravity reservoirs:

1. Devonian reservoir (41 API Oil):
   - Initial production (April 2008): 2.5 m^3/day oil (16 BOPD) and 29.2 m^3/day water (184 BWPD), with a 92% water cut.

- Posttreatment production: 5.1 m^3/day oil (32 BOPD) and 29.1 m^3/day water (183 BWPD), with an 85% water cut. A pilot project was later launched in this reservoir.

2. Sparky reservoir (16 API Oil):
   - Initial production (August 2010): 1.4 m^3/day oil (9 BOPD) and 22.9 m^3/day water (144 BWPD), with a 94% water cut.
   - Posttreatment production: 9.0 m^3/day oil (57 BOPD) and 51.0 m^3/day water (321 BWPD), with an 85% water cut. Following this, a full-field application was executed.
3. Upper Topanga reservoir (16 API Oil):
   - Initial production: 28.7 m^3/day oil (181 BOPD) and 177 m^3/day water (1113 BWPD), with an 86% water cut.
   - Posttreatment production: 42.2 m^3/day oil (266 BOPD) and 169 m^3/day water (1065 BWPD), with an 80% water cut. Although initial responses were temporary, subsequent treatments improved production significantly.

These results validate that OOR can markedly enhance oil production across a range of oil gravities and reservoir conditions.

Historically, MEOR carried risks of reservoir plugging due to biomass buildup when microbial activity was stimulated, with documented issues in fields like East Beverly Hills and San Vincente. However, the latest generation of OOR technology has effectively eliminated these risks. Of the 183 OOR applications, none have shown signs of perforation or formation damage.

This improvement is attributed to two primary modifications in contemporary OOR practices:

1. *No microbe Injection*: Current methods do not involve injecting microbes, which previously could clog small pore spaces and lead to plugging.
2. *Utilization of Nonglucose Nutrients*: Today's OOR uses environmentally benign nutrients instead of glucose, which can promote uncontrolled microbial growth and potential plugging. Controlled growth of specific microbes with these nutrients helps prevent such complications.

Controlled microbial growth has also been studied for optimizing waterflooding techniques by adjusting the sweep pattern, further enhancing the application of this technology.

OOR has proven to be highly effective even under severe environmental conditions, withstanding temperatures up to 93 °C (200 °F) and salinities as high as 142,000 ppm TDS. A notable example is the Hauser formation in California, where OOR was applied in a reservoir experiencing temperatures between 88 and 93 °C (190–200 °F).

1. *Pretreatment Production*: Initially, the target well produced approximately 4.4 m^3/day of oil (28 BOPD) and 30.3 m^3/day of water (91 BWPD).
2. *Peak Posttreatment Production*: The well achieved a peak of 17.8 m^3/day of oil (112 BOPD) and 11.4 m^3/day of water (72 BWPD).
3. *Three-month Average Posttreatment*: The production stabilized at 12.5 m^3/day of oil (79 BOPD) and 30.3 m^3/day of water (91 BWPD).

This treatment alone resulted in an additional 4,500 barrels of oil. Given the field's limited number of injectors, a comprehensive OOR strategy was implemented. In another instance, the Devonian reservoir with water salinity reaching 142,600 ppm TDS had limited microbial diversity but still demonstrated significant improvements in flow characteristics following successful nutrient application.

Concerns about the efficacy of OOR in dual-porosity reservoirs, especially regarding nutrient penetration in the matrix versus remaining in high-permeability channels, have been addressed with successful applications. One such case is the Sparky reservoir in Alberta, Canada, known for its distinctive "wormholes" or high-permeability channels that facilitate rapid communication between injector and producer wells, typically taking only hours instead of months as in more homogeneous reservoirs.

- **Nutrient Test** (January 2009):
- **Initial Production**: The well started at 1.3 m^3/day of oil (8 BOPD) and 14.4 m^3/day of water (91 BWPD), with a 92% water cut.
- **Peak Production**: It reached 6.2 m^3/day of oil (39 BOPD) and 14.8 m^3/day of water (93 BWPD), with a 71% water cut.

This test confirmed that nutrients could prompt microbial growth and enhance oil production even in highly permeable channels. The treatment also positively affected three 20-acre offset wells, with average production increasing from 1.2 m^3/day of oil (8 BOPD) and 20 m^3/day of water (126 BWPD), with a 95% water cut, to 2.6 m^3/day of oil (16 BOPD) and 17 m^3/day of water (107 BWPD), with an 88% water cut. Overall, nutrient application raised oil output from an average of 4.8 m^3/day (40 BOPD) to 10.6 m^3/day (89 BOPD) in the well and its offsets. These instances demonstrate OOR's capability to effectively increase oil production in dual-porosity reservoirs by successfully infiltrating both high-permeability channels and the reservoir matrix.

In March 2009, nutrient treatments were carried out in the Upper Topanga Reservoir off the coast of California. The injector well E-20S underwent treatments on three separate occasions: March 5, 2009, December 29, 2009, and January 25, 2010. These interventions not only boosted oil output in nearby offset wells but also significantly reduced the levels of hydrogen sulfide (H_2S) produced. The data show a decline in H_2S concentrations in all wells except for well E-13L. This decrease is attributed to the nutrient-stimulated microbes, which outcompeted sulfate-reducing bacteria (SRB) for available nutrients. By inhibiting SRB growth, the production of hydrogen sulfide—a byproduct of SRB activity—is reduced. This competitive mechanism effectively helps mitigate the souring of the reservoir, thereby enhancing the reservoir's overall health and safety. These results underline that nutrient treatments in OOR not only augment oil production but also play a critical role in reducing reservoir souring, offering a dual advantage in reservoir management.

In November 2010, OOR was implemented in two producing wells in the Big Wells Field, Texas, characterized by the San Miguel sandstone reservoir with an average permeability of 7.5 mD. However, sidewall core analyses indicated that permeability might reach up to 40 mD. Following the treatment, both wells showed increased oil production and decreased water cuts, prompting the collection of produced fluid samples. Remarkably, Well B-17 initially had no water production, which made collecting a water sample unfeasible. Before the treatment, Well B-17 was producing at a rate of 11 BOPD with a 50% water cut. Surprisingly, just two days after resuming production posttreatment, the well's output increased to 21 BOPD, and the water cut fell to 46%, showing 18 BWPD. This boost was notable, especially since a similar two-day shutdown in January 2010 hadn't resulted in any improvement. However, production eventually reverted back to 11 BOPD and 11 BWPD, maintaining a 50% water cut until March. At that point, it was reported that the well ceased producing water. On March 11, 2011, Well B-17's production stabilized at 13 BOPD and 2 BWPD, with a dramatically reduced water cut of 13%. Since then, the water cut has gradually increased again.

These observations confirm that OOR can be successfully applied in reservoirs with permeabilities as restrictive as 40 mD, effectively enhancing oil recovery in tight formations.

Despite positive microbial reactions, 9 out of 41 treated producing wells did not show an increase in oil production. The reasons behind this lack of incremental oil production in these specific instances require additional exploration. For example, the Upper Shaunavon reservoir underwent three successful nutrient treatments, but the fourth application failed to enhance oil production, an outcome that remains unexplained. One potential reason might be the minimal residual oil available in the treated formation; low residual oil limits the potential for enhancing production.

Studies such as those by Romero et al. (2009) suggest that residual oil saturation in areas swept by water can drop below 15%. This theory is supported by data from two different Mannville reservoirs within the same field, which includes four separate pools. Despite nutrient treatments across all pools, only two exhibited increases in oil production. For instance, one successful well initially producing about 1.25 m^3/day (7.9 BOPD) with a 93% water cut saw production peak at 4.4 m^3/day (27.7 BOPD) with a 76% water cut, then averaged 2.9 m^3/day (18 BOPD) with an 87% water cut over the next month. After some mechanical issues, production eventually stabilized at 1.8 m^3/day (11.3 BOPD) with an 89% water cut, yielding an estimated 167 m^3 (1,050 barrels) of incremental oil. Conversely, another well that did not respond as expected initially produced 0.78 m^3/day (4.9 BOPD) with a 98% water cut; posttreatment, production dropped to 0.31 m^3/day (2 BOPD) and later stabilized at roughly pretreatment levels.

Interestingly, the unsuccessful applications occurred in pools that had previously been treated with tertiary alkaline polymer (AP) flood EOR methods, which had concluded five years before the nutrient treatments. While the AP flood temporarily increased the pH, it was anticipated that pH levels would have returned to normal by the time of the nutrient testing in May 2010. Although there's no direct correlation yet established between the lack of incremental oil and the residual effects of the AP flood, this aspect should be closely monitored in future trials.

Despite the overall positive microbial responses, the absence of a noticeable increase in oil might be due to the lack of movable oil in the treated zones. High water cut producers may show less responsiveness to this recovery method, though no definitive quantitative link has been established to support this hypothesis. However, two reservoirs with high water cuts did experience successful applications, suggesting that other factors may also play a role in the outcomes.

In summary, over 180 nutrient applications across various reservoirs have underscored the effectiveness of OOR. The method has proven to be a valuable adjunct to secondary recovery techniques, utilizing existing infrastructure without necessitating significant new capital investments. OOR has been successful across various oil gravities (16 API to 41 API) and in reservoirs with higher temperatures and salinities. Limited experiences in dual-porosity systems also suggest potential for OOR applications. Beyond enhancing oil recovery, OOR can also reduce H_2S levels, and given its low risk of causing reservoir damage, it remains a viable option when microbial and reservoir conditions are properly assessed before treatment. A thorough scientific evaluation is critical to maximizing the benefits of OOR, ensuring that each application is customized to the specific conditions of the reservoir.

Bibliography

Akintunji, F., Marcotte, B., Sheehy, A., & Govreau, B. (2012, April 14–18). A Texas MEOR application shows outstanding production improvement due to oil release effects on relative permeability. *Presented at Improved Oil Recovery Symposium. Tulsa, OK.*

Al-Anazi, H.A., & Sharma, M.M. (2002, February). Use of a pH sensitive polymer for conformance control. *Paper SPE-73782-MS Presented at the SPE International Symposium and Exhibition on Formation Damage Control.*

Al-Assi, A.A., Willhite, G.P., Green, D.W., & McCool, C.S. (2009). Formation and propagation of gel aggregates using partially hydrolyzed polyacrylamide and aluminum citrate. *SPE Journal, 14*(3), 450–461.

Alvarado, V., & Manrik, E. (2011). *Methods for Enhancing Oil Recovery: Planning and Application Strategies*. Premium Engineering LLC.

Anderson, W.G. (1987). Wettability literature survey part 5: The effects of wettability on relative permeability. *Journal of Petroleum Technology, 39*(11), 1–453.

Austad, T. (2008). Smart water for enhanced oil recovery: A comparison of mechanisms in carbonates and sandstones. *Presentation at the FORCE Seminar on Low Salinity. Stavanger, Norway.*

Austad, T., RezaeiDoust, A., & Puntervold, T. (2010, April). Chemical mechanism of low salinity water flooding in sandstone reservoirs. *Paper SPE-129767-MS Presented at the SPE Improved Oil Recovery Symposium.*

Ayirala, S.C., Uehara-Nagamine, E., Matzakos, A.N., et al. (2010, April). A designer water process for offshore low salinity and polymer flooding applications. *Paper SPE-129926-MS Presented at the SPE Improved Oil Recovery Symposium.*

Bagirov, M.K., & Mamedov, T.M. (2001). *Enhanced Oil Recovery using Microfoam Systems*. Sabah.

Bai, B., Li, L., Liu, Y., et al. (2007). Conformance control by preformed particle gel: Factors affecting its properties and applications. *SPE Reservoir Evaluation and Engineering, 10*(4), 415–421.

Bai, B., Liu, Y., Coste, J.P., et al. (2007). Preformed particle gel for conformance control: Transport mechanism through porous media. *SPE Reservoir Evaluation & Engineering, 10*(2), 176–184.

Bai, B., Wei, M., & Liu, Y. (2013, March). Field and lab experience with a successful preformed particle gel conformance control technology. *Paper SPE-164511-MS Presented at the SPE Production and Operations Symposium.*

Betelin, V.B., Yudin, V.A., Afanaskin, I.V., et al. (2015). Creation of a domestic thermal hydraulic simulator is a necessary stage in the development of unconventional hydrocarbon deposits in Russia. FGU FSC NIISI RAS.

Bogomolov, A.I., Gaile, A.A., Gromova, V.V., et al. (1981). *Chemistry of Oil and Gas.* Chemistry.

Boxerman, A.A., Savelyev, V.A., Jafarov, I.S., et al. (2003). Thermal gas impact: An innovative technology for developing deposits in Siberia. *Proceedings of the XII European Symposium on Enhanced Oil Recovery. Kazan.*

Boxerman, A.A., & Yambaev, M.F. (2003). Method of injection and in-situ transformation of air in light oil fields. *Collection of Reports of the XII European Symposium on Enhanced Oil Recovery. Kazan.*

Castro, R., Maya, G., Sandoval, J., et al. (2013, July). Colloidal dispersion gels (CDG) in Dina Cretaceous field: From pilot design to field implementation and performance. *Paper SPE-165273-MS Presented at the SPE Enhanced Oil Recovery Conference. Kuala Lumpur, Malaysia.*

Chang, H.L., Sui, X., Xiao, L., et al. (2004, April). Successful field pilot of in-depth colloidal dispersion gel (CDG) technology in Daqing oil field. *Paper SPE-89460-MS Presented at the SPE/DOE Fourteenth Symposium on Improved Oil Recovery. Tulsa, Oklahoma.*

Chauveteau, G., Tabary, R., Le Bon, C., et al. (2003, May). In-depth permeability control by adsorption of soft size-controlled microgels. *Paper SPE-82228-MS Presented at the SPE European Formation Damage Conference.*

Cheung, S., Ng, R., Frampton, H., et al. (2007, March). A swelling polymer for in-depth profile modification: Update on field applications. *Paper Presented at the SPE Applied Technology Workshop of Chemical Methods of Reducing Water Production. San Antonio, Texas.*

Coste, J.P., Liu, Y., Bai, B., et al. (2000, April). In-depth fluid diversion by pre-gelled particles: Laboratory study and pilot testing. *Paper SPE-59362-MS Presented at the SPE/DOE Improved Oil Recovery Symposium. Tulsa, Oklahoma.*

De Bons, F.E., & Braun, R.W. (1995, May). Polymer flooding: Still a viable IOR technique. *Paper Presented at the IOR 1995-8th European Symposium on Improved Oil Recovery.*

Denisov, E.T., & Shestakov, A.F. (2008). Reactions of alkoxy and peroxy radicals with carbon monoxide. *Kinetics and Catalysis, 49*(1), 5–14.

Diaz, D., Somaruga, C., Norman, C., & Romero, J. (2008, April). Colloidal dispersion gels improve oil recovery in a heterogeneous Argentina waterflood. *Paper SPE-113320-MS Presented at the SPE/DOE Improved Oil Recovery Symposium. Tulsa, Oklahoma.*

Elkarsani, K.S.M. (2013). Development of a new cost-effective polymer gel for water control in oil and gas wells (Doctoral dissertation, King Fahd University of Petroleum and Minerals, Saudi Arabia).

Emmanuel, N.M., Denisov, E.T., & Mayzus, Z.K. (1965). *Chain Reactions of Hydrocarbon Oxidation in the Liquid Phase.* Nauka.

Fielding, R.C. Jr., Gibbons, D.H., & Legrand, F.P. (1994, April). In-depth drive fluid diversion using an evolution of colloidal dispersion gels and new bulk gels: An operational case history of North Rainbow Ranch unit. *Paper SPE-27773-MS Presented at the SPE/DOE Ninth Symposium on Improved Oil Recovery. Tulsa, Oklahoma.*

Frampton, H., Denyer, P., Ohms, D.H., et al. (2009, April). Sweep improvement from the lab to the field. *Paper Presented at the IOR 2009-15th European Symposium on Improved Oil Recovery.*

Frampton, H., Morgan, J.C., Cheung, S.K., et al. (2004, April). Development of a novel waterflood conformance control system. *Paper SPE-89391-MS Presented at the SPE/DOE Symposium on Improved Oil Recovery. Tulsa, Oklahoma.*

Gurevich, I.L. (1972). *Oil and Gas Processing Technology*. Chemistry.

Gutierrez, D., Miller, R.J., Taylor, A.R., et al. (2009). Buffalo field high-pressure-air-injection projects 1977 to 2007: Technical performance and operational challenges. *SPE Reservoir Evaluation & Engineering, 12*(4).

Havemann, G.D., Clement, B.G., et al. (2015). New microbial method promises in EOR. *Journal of Petroleum Technology, 67*(03), 32–35.

Huh, C., Choi, S.K., & Sharma, M.M. (2005, October). A rheological model for pH-sensitive ionic polymer solutions for optimal mobility control applications. *Paper Presented at the SPE Annual Technical Conference and Exhibition.*

Husband, M.E., Ohms, D.S., & Frampton, H., et al. (2010, April). Results of a three-well waterflood sweep improvement trial in the Prudhoe Bay field using a thermally activated particle system. *Paper SPE-129967-MS Presented at the SPE Improved Oil Recovery Symposium.*

Khlebnikov, V.N., Zobov, P.M., Antonov, S.V., & Ruzanova, Y.F. (2008). Study of the thermal gas method of oil production: Kinetic patterns of auto-oxidation of oil from Jurassic formations. *Bashkir Chemical Journal, 15*(4), 105–110.

Kumar, V.K., Fassihi, M.R., & Yannimaras, D.V. (1995). Case history and appraisal of the Medicine Pole Hills Unit air-injection project. *SPE Reservoir Engineering, 10*(3), 198–202.

Lager, A. (2006). BP low salinity patent application (hydrocarbon recovery process). PCT/GB2007/003337-WO2008/029124.

Larkin, R.J., & Creel, P.G. (2008, April). Methodologies and solutions to remediate inner-well communication problems on the SACROC CO_2 EOR project: A case study. *Paper SPE-113305-MS Presented at the SPE Symposium on Improved Oil Recovery.*

Lee, S.Y., Webb, K.J., & Collins, I.R., et al. (2010, April). Low salinity oil recovery–increasing understanding of the underlying mechanisms of double layer expansion. *Paper SPE-129722-MS Presented at the SPE Improved Oil Recovery Symposium.*

Li, X., Liu, J., Pu, J., et al. (2016). Accelerated oxidation during air flooding in a low-temperature reservoir. *Petroleum Science and Technology, 35*(1), 86–91.

Ligthelm, D.J., Gronsveld, J., & Hofman, J., et al. (2009, June). Novel waterflooding strategy by manipulation of injection brine composition. *Paper SPE-119835-MS Presented at the SPE EUROPEC/EAGE Conference and Exhibition.*

Liu, Y., Bai, B., & Wang, Y. (2010). Applied technologies and prospects of conformance control treatments in China. *Oil & Gas Science and Technology–Revue d'IFP Energies Nouvelles, 65*(6), 859–878.

Liu, P.G., Pu, W.F., Li, Y.B., et al. (2016). Low-temperature isothermal oxidation of crude oil. *Petroleum Science & Technology, 34*, 838–844.

Liu, P.G., Pu, W.F., & Ni, J.H. (2016). Catalytic effect analysis of clay minerals on low-temperature oxidation of crude oil through combined thermal analysis methods. *Petroleum Science & Technology, 34*, 343–349.

Lu, X., Song, K., Niu, J., & Chen, F. (2000, April). Performance and evaluation methods of colloidal dispersion gels in the Daqing oil field. *Paper SPE-59466-MS Presented at the SPE Asia Pacific Conference on Integrated Modeling for Asset Management. Yokohama, Japan.*

Mack, J.C., & Smith, J.E. (1994, April). In-depth colloidal dispersion gels improve oil recovery efficiency. *Paper SPE-27780-MS Presented at the SPE/DOE Ninth Symposium on Improved Oil Recovery. Tulsa, Oklahoma.*

Maharramov, A.M., Akhmedova, R.A., & Akhmedova, N.F. (2009). *Petrochemistry and Oil Refining*. Baki Universiteti.

Manrique, E., Thomas, C., Ravikiran, R., et al. (2010, April). EOR: Current status and opportunities. *Paper SPE-130113-MS Presented at the SPE Improved Oil Recovery Symposium. Tulsa, Oklahoma.*

McGuire, P.L., Chatham, J.R., Paskvan, F.K., et al. (2005, April). Low salinity oil recovery: An exciting new EOR opportunity for Alaska's North Slope. *Paper SPE-93903-MS Presented at the SPE Western Regional Meeting.*

Menconi, F., Giaccaglia, F., Ramirez, J., & Berto, C. (2013, April). Inyección de geles en el yacimiento El Tordillo: Desde los pilotos hasta la masificación. *Petrotecnia, 2*, 42–55.

Morrow, N.R., Tang, G.Q., Valat, M., & Xie, X. (1998). Prospects of improved oil recovery related to wettability and brine composition. *Journal of Petroleum Science and Engineering, 20*(3–4), 267–276.

Mustoni, J.L., Denyer, P., & Norman, C. (2010, April). Deep conformance control by a novel thermally activated particle system to improve sweep efficiency in mature waterfloods of the San Jorge Basin. *Paper SPE-129732-MS Presented at the SPE Improved Oil Recovery Symposium.*

Ohms, D., McLeod, J.D., Frampton, H., et al. (2010). Incremental-oil success from waterflood sweep improvement in Alaska. *SPE Production & Operations, 25*(3), 247–254.

Paez Yanez, P.A., Mustoni, J.L., Frampton, H., et al. (2007, April). New attempt in improving sweep efficiency at the mature Koluel Kaike and Piedra Clavada waterflooding projects of the San Jorge Basin in Argentina. *Paper SPE-107923-MS Presented at the SPE Latin American & Caribbean Petroleum Engineering Conference.*

Panfilov, M.T. (2000). Analysis of the results of laboratory research and pilot industrial implementation of water-gas impact in the fields of Western Siberia. *Proceedings of the Conference "Severgeoelletech"*. Ukhta State Technical University.

Pascual, M., Crosta, D., & Coombe, D. (2005, June). Air injection into a mature waterflooded light oil reservoir: Laboratory and simulation results for Barrancas Field, Argentina. *Paper SPE-94092-MS Presented at the SPE Europec / EAGE Annual Conference. Madrid, Spain.*

Peirce, J.W., Hutcherson, M.R., Brice, B.W., et al. (2014, April). An overview of conformance control efforts for the West Sak Field on the North Slope of Alaska. *Paper SPE-169073-MS Presented at the SPE IOR 2015–18th European Symposium on Improved Oil Recovery.*

Portwood, J.T. (1995, April 2–4). A commercial microbial enhanced oil recovery technology: Evaluation of 322 projects. *Paper SPE-29518-MS Presented at the SPE Production Operations Symposium. Oklahoma City, Oklahoma.*

Pritchett, J., Frampton, H., Brinkman, J., et al. (2003, October). Field application of a new in-depth waterflood conformance improvement tool. *Paper SPE-84897-MS Presented at the SPE International Improved Oil Recovery Conference in Asia Pacific.*

Pyziak, D., & Smith, D. (2007, November). Update on Anton Irish conformance effort. *Paper Presented at the 6th International Conference on Production Optimization–Reservoir Conformance-Profile Control–Water and Gas Shut-off. Houston, TX.*

Ranganathan, R., Lewis, R., McCool, C.S., et al. (1997, February). An experimental study of the *in situ* gelation behavior of a polyacrylamide/aluminum citrate "colloidal dispersion" gel in a porous medium and its aggregate growth during gelation reaction. *Paper SPE-37220-MS Presented at the International Symposium on Oilfield Chemistry. Houston, Texas.*

Rivet, S.M. (2009). Coreflooding oil displacements with low salinity brine (MS thesis). University of Texas at Austin.

Romero, C., Aubertin, F., Cassou, E., et al. (2009, April 27–29). Single-well chemical tracer tests (SWTT) experience in the mature Handil field: Evaluating stakes before launching an EOR project. *Presented at the EAGA 15th European Symposium on Improved Oil Recovery.*

Saracov, T.H., Ahmadov, E.A., & Aghazada, E.H. (1997). *Results and Future Prospects of Air Injection into Layers at the Qala Field* [Manuscript in Azerbaijani] (pp. 115–118). Republic Oil Industry Problem-Solving Methods (AzNSETLI Scientific Works Collection), Baku.

Scheidegger, A.E. (1958). The physics of flow through porous media. *Soil Science, 86*(6), 355.

Seccombe, J.C., Lager, A., Webb, K., et al. (2008, April). Improving waterflood recovery: LoSal™ EOR field evaluation. *Paper SPE-113480-MS Presented at the SPE/DOE Improved Oil Recovery Symposium, Tulsa, Oklahoma.*

Seright, R.S. (1995). *Propagation of an Aluminum-Citrate-HPAM 'Colloidal-Dispersion' Gel through Berea Sandstone.* New Mexico PRRC.

Seright, R.S., & Liang, J. (1994, April). A survey of field applications of gel treatments for water shutoff. *Paper SPE-26991-MS Presented at the III Latin American/Caribbean Petroleum Engineering Conference. Buenos Aires, Argentina.*

Sharma, M.M., & Filoco, P.R. (2000). Effect of brine salinity and crude-oil properties on oil recovery and residual saturations. *SPE Journal, 5*(3), 293–300.

Sheng, J.J. (2011). *Modern Chemical Enhanced Oil Recovery: Theory and Practice.* Elsevier.

Sheng, J.J. (2013). *Enhanced Oil Recovery Field Case Studies* (1st ed.). Gulf Professional Publishing.

Smith, D.D., Giraud, M.J., Kemp, C., et al. (2006, September). The successful evolution of Anton Irish conformance efforts. *Paper SPE-103044-MS Presented at the SPE Annual Technical Conference and Exhibition.*

Sorbie, K.S., & Collins, I. (2010, April). A proposed pore-scale mechanism for how low salinity waterflooding works. *Paper SPE-129833-MS Presented at the SPE Improved Oil Recovery Symposium. Tulsa, Oklahoma.*

Spildo, K., Skauge, A., & Skauge, T. (2010, April). Propagation of colloidal dispersion gels (CDG) in laboratory corefloods. *Paper SPE-129927-MS Presented at the SPE Improved Oil Recovery Symposium. Tulsa, Oklahoma.*

Tang, G.Q., & Morrow, N.R. (1999). Oil recovery by waterflooding and imbibition – invading brine cation valency and salinity. *Paper SCA-9911 Presented at the International Symposium of the Society of Core Analysis. Golden, Colorado.*

Trujillo, P.M.L., Mercado, S.D.P., Maya, G.A., et al. (2010, December). Selection methodology for screening evaluation of enhanced-oil-recovery methods. *Paper SPE-139222-MS Presented at the SPE Latin American and Caribbean Petroleum Engineering Conference.*

Vasilievsky, V.N., & Petrov, A.I. (1973). *Study of Oil Reservoirs and Wells*. Nedra.

Verbeek, P., & Matzakos, A. (2009, January). Water treatment for flooding with lower salinity than formation water. *Paper Presented at the Tekna Produced Water Conference. Stavanger.*

Vledder, P., Gonzalez, I.E., Carrera Fonseca, J.C., et al. (2010, April). Low salinity water flooding: proof of wettability alteration on a field-wide scale. *Paper SPE-129564-MS Presented at the SPE Improved Oil Recovery Symposium. Tulsa, Oklahoma.*

Wang, C.J., Liu, H.Q., Pang, Z.X., et al. (2015). Enhance oil recovery for steam flooding: low-temperature oxidative decomposition of heavy oil with air injection. *Energy & Fuels, 29*(9), 6242–6249.

Wang, J., Wang, T., Feng, C., et al. (2015). The catalytic effect of organometallic additives on low-temperature oxidation of light crude oil during air flooding. *Petroleum Science & Technology, 33*(9), 1118–1125.

Yang, J., Li, X.F., Chen, Z.X., et al. (2016). Combined steam-air flooding studies: experiments, numerical simulation, and field test in the Qi-40 Block. *Energy & Fuels, 30*(3), 2060–2065.

Zahner, R.L., Tapper, S.J., Marcotte, B.W.G., et al. (2011, July 19–21). What has been learned from a hundred MEOR applications. *Presented at the SPE Enhanced Oil Recovery Conference. Kuala Lumpur, Malaysia.*

Zaitoun, A., Tabary, R., & Rousseau, D. (2007, March). Using microgels to shutoff water in a gas storage well. *Paper SPE-106042-MS Presented at the SPE International Symposium on Oilfield Chemistry. Houston, Texas.*

Zhang, Y., Xie, X., & Morrow, N.R. (2007, November). Waterflood performance by injection of brine with different salinity for reservoir cores. *Paper SPE-109849-MS Presented at the SPE Annual Technical Conference and Exhibition. Anaheim, California.*

Zhidong, G., Chang, H.L., & Chengyong, F., et al. (2011, July). Comparison of oil displacement characteristics between CDG and polymer flooding in the Daqing oilfield. *Paper SPE-144119-MS Presented at the SPE Enhanced Oil Recovery Conference. Kuala Lumpur, Malaysia.*

Zohuriaan-Mehr, M.J., & Kabiri, K. (2008). Superabsorbent polymer materials: A review. *Iranian Polymer Journal, 17*(6), 451–462.

10

Practical Implementation of Enhanced Oil Recovery (EOR)

10.1 Screening Assessment

The initial step to the application of a particular enhanced oil recovery (EOR) in a field is the selection of the most suitable method of increasing oil recovery for a given specific field—screening. Screening studies are designed to assess the feasibility of using an EOR method based on a limited number of reservoir and fluid characteristics, usually critical. Screening is a comparison of average reservoir characteristics with tabular applicability criteria previously developed on the basis of global industrial experience. Most often, to select methods for increasing oil recovery, conventional screening is used, i.e., analysis of geological and physical properties according to the "suitable or not suitable" principle. The most applicable today, due to its simplicity and lack of need for additional software, is screening carried out using Taber tables (see Table 10.1). He analyzed the selection criteria for all the main EOR; the results of the analysis were summarized in tables, called Taber tables in the literature. He identified 10 criteria, each of which is compared with the corresponding values for the field in question. A score is assigned from 0 to 1. A score of 1 is assigned when the property is within the range, and zero when it is not within the range.

When a parameter matches only part of the range, a score is assigned proportional to the match rate.

Figure 10.1 shows the scoring procedure. Using field A as an example, we will consider the main screening criteria for conditional EOR. All criteria, with the exception of formation permeability, are within the specified screening ranges and are assigned a value of 1. Permeability as a criterion is assigned a proportional score of 0.6666. Next, find the arithmetic mean of the value of the assigned points, which is 0.9524. Thus, the applicability of this EOR method is 0.9524 points, i.e., quite high.

Methods for Enhanced Oil Recovery: Fundamentals and Practice, First Edition.
Baghir A. Suleimanov and Elchin F. Veliyev.

Table 10.1 Summary of screening criteria for EOR methods.

		Oil properties			Reservoir characteristics					
	EOR Method	Gravity (°API)	Viscosity (cP)	Composition	Oil Saturation (% PV)	Formation Type	Net Thickness (ft)	Average Permeability (mD)	Depth (ft)	Temperature (°F)
Gas injection methods (miscible)										
1	Nitrogen and flue gas	>35 ↗ 48 ↗	<0.4 ↘ 0.2 ↘	High percent of C_1 to C_7	>40 ↗ 75 ↗	Sandstone or carbonate	Thin unless dipping	NC	>6,000	NC
2	Hydrocarbon	>23 ↗ 41 ↗	<3 ↘ 0.5 ↘	High percent of C_2 to C_7	>30 ↗ 80 ↗	Sandstone or carbonate	Thin unless dipping	NC	>4,000	NC
3	CO_2	>22 ↗ 36 ↗	<10 ↘ 1.5 ↘	High percent of C_5 to C_{12}	>20 ↗ 55 ↗	Sandstone or carbonate	Wide range	NC	>2,500	NC
1–3	Immiscible gases	>12	<600	NC	>35 ↗ 70 ↗	NC	NC if dipping and/or good vertical permeability	NC	>1,800	NC
(Enhanced) Waterflooding										
4	Micellar/ Polymer, ASP, and Alkaline Flooding	>20 ↗ 35 ↗	<35 ↘ 13 ↘	Light, intermediate, some organic acids for alkaline floods	>35 ↗ 53 ↗	Sandstone preferred	NC	>10 ↗ 450 ↗	>9,000 ↘ 3,250	>200 ↘ 80

5	Polymer Flooding	>15	<150, >10	NC	>50 ↗ 80 ↗	Sandstone preferred	NC	>10 ↗ 800 ↗[a]	<9,000	>200 ↘ 140
Thermal/Mechanical										
6	Combustion	>10 ↗ 16 →?	<5,000 ↓ 1,200	Some asphaltic components	>50 ↗ 72 ↗	High-porosity sand/ sandstone	>10	>50[b]	<11,500 ↘ 3,500	>100 ↗ 135
7	Steam	>8 to 13.5 → ?	<200,000 ↓ 4,700	NC	>40 ↗ 66 ↗	High-porosity sand/ sandstone	>20	>200 ↗ 2,540 ↗[c]	<4,500 ↘ 1,500	NC
—	Surface mining	7 to 11	Zero cold flow	NC	>8 wt% sand	Mineable tar sand	>10[d]	NC	>3 : 1 overburden to sand ratio	NC

NC = not critical.
Underlined values represent the approximate mean or average for current field projects.
a. >3 mD from some carbonate reservoirs if the intent is to sweep only the fracture system.
b. Transmissibility >20 mD-ft/cP.
c. Transmissibility >50 mD-ft/cP.
d. See depth.

	Fluid and reservoir properties. Field under study (field a)	Screening criteria Taber-Seright (1997)	Screening results
Viscosity, cP	500	<100,000	1
Density, °API	14	8 – 25	1
Current oil saturation	0, 6	>0, 4	1
Reservoir thickness, ft	120	>20	1
Permeability, mD	150 – 300	>200	0.6666
Porosity	0.38 – 0.52	NC	1
Depth, ft	2,900 – 3,200	<5,000	1
Reservoir pressure, MPa	4,14	NK	
Lithology	Sandstone	NK	
			0.9524

Score for each property

Score of the method

Figure 10.1 Scoring procedure.

10.2 Phase Behavior of Formation Fluids and Core Analysis

To carry out geological modeling, it is necessary to know the physicochemical, reservoir, and filtration properties of the reservoir in which it is planned to use one or another EOR. In this case, along with geophysical research, it is necessary to conduct laboratory experiments with formation fluids and core material at formation pressure and temperature.

10.2.1 Study of the Phase Behavior of Formation Fluids

The term "PVT properties" is an acronym for: **pressure** (pressure), **volume** (volume), and **temperature** (temperature). These are the physical properties of substances, particularly formation fluids. To determine them, thermodynamic studies are carried out.

The properties of formation fluids are necessary for:

- calculation of oil and gas reserves,
- well production forecast,
- field development design, and
- hydrodynamic modeling of the field.

According to the law of Darcy, it is impossible to correctly calculate the productivity of a well without knowing the properties of the fluids.

The main parameters determined during PVT analysis for reservoir oil:

- oil saturation pressure with gas using contact and differential degassing methods,
- density,
- viscosity, and
- paraffin crystallization temperature.

The main parameters determined during PVT analysis for reservoir gas (gas condensate):

- pressure of the onset of retrograde condensation,
- density of gas and gas condensate, and
- viscosity of reservoir gas.

The main manufacturers of installations for studying the PVT properties of formation fluids are:

1. Chandler Engineering
2. Vinci Technologies

10.2.2 Core Analysis

To determine permeability in reservoir conditions, it is necessary to take into account several basic indicators: lithostatic or rock pressure, pore pressure, and temperature. To create these conditions, automated installations for modeling reservoir conditions are used. The units are designed to conduct core studies under conditions simulating reservoir conditions in a wide range of pressures and temperatures.

All core studies can be divided into two large groups: standard and special analysis.

Standard analysis is carried out in order to determine the parameters necessary for calculating geological reserves of oil and gas: oil and water saturation, porosity, permeability, pore size distribution, and the nature of the distribution of saturation over the oil–water contact.

Special core studies are carried out in order to obtain reliable initial data for hydrodynamic modeling of fields: relative phase permeability during two- and three-phase filtration; coefficients of oil displacement by water, gas, and various chemicals.

The main manufacturers of core analysis equipment are:

1. Core Laboratories
2. Coretest Systems
3. Vinci Technologies

The more detailed information on laboratory evaluation of cores is described in Chapter 11.

10.3 EOR Simulation

Reservoir modeling includes a description of the physical properties of reservoirs and the fluids saturating them, consideration of the technological process for developing productive formations, and the creation of a mathematical model of the reservoir, through which it is possible to predict the behavior of the reservoir under various operating conditions.

Field modeling is important not only for the purpose of saving time and forecasting accuracy, but also for a more in-depth understanding of the structure of the field reservoir, capacitance-filtration parameters, and filtration properties of fluids, and monitoring changes in these parameters during the development process.

With the classical approach, calculations are carried out using averaged parameters, since it is impossible to take into account the dynamics of changes in parameters that affect development indicators. Using the capabilities of modern modeling programs, the deposit can be divided into millions of small cells. This allows you to study the field in more detail by applying filtration equations to each small cell.

Hydrocarbon reservoir modeling mainly consists of two stages: static and dynamic.

A static model is a geological model of a field. A geological model is a framework of a field, divided into millions of cells, where each cell contains all the petrophysical and reservoir-filtration parameters of a given field.

The dynamic model is a hydrodynamic model of the field. The hydrodynamic model is loaded with data on the three-dimensional (3D) geological model, perforation, production, reservoir stimulation measures, and other dynamic data necessary to reconstruct the development history (history matching) and predict its indicators.

10.3.1 Geological Modeling

Currently, the main software packages for creating geological 3D models of oil and gas fields are DecisionSpase and Geographix (Landmark), IRAP RMS (Roxar), Petrel (Schlumberger), Gocad (Paradigm), etc.

The 3D model is built on the basis of seismic data and the results of detailed correlation of well sections interconnected (Figures 10.2 and 10.3).

The traditional technology of geological 3D modeling is presented in the form of the following main stages:

1. Collection, analysis, and preparation of necessary information, data loading (import and export)
2. Correlation of formations based on well data
3. Interpretation of seismic data (identification of faults, horizon tracing and mapping, attribute analysis, etc.)
4. Building and editing maps
5. Construction of a model of tectonic disturbances
6. Structural modeling (creating a frame)
7. Creating a grid (3D grid), transferring well data to the grid
8. Facies (lithological) modeling

9. Petrophysical modeling
10. Calculation of hydrocarbon reserves
11. Planning for drilling new wells
12. Analysis of uncertainties and risks

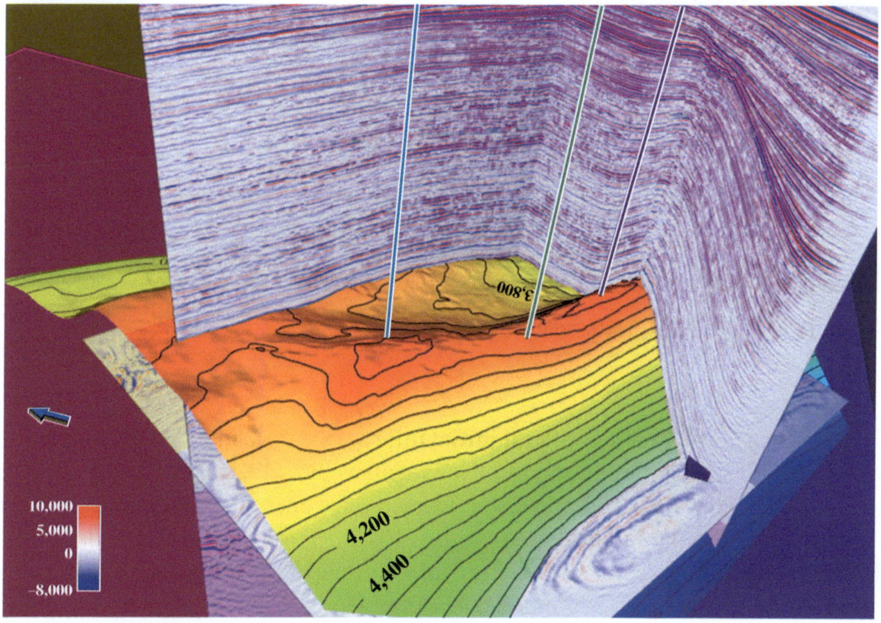

Figure 10.2 Seismic profile of the field.

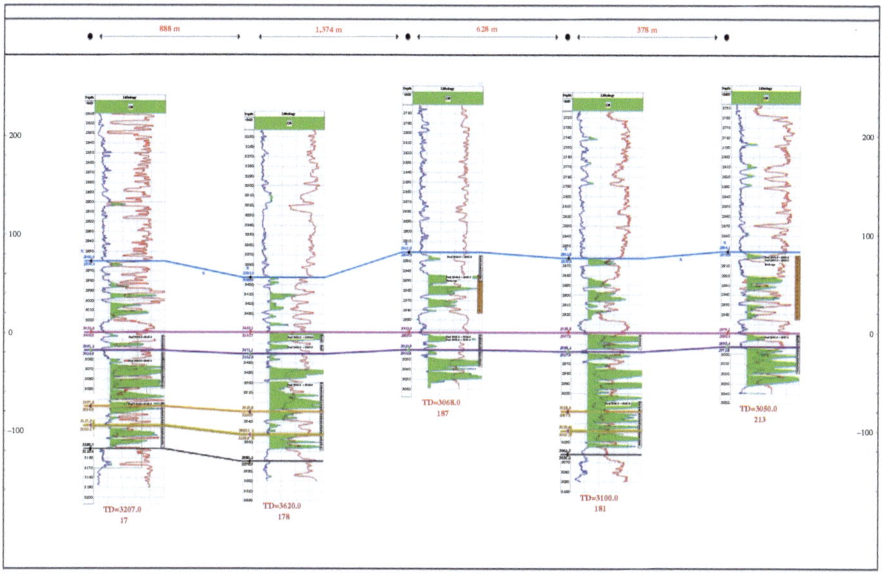

Figure 10.3 Log Correlation.

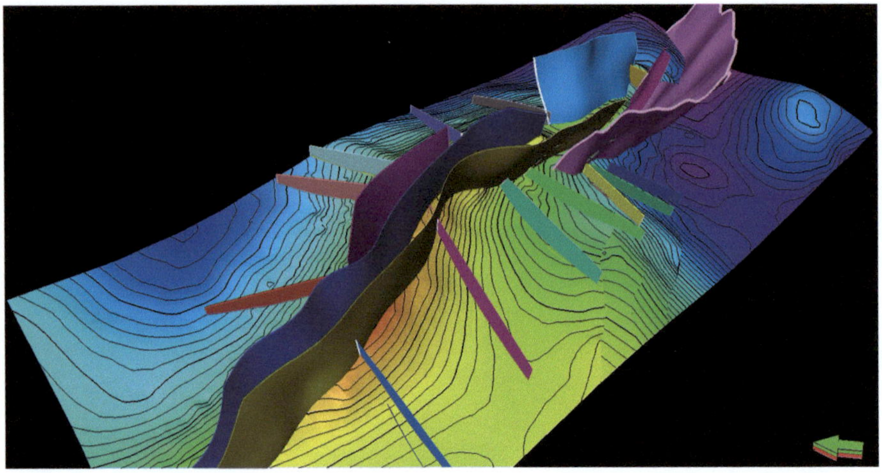

Figure 10.4 Structural-stratigraphic framework of the horizon.

To create a 3D model of the field, initial information is loaded: wellhead coordinates, altitudes, inclinometry, and well logs of all wells drilled in the structure. After loading the initial data and creating a working design, a structural-stratigraphic framework of the model is created (Figure 10.4). To do this, using the data from reference seismic horizons as a trend, well correlation is first performed (strata breakdowns in the wells are indicated). As a result, a model of tectonic disturbances is created. Detailed section correlation is carried out based on a series of transverse and longitudinal profiles covering the entire structure.

After creating a structural-stratigraphic framework of the horizons of a given field, in order to obtain structural maps of the roof and bottom of the formation, external and internal oil-bearing contours, and a structural model, is created that includes all tectonic disturbances.

When studying sedimentation conditions, the principle of sequential stratigraphy is taken into account, which makes it possible to predict the order of study of stratigraphic units, including the sequence in the process of sedimentation, to create a general chronostratigraphic system of cycles in the process of formation of genetically related strata. These sedimentary units are defined on the basis of internal strata geometry and evolve in response to changing shelf environments.

Within the framework of the structural-stratigraphic framework, taking into account the patterns of sedimentation, thin "slicing" of layers is performed for each layer, thus creating a 3D grid taking into account the fault model (3D grid). A 3D grid is a cellular frame within which all the main stages of geological modeling take place.

The main difference between a 3D grid and a two-dimensional grid is that each cell of a 3D grid occupies a certain volume in space, while a cell of a two-dimensional grid is characterized only by area. A correctly constructed 3D grid is the basis for constructing a correct geological model. The vertical dimensions of the model cells are selected taking into account the differentiation of the section by filtration and

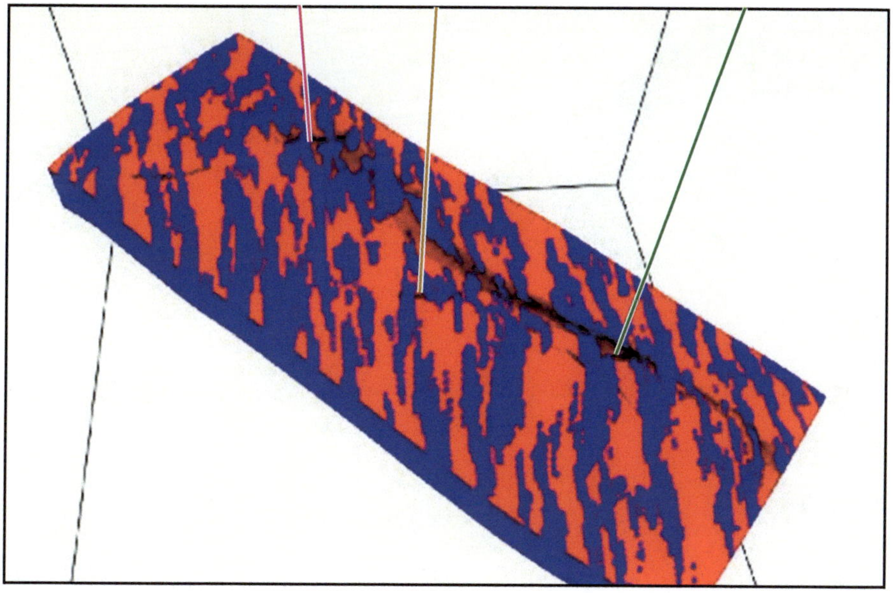

Figure 10.5 Reservoir facies model.

capacitance properties. The vertical resolution of the grid is determined by the number of layers, which are selected in such a way as to most adequately restore the distribution field of reservoirs without missing a single layer. The horizontal grid increment is selected taking into account the placement of wells and the size of the field.

The results of the interpretation of logging curves of facies, lithology, porosity, and oil saturation are transferred to the grid cells along the well trajectories.

As a result of the modeling, based on the probability of occurrence of a particular facies, each parameter cell is assigned a code (number) of the corresponding (reservoir or nonreservoir) facies (Figure 10.5). Additionally, a critical part of indicator modeling is the determination of variograms for each facies. Variogram analysis works best when a large number of wells are being evaluated. At this stage, modeling of reservoir properties of deposits is performed separately for each lithotype of rocks (sandy silt and clay), which makes it possible to identify clear boundaries during the transition from one type of rock to another.

Then a petrophysical model of the field is built (Figure 10.6), which is based on the results of the lithological modeling stage and allows us to obtain consistent 3D models of reservoir properties and oil and gas saturation of rocks, reflecting (Figure 10.7):

- well data,
- correlations between parameters (porosity, oil saturation),
- spatial heterogeneity of reservoir properties,
- lithological heterogeneity of the reservoir layer, and
- geological patterns.

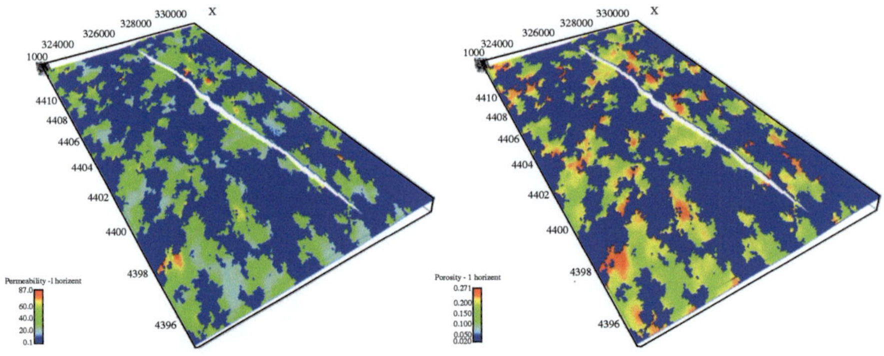

Figure 10.6 Petrophysical modeling.

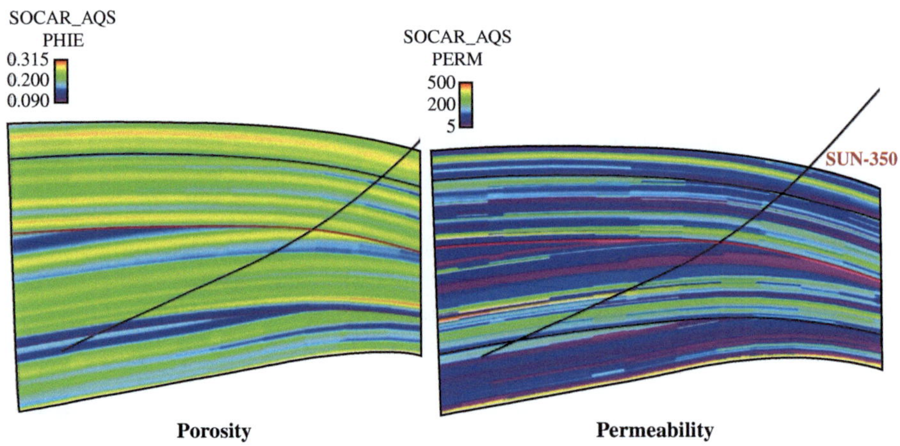

Figure 10.7 Distribution of formation flow-capacity properties of a reservoir over the area and depth of the well.

As a result, the resulting model of the spatial distribution of reservoir properties (porosity, permeability, oil saturation) is one of the key components of the 3D computer model of the field. It is important both for the correct assessment of initial geological reserves and for obtaining correct results in hydrodynamic modeling.

The main result of the created geological 3D model is the calculation of the initial geological reserves of oil and gas.

An important circumstance confirming the high quality of the creation of this geological model is the correspondence of oil and gas reserves obtained through geological 3D modeling with those available on the balance sheet of the state commission for mineral reserves.

Next, based on the constructed geological model, a hydrodynamic model of the field is created, which makes it possible to develop an optimal field development strategy.

The main problem when drawing up design documents is caused by discrepancies between the static (geological) and dynamic (hydrodynamic) models. It is typical that experts require that the static and dynamic models differ from each other by no more than 3%.

10.3.2 Hydrodynamic Modeling

The main software packages for creating hydrodynamic models of oil and gas fields are Eclipse (Schlumberger), Nexus (Landmark), TempestMore (Roxar), VIP (Landmark), etc.

To create a hydrodynamic model of the horizon, first of all, the constructed 3D geological model of the horizon is loaded into a hydrodynamic simulator (Figure 10.8). However, the multimillion-dollar "grid" constructed in the geological model is not applicable for hydrodynamic calculations, since in the hydrodynamic model, along with determining the dependences of the physicochemical and thermodynamic properties of fluids on pressure, functions of changes in phase permeability, and the filtration equation is solved in each cell, which in turn leads to an increase in calculation time. Therefore, the geological model is upscaled

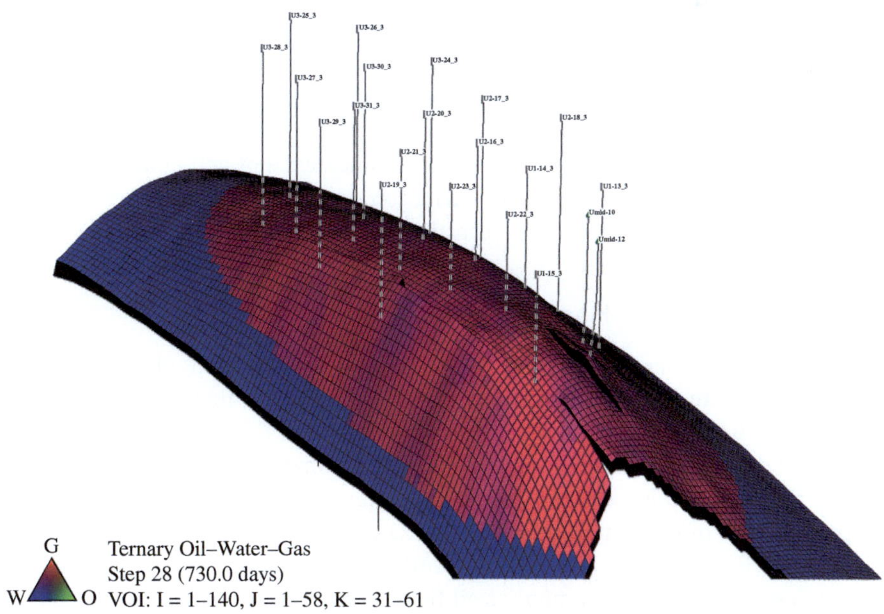

Figure 10.8 Saturation distribution in the hydrodynamic model.

when loading it into the hydrodynamic model, and a more accurate geological grid is replaced with an equally accurate one with a larger cell size.

In the process, a geological model containing information about the coordinates of wells, altitudes, inclinometry, depths of the water-water contact, depths of the roof and bottom of the layers, and cubes of reservoir capacity-filtration parameters are loaded into the hydrodynamic model. In addition to this information, all dates and depth intervals of perforations and cement bridges of production and injection wells, data on oil, gas, and water production, data on water injection, and the dependence of the physicochemical and thermodynamic properties of fluids are loaded into the hydrodynamic model.

It is known that during development, reservoir pressure gradually decreases. Since the physicochemical and thermodynamic properties of reservoir products depend on pressure, changes in pressure during the development process cause changes in thermodynamic properties. On the other hand, calculations carried out within the framework of a hydrodynamic model are carried out not only in accordance with the value of the initial reservoir pressure and the initial thermodynamic properties, but also in accordance with both the pressure value over time and the thermodynamic properties of the reservoir product at these pressures. These data are determined based on PVT analysis of reservoir products in laboratory conditions.

After this, the development history is restored, i.e., the process of adapting the parameters included in the new model with actual data is carried out. Since pressure values and production data are considered more accurate and informative, the adaptation of these parameters is taken as the basis.

The most important parameter for the complete adaptation of the hydrodynamic model is the relative phase permeability values. The values of this parameter are important for accurately representing ongoing processes and predicting development performance. Relative phase permeability values are determined based on core processing data.

The process of model adaptation, i.e., the coincidence of actual and calculated values, means a more accurate representation of the processes occurring in the reservoir and directly affects the forecasting of development indicators.

10.3.3 Hydrodynamic Modeling of EOR

Additional development of fields at a late stage of development, characterized by a complex structure and a significant proportion of residual reserves, is impossible without the use of EOR. As is known, oil production in fields can be significantly increased due to modern methods of influencing formations.

Planning the application of EOR is a complex task that requires a systematic approach. One of the important stages in EOR planning is the selection of methods and preliminary data analysis. Currently, the methodology for selecting EOR when designing the development of oil fields is not a fully formalized procedure.

At the same time, choosing an effective technology for increasing oil recovery for specific geological, physical, and economic development conditions is one of the most difficult tasks for a reservoir engineer.

Before carrying out pilot implementation, it is advisable to have estimated data on the effect of the activity to know how much the oil recovery factor and other characteristics will change. Therefore, today calculations and analysis using software are an integral stage in the development of oil and gas fields. In particular, hydrodynamic modeling is widely used to assess and predict the effectiveness of measures to enhance oil recovery.

For these purposes, production and service companies conduct full-scale hydrodynamic modeling using one of the software packages.

Eclipse (Schlumberger) is used to create hydrodynamic models of oil and gas fields, allows you to calculate various development options, and also analyzes the effectiveness of geological and technical measures and EOR methods.

The Eclipse simulator supports a wide range of EOR techniques, including thermal stimulation, chemical EOR, and miscible and immiscible gas displacement.

Eclipse modules have a wide range of functionality, are easy to use, and have a fairly simple interface.

INTERSECT EOR (Schlumberger) allows you to simulate chemical EOR with high accuracy, which allows you to use this module in fields with different geological conditions and physical properties of saturating fluids. The module allows you to simulate various EOR, including the injection of polymers and various gases.

Tempest More (Roxar) is an efficient and full-featured filtration modeling system with unique capabilities for creating multivariate models. These include convenient 3D visualization, non-isothermal models, dual porosity/permeability models, compositional modeling, gas desorption accounting, reproduction of PVT experiments, well hydraulics calculations, economic assessment, and much more. Tempest More PC contains automated adaptations of geological and hydrodynamic models. Therefore, Tempest users can not only create and adapt one version (implementation) of the filtration model based on the development history, but also obtain a representative multivariate reservoir model, which can then be used for evaluation and optimization under conditions of uncertainty. The Tempest More PC takes into account the properties of the reservoir formation (heterogeneity, fracturing), the filtration features of formation fluids (non-Newtonian properties of high-viscosity oils, the formation of a "plug" in the near zone of gas condensate wells), and also the use of various methods of increasing oil recovery.

As mentioned above, before carrying out pilot testing of any EOR technology in a field, it cannot be done without its preliminary analysis on appropriate hydrodynamic simulators. The use of appropriate simulators is a necessary and very important step in the development of oil and gas fields.

10.4 Implementation of EOR

Once hydrodynamic modelling has been carried out and positive results from the proposed EOR at the pilot site have been obtained, the project will proceed to implementation.

Typically, a pilot area with an injection well (usually a 5–7 point pattern) with surrounding production wells is considered for implementation (see Figure 10.9).

In order to control and evaluate the process and determine the mechanism, a number of necessary studies are carried out at the wells before and after the start of the activity. At the injection well, before the start of the event, an injectivity profile (PLT study) is determined to assess the vertical sweep efficiency.

In order to monitor the ongoing activity, the injectivity profile of the injection well is determined after a certain time or periodically (see Figure 10.10).

An assessment of the impact area of a given area can be made based on the distribution of streamlines shown in Figure 10.11.

As can be seen from Figure 10.11, before the intervention, the injected water moved in one direction (a), and after the intervention, the flow lines had a uniform distribution over the area (b).

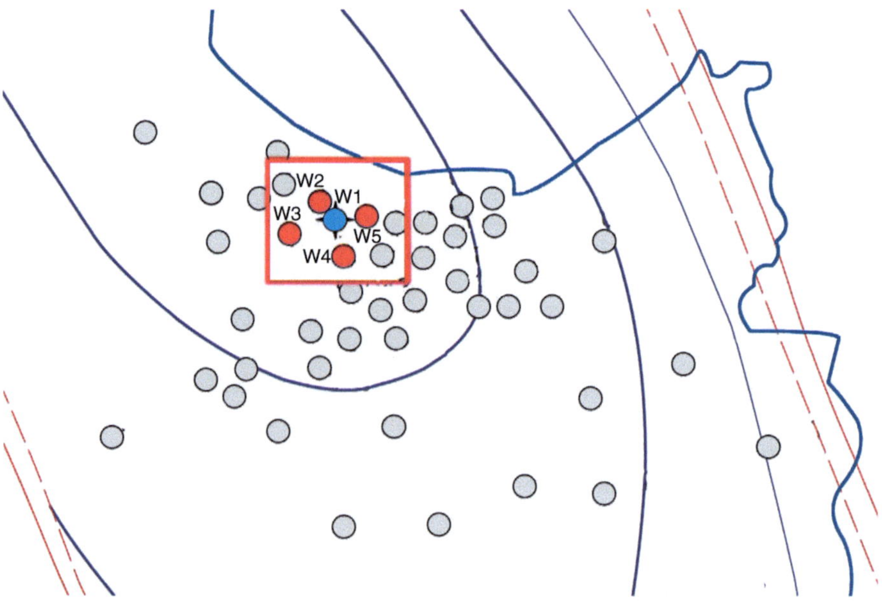

Figure 10.9 Field development map with a highlighted pilot area.

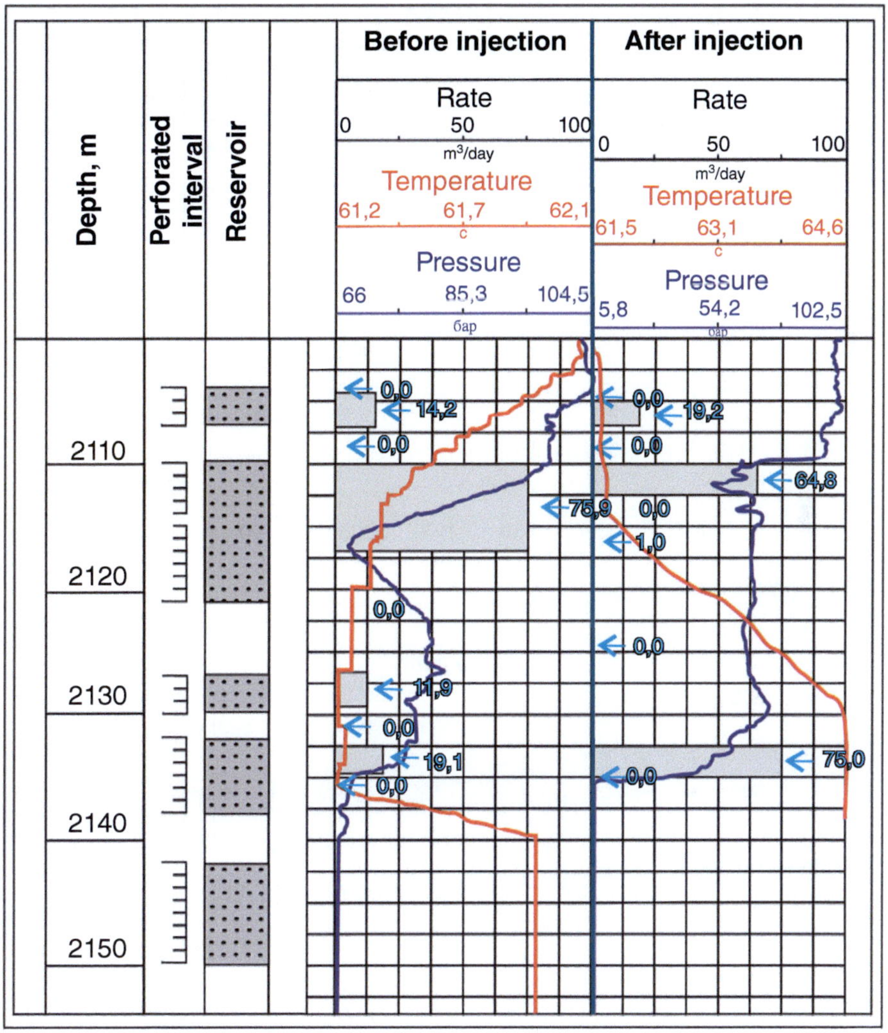

Figure 10.10 Injection well injectivity profile before and after events.

To determine additional oil production, curves of the dynamics of oil and water production before and after the impact are plotted (see Figure 10.12) and additional oil production is determined (see Figure 10.13).

If positive results are obtained from the stimulation technology at the pilot site, i.e., if the additional production is at least 10% of oil production before the event, the technology is recommended for widespread implementation in the field.

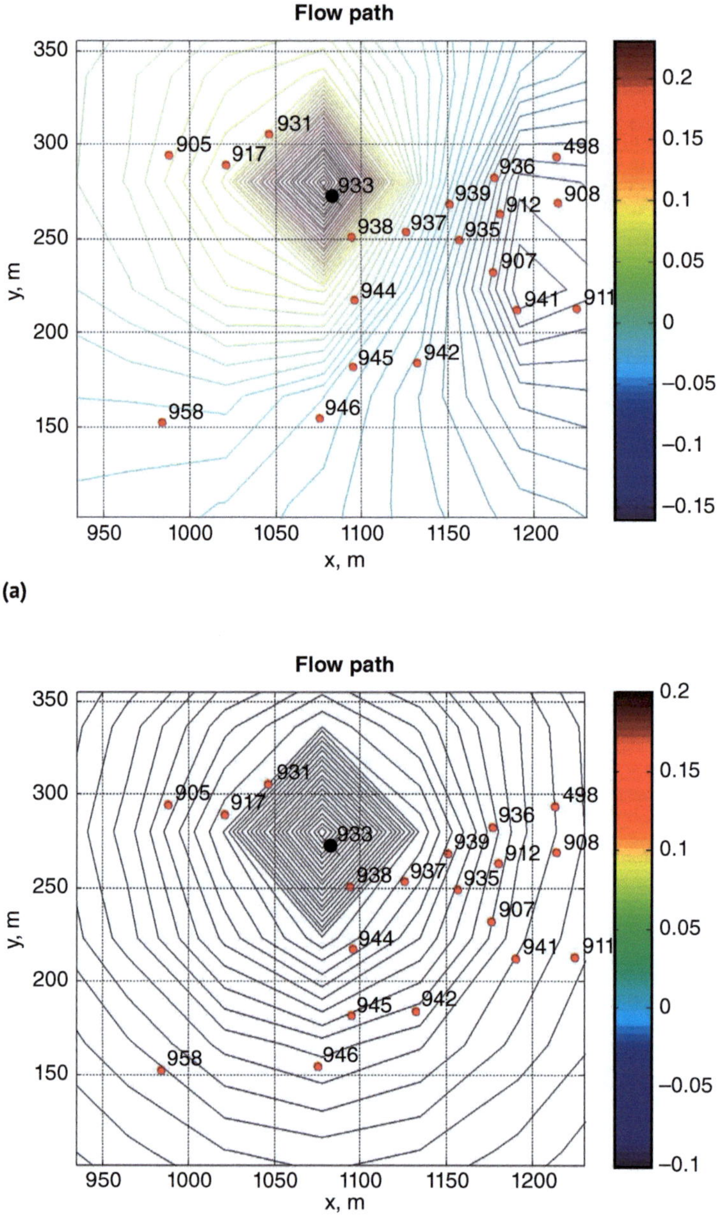

Figure 10.11 Distribution of streamlines: (a) before the start of impact on the formation; (b) after impact on the formation.

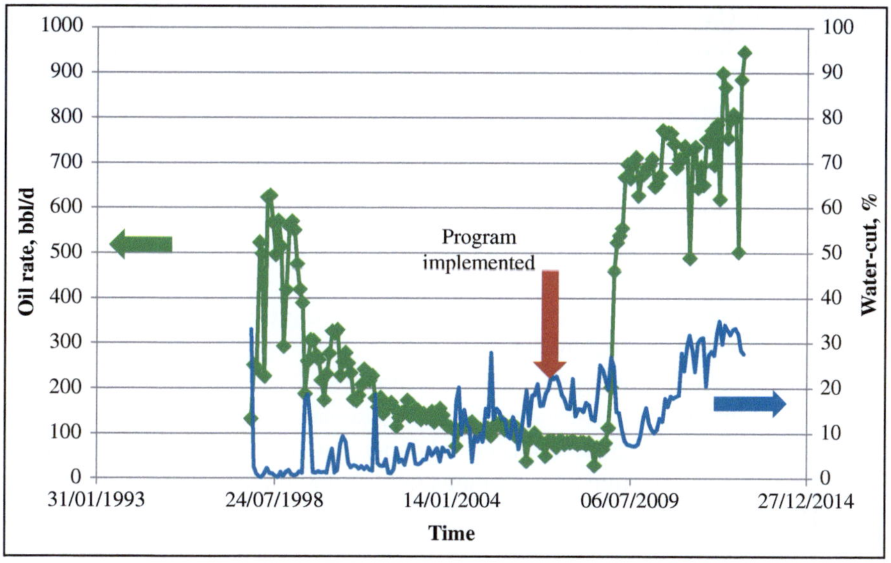

Figure 10.12 Dynamics of oil flow rate and water cut of products.

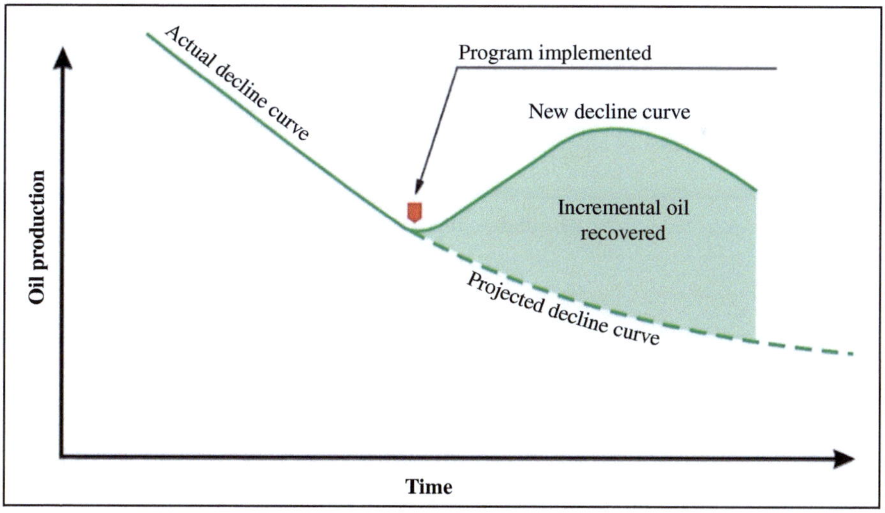

Figure 10.13 Determination of additional oil production.

10.5 Technology Readiness Level

Technology readiness levels (TRLs) were first developed (by NASA) in the USA and taken as the basis for grading the stages of development of new technologies by many departments and organizations both in the USA and in a number of other countries. The classification used reflects the state of research programs depending on the current level of technology readiness, which makes it easier for developers and customers to control the progress of research work and select the technologies that are most ready for industrial implementation.

There are nine levels of technology readiness. From the first to the sixth levels, this is the development of technology, which is carried out as part of research and development work. From the seventh level and above, development work begins, or demonstration of the functionality of technologies on real devices under development.

For the oil industry, the TRL approach to assessing the maturity of software technology was first applied by TOTAL.

In this approach, technology maturity has nine levels.

Levels 1–3 represent yourself stage "Research" (Research: Basic principles, Concept and application formulation, Concept validation), 4–6 represent "Development" (Development: Experimental pilot, Demonstration pilot, Industrial pilot), and grades 7–9—"Deployment" (Deployment: First implementation, a few records of implementation, Extensive implementation).

Many oil companies use this method to assess the maturity of oil and gas production technologies. Table 10.2 shows the TRL for various EOR technologies.

The form shown in Figure 10.14 can be used for easy comparison of technologies.

Table 10.2 Technology readiness level (TRL) for various enhanced oil recovery technologies.

Stage	TRL	
R&D	TRL 1–3	Injection of colloidal dispersed gels (CDG)
		Injection of preformed particle gels (PPG)
		In situ upgrading (heating/catalytic)
		Injection of delayed action polymers (DAP)
		Hybrid processes
Demonstration-optimization	TRL 4–8	High pressure air injection (HPAI)
		Injection of thermoactive polymers (TAP)
		Injection of low salinity water
		Water alternate gas (WAG) injection
		Microbial EOR
		Alkaline polymer (AP) flooding
		Steam assisted drainage (SAGD)
Well established EOR processes	TRL 9	Steam injection
		Chemical methods
		Injection of hydrocarbon gases
		Miscible flooding of CO_2

Company name	Polymer flooding		
Date		Theme	Direction
Current TRL	8	SPA/s	
Technology description			
Implementation experience			
Company position			
References			

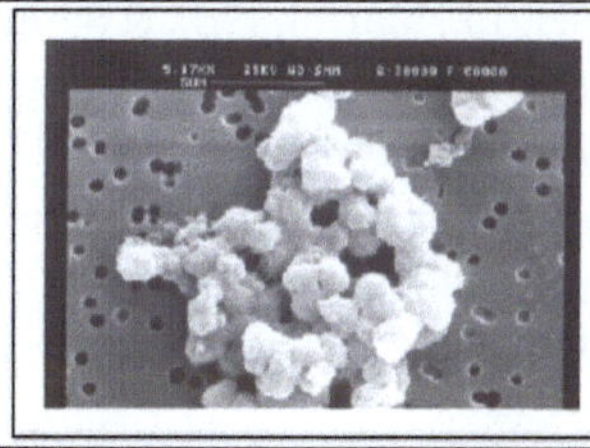

Advantages of the technology		
Production Increase		
Increase in reserves		
Decrease in capital expenditure		
Reduced operating costs		
Reduced risks		

Potential applications	Continuance

Risks	Minimisation of unforeseen circumstances

Figure 10.14 EOR implementation form.

Bibliography

Al Arfi, S.A., Mohamed, O.Y., Keshka, A.A., et al. (2009, March 15–18). Inflow control device: An innovative completion solution from "extended wellbore to extended well life cycle". *SPE Middle East Oil and Gas Show and Conference, Manama, Bahrain.* https://doi.org/10.2118/119599-MS

Chandler Engineering. (n.d.). Retrieved October 8, 2024, from https://www.chandlereng.com/

Core Laboratories. (n.d.). Retrieved October 8, 2024, from https://www.corelab.com/

Coretest Systems. (n.d.). Retrieved October 8, 2024, from http://www.coretest.com/

Irap RMS 2009.1. (2009). *User Guide*. Stavanger, Norway.

Kjeilen-Eilertsen, G., Merzi, T., Burgos, M., & Brönner, U. (2016, April 11–13). TRL (technology readiness level) assessment of DREAM (dose-related risk and effects assessment model) to qualify its use for modeling of produced water and drilling discharges. *SPE International Conference and Exhibition on Health, Safety, Security, Environment, and Social Responsibility, Stavanger, Norway*. https://doi.org/10.2118/119599-MS

Romero-Zerón, L. (Ed.). (2012). *Introduction to Enhanced Oil Recovery (EOR) Processes and Bioremediation of Oil-Contaminated Sites.* IntechOpen.

Shandrygin, A.N., & Lutfullin, A. (2008, September 21–24). Current status of enhanced recovery techniques in the fields of Russia. *SPE Annual Technical Conference and Exhibition, Denver, Colorado*. https://doi.org/10.2118/115712-MS

Shokir, E.M., & Sayyoch, M.H. (2002, August 5–7). Selection and evaluation EOR method using artificial intelligence. *SPE Annual International Conference and Exhibition, Abuja, Nigeria*. https://doi.org/10.2118/79163-MS

Taber, J.J., Martin, F.D., & Seright, R.S. (1997). EOR screening criteria revisited: Part 1: Introduction to screening criteria and enhanced recovery field projects. *SPE Reservoir Engineering*, *12*(3), 189–198. https://doi.org/10.2118/35385-PA

Vinci Technologies. (n.d.). Retrieved October 8, 2024, from http://www.vinci-technologies.com/

Zakirov, E.S. (2007). *Upscaling in 3D Computer Modeling*. Book and Business.

Zakrevsky, K.E. (2011). *Geological 3D Modelling*. EAGE Publications.

11

Laboratory Evaluation of Oil-bearing Rock Properties

11.1 Introduction

Understanding the properties of oil-bearing rocks is critical for optimizing EOR techniques. This chapter offers a comprehensive overview of the physical and chemical characteristics of reservoir rocks, including their granulometric composition, density, carbonate content, porosity, and permeability. Accurate characterization of these properties enables engineers to predict the behavior of oil reservoirs under various EOR methods and adjust extraction techniques accordingly. The chapter delves into the critical processes involved in measuring porosity, permeability, wettability, and IFT, all of which directly impact oil recovery efficiency. Given the complexity of reservoirs and the need for tailored EOR approaches, this chapter is essential for providing a scientific foundation for improving recovery rates, especially in mature fields where conventional methods have become ineffective.

11.2 Granulometric Composition of Rock

The granulometric (mechanical) composition of a rock refers to the quantitative distribution of particles of different sizes, expressed as a percentage by weight. It is one of the key characteristics of rock.

The most common methods of determining the mechanical composition of rocks are the sieve method and the sedimentation method. The sieve method, which involves separating the rock into fractions using sieves, is very simple, but its application is limited by the minimum mesh sizes and the tendency for very small particles to stick together. The smallest sieve mesh size is 0.053 mm. Finer particles, such as silty and clayey materials, which play an important role in the properties of the rock, are determined using sedimentation methods. Sieve analysis provides the "mechanical grain size," i.e., the absolute size of the grains, while sedimentation gives the "hydraulic grain size," characterized by the settling velocity of particles in water.

The mechanical composition of the rock can provide insights into the geological and paleogeographical conditions of sediment deposition in the reservoir. The

Methods for Enhanced Oil Recovery: Fundamentals and Practice, First Edition.
Baghir A. Suleimanov and Elchin F. Veliyev.

granulometric composition of the rock determines the amount of oil remaining in the reservoir after its development, either as films covering the surface of the grains or as oil retained by capillary forces.

Granulometric analysis of sands is used in oilfield practice. For example, based on mechanical analysis, filters installed at the wellbore are selected during oil field development to prevent sand from entering the well.

In laboratory conditions, a set of stamped wire or silk sieves is usually used. Stamped sieves have openings of 10, 7, 5, 3, 2, 1, 0.5, and 0.25 mm.

Sedimentation separation of particles into fractions occurs due to the different settling velocities of grains of varying sizes in a viscous liquid. According to Stokes' law, the settling velocity of spherical particles in a liquid is calculated as:

$$V = \frac{g \times d^2}{18 \times \upsilon} \times \left(\frac{\rho_{rock}}{\rho_{liq}} - 1 \right) \tag{11.1}$$

where:

- g is the acceleration due to gravity,
- d is the diameter of the particles,
- υ is the kinematic viscosity,
- ρ_{liq} is the density of the liquid, and
- ρ_{rock} is the density of the rock particles.

11.3 Determining the Density of Rocks

Density is defined as the mass per unit volume of a substance:

$$\rho = \frac{M}{V} \tag{11.2}$$

where ρ is the density (g/cm^3), M is the mass of the substance (g), and V is the volume of the substance (cm^3).

In rock studies, two main concepts are used: apparent and true density. Apparent density ρ_{app}, also referred to as bulk density or sample density, is calculated using the formula:

$$\rho_{app} = \frac{M}{V_{sample}} \tag{11.3}$$

where M is the mass of the dry, extracted rock sample, and V_{sample} is the volume of the rock sample, consisting of pore volume (V_{pore}) and solid phase volume (V_{solid}):

$$V_{sample} = V_{pore} + V_{solid} \tag{11.4}$$

True density (ρ_{true}), also known as mineral density or grain density, is determined as:

$$\rho_{true} = \frac{M}{V_{solid}} \tag{11.5}$$

To determine the density of a dry rock, its mass is measured by weighing. The volume of the rock sample is determined using methods such as liquid weighing: porosimetric and geometric methods.

11.3.1 Methods of Liquid Weighing

These methods are divided into the saturation method and Melcher's method:

1. **Saturation method:** Based on saturating the sample with liquid and immersing it in the same liquid. The volume of the liquid-saturated sample is determined by hydrostatic weighing, where the sample is weighed in air and in the same saturation liquid.

   Let's consider the following weights:

   - P_1—weight of the dry extracted sample in air (g)
   - P_2—weight of the liquid-saturated sample in the same liquid (g)
   - P_3—weight of the liquid-saturated sample in air (g)

   The forces acting on the sample submerged in liquid are the gravitational force (P_g) directed downward, equal to the weight P_3, and the buoyant force (P_{Arch}) directed upward. The resultant force, directed downward, equals P_2 (Figure 11.1). Thus, we have the vector equation:

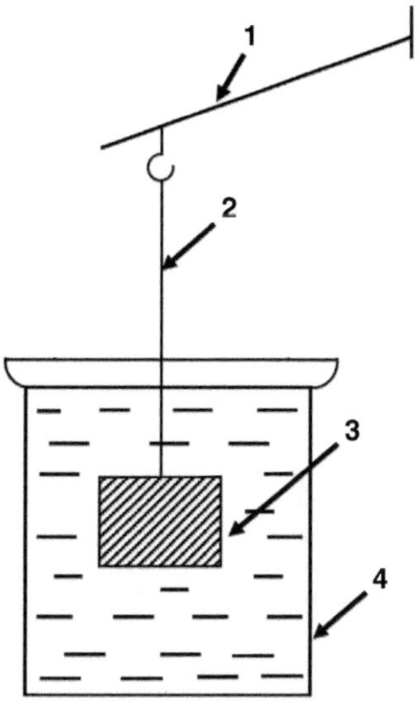

Figure 11.1 Weighing a saturated sample in a liquid: 1—scale; 2—wire; 3—sample; and 4—container with liquid.

$$P_g + P_{Arch} + P_2 = 0$$

and the scalar equation:

$$P_2 = P_3 - P_{Arch} \tag{11.6}$$

According to Archimedes' principle, the buoyant force (P_{Arch}) is equal to the weight of the displaced liquid (P_{liquid}):

$$P_{Arch} = P_{liquid} = V_{liquid} \times \gamma_{liquid} = V_{sample} \times \gamma_{liquid} \tag{11.7}$$

where γ_{liquid} is the specific weight of the liquid.

Using this, we can rewrite Eq. (11.6) as:

$$P_2 = P_3 - V_{sample}\gamma_{liquid}$$

which gives:

$$V_{sample} = \frac{P_3 - P_2}{\gamma_{liquid}} \tag{11.8}$$

Now, using the relations:

- $P = mg$ and
- $\gamma_{liquid} = \rho_{liquid} \times g$

where ρ_{liquid} is the density of the liquid and g is the acceleration due to gravity, we get:

$$V_{sample} = \frac{m_3 - m_2}{\rho_{liquid}} \tag{11.9}$$

If we denote m_1 as the mass of the dry, extracted sample (g), we obtain the standard formula from GOST 26450.1–85 for determining bulk (apparent) density:

$$\rho_{app} = \frac{m_1 \rho_{app}}{m_3 - m_2} \tag{11.10}$$

2. **Melcher's method (Paraffin coating method):** This method is used to determine the volume of a rock sample of any geometric shape by first covering the sample with a thin layer of paraffin and then weighing it in a liquid.

   A sample with mass M_1 is coated with paraffin, which is known as density $\rho_{paraffin}$. The paraffinization is done by immersing the sample in molten paraffin, first one half, then the other. The paraffin-coated sample is then weighed again ($\grave{M_1}$).

   The volume of the paraffin layer is calculated as:

$$V_{paraffin} = \frac{\grave{M_1} - M_0}{\rho_{paraffin}} \tag{11.11}$$

   where $V_{paraffin}$ is the volume of paraffin, $\rho_{paraffin}$ is the density of paraffin (900 kg/m^3), $\grave{M_1}$ is the mass of the rock sample with paraffin, and M_0 is the mass of the dry rock sample.

   The paraffin-coated sample is weighed in liquid, giving a weight M_2.

   The volume of the paraffin-coated rock sample is:

$$V_{sample\ with\ paraffin} = \frac{M_1 - (M_2 - a)}{\rho_{liquid}} \tag{11.12}$$

   where $V_{sample\ with\ paraffin}$ is the volume of the rock sample with paraffin, M_2 is the weight of the paraffin-coated sample in liquid, a is the weight of the wire in the liquid, and ρ_{liquid} is the liquid's density.

   The rock sample's volume is then determined using:

$$V_{sample} = V_{sample\ with\ paraffin} - V_{paraffin} \tag{11.13}$$

11.3.2 Porosimeter Method

Figure 11.2 shows a diagram of the porosimeter.

While the rock sample is being saturated, the porosimeter is filled with liquid. The instrument is tightly sealed with a cup, inverted, and after 4–7 minutes in an upright position, the liquid level is read from the scale (V_1).

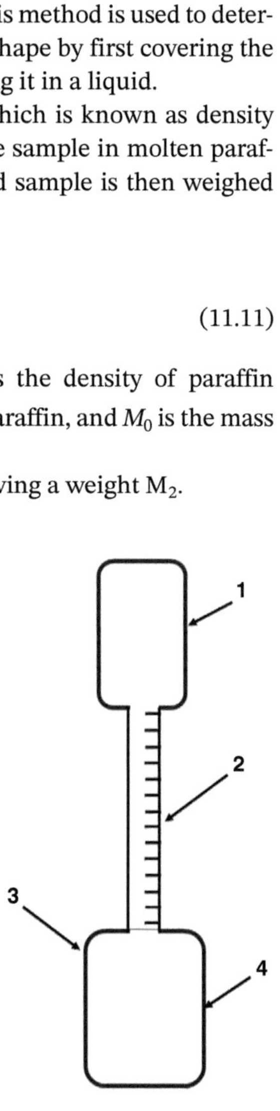

Figure 11.2 Porosimeter: 1—chamber; 2—scale; 3—fitted lid; and 4—glass.

Next, the saturated sample is placed into the cup, the instrument is sealed again, inverted, and set in place. A second reading of the liquid level is taken from the scale (V_2), similar to the first case.

The volume of the sample is calculated as the difference between the two liquid levels:

$$V_{sample} = V_2 - V_1 \tag{11.14}$$

With the sample's mass known before saturation (M), the apparent (bulk) density is calculated using Formula (11.3).

11.3.3 Geometric Method

If a rock sample has a regular geometric shape, such as a cylinder or cube, the geometric method can be applied to determine its volume. This method is based on measuring the geometric dimensions of the sample—the arithmetic mean values of the diameter (d_{avg}) and length (l_{ong}):

$$d_{avg} = \frac{\sum_i d_i}{n} \text{ and } l_{avg} = \frac{\sum_i l_i}{n}$$

where i is the number of measurements. Using these average values, the volume of the sample is calculated as:

$$V_{sample} = \frac{\pi d_{avg}^2 l_{avg}}{4} \tag{11.15}$$

This method is suitable for porous rocks in its pure form.

When determining the **true grain density** (mineralogical density) of loose rock materials, the volume of the solid phase of the rock sample (core) must be known. The following methods are used to determine the volume of the solid phase of a rock sample: volumetric, pycnometric, flotation, hydrostatic weighing, and others. The most commonly used methods are the volumetric and pycnometric methods.

1. **Volumetric Method**
   The volumetric method is based on measuring the change in the liquid volume when crushed rock particles are immersed in the liquid. One instrument used for determining the density of loose materials is the porosimeter (see Figure 11.2).
2. **Pycnometric Method**
   The weight method is based on the change in weight of a standard volume with and without a rock sample. The simplest instrument for this method is the **pycnometer**—a glass vessel with a narrow neck, calibrated for a specific volume, which is indicated by a mark on the neck (Figure 11.3).
   In Figure 11.3, the following symbols are used:

   - M—mass of the empty pycnometer,
   - M_1—mass of the pycnometer filled with liquid up to the mark,
   - M_2—mass of the pycnometer with a rock sample,
   - M_3—mass of the pycnometer with the rock sample and liquid filled to the mark.

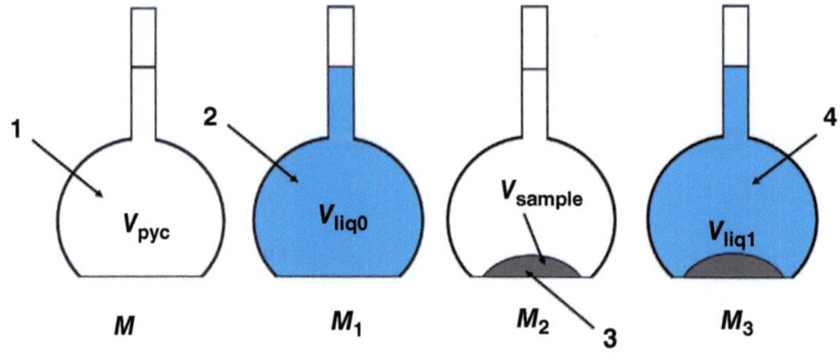

Figure 11.3 Determination of the density of the solid phase of the rock using a pycnometer: 1—dry pycnometer; 2—liquid volume to the mark in the pycnometer; 3—sample volume; 4—rock sample with liquid filled to the mark in the pycnometer.

The mineralogical density of the rock is calculated as follows:

$$\rho_{min} = \frac{M_n}{V_n} = \frac{M_2 - M}{V_k - V_{liq,n}} \tag{11.16}$$

where:

- M_n is the mass of the rock sample,
- V_k is the volume of the pycnometer up to the mark, and
- $V_{liq,n}$ is the volume of liquid in the pycnometer with the rock sample.

The equation for mineralogical density can also be written as:

$$\rho_{min} = \frac{(M_2 - M)\rho_{liq}}{(M_1 - M) - (M_3 - M_2)} \tag{11.17}$$

where:

- ρ_{liq} is the density of the liquid.

3. **Apparent Mineralogical Density**
There is also the concept of apparent mineralogical density, where the volume of the solid phase is determined by subtracting the volume of open pores from the total sample volume, assuming no closed pores:

$$V_{solid} = V_{sample} - V_{open\ pores} = \frac{M_3 - M_2}{\rho_{liq}} - \frac{M_3 - M_1}{\rho_{liq}} = \frac{M_1 - M_2}{\rho_{liq}} \tag{11.18}$$

The apparent mineralogical density is then calculated as:

$$\rho_{apparent\ min.} = \frac{M_1 \rho_{liq}}{M_1 - M_2} \tag{11.19}$$

11.3.4 Procedure for Determination of the Apparent Density of Rock

The objective of this test is to determine the apparent density of rock using hydrostatic weighing and geometric methods.

Equipment and Instruments:

- Installation for vacuum saturation of rock samples.
- Laboratory scales, specifically designed for weighing saturated core samples in liquid.

Steps of the Procedure:

1. Determine geometric dimensions:
   - Measure the average geometric dimensions of the cylindrical sample: diameter (d_{avg}) and length (l_{avg}).
2. Determine sample mass:
   - Weigh the extracted and dried rock sample using laboratory scales with a precision of 0.001 g. Record the data in Table 11.1.
3. Saturate the sample:
   - Place the sample in a vacuum chamber to saturate it with the liquid (refer to Figure 11.4).
4. Weigh saturated sample in liquid:
   - Weigh the saturated sample while it is still submerged in the liquid using the laboratory scales. Record the data in the table. It is essential to keep the samples submerged to avoid evaporation of the liquid absorbed in the sample.

Table 11.1 Physical and dimensional properties of rock samples for density determination.

No.	Name of quantity	Symbol	Unit of measurement	Core 1	Core 2
1	Diameter of rock sample	D	cm		
2	Length of rock sample	L	cm		
3	Mass of dry sample	M_0	g		
4	Mass of saturated sample in liquid	M_2	g		
5	Mass of saturated sample in air	M_1	g		
6	Density of liquid	ρ_{liq}	g/cm^3		
7	Volume of sample	V_{sample}	cm^3		
	• using Archimedes' principle				
	• by geometric measurements				
8	Apparent density of rock	$\rho_{apparent}$	g/cm^3		
	• using Archimedes' principle				
	• by geometric measurements				

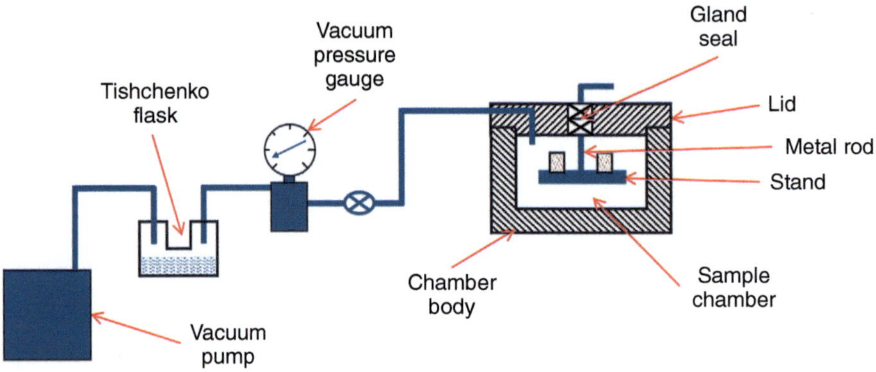

Figure 11.4 Vacuum system for saturating rock samples with liquid.

5. Weigh saturated sample in air:
   - Weigh the saturated sample in air, documenting the data in the table. Remove excess liquid from the samples using filter paper that has been saturated with the same liquid.

Data Processing:

1. Calculate the volume of the sample:
   - Utilize Archimedes' principle to determine the volume, applying the appropriate Formula (11.9).
   - Calculate the volume through geometric measurements using Formulas (11.11) and (11.12).
2. Calculate the apparent density of the rock sample:
   - Employ Formula (11.3) to calculate the apparent density for both methods of volume determination.
3. Compare obtained density values:
   - Assess and compare the density values obtained to identify the type of rock. The average density values for various rocks are as follows:
     (i) Porous sandstone—2.15 g/cm^3
     (ii) Dense sandstone—2.48 g/cm^3
     (iii) Porous limestone—2.20 g/cm^3
     (iv) Dense limestone—2.525 g/cm^3

11.3.5 Procedure for Determination of True Density of Rocks by Pycnometric Method

The objective of this test is to determine the true density of a rock sample using a glass pycnometer. The true density of rocks is defined as the mass of a unit volume of the solid phase of the rock, which is the ratio of the mass of the minerals to their volume. The density of minerals can be determined through various methods, with the pycnometric method being one of the most widely used.

Equipment and Instruments:

- Glass pycnometer.

Steps of the Procedure:

1. Determine the mass of the empty pycnometer:
   - Measure and record the mass of the dry pycnometer M_0 in the data Table 11.2.
2. Fill the pycnometer with liquid:
   - Pour the liquid (either purified kerosene or a saline solution) into the pycnometer up to the mark on the neck and determine the mass of the pycnometer with liquid M_1.
3. Empty and clean the pycnometer:
   - Drain the liquid from the pycnometer, wash it thoroughly, and dry it.
4. Prepare the rock sample:
   - Crush the rock sample and add the sample into the dry pycnometer. Determine and record the mass of the pycnometer with the rock sample M_2.
5. Add liquid and mix:
   - Add liquid to the pycnometer with the rock sample until it is two-thirds full. Mix the contents to remove any air bubbles. Then, fill the pycnometer to the mark on the neck and determine the mass of the pycnometer with the rock sample and liquid M_3.

Table 11.2 Measurement data for determining rock sample density using a pycnometer.

No.	Parameters	Unit of measurement	Value
1	Mass of dry pycnometer	kg	
2	Mass of pycnometer with liquid	kg	
3	Liquid number of pycnometer		
4	Mass of pycnometer with sample	kg	
5	Mass of rock sample	kg	
6	Mass of pycnometer with sample	kg	
7	Mass of pycnometer with sample and liquid	kg	
8	Density of liquid	kg/cm^3	
9	Temperature of experiment	°C	
10	Volume of liquid in pycnometer without rock sample	m^3	
11	Volume of liquid in pycnometer with rock sample	m^3	
12	Volume of rock sample	m^3	
13	True density of minerals in a rock sample	kg/m^3	

Data Processing:

- Calculate the mass of the sample in the pycnometer:

$$M_{sample} = M_2 - M_0 \tag{11.20}$$

- Calculate the mass of liquid in the pycnometer without the rock sample:

$$M_{liquid} = M_1 - M_0 \tag{11.21}$$

- Determine the density of the liquid using a hydrometer (ρ_{liquid}).
- Measure the temperature of the liquid in the pycnometer (T).
- Calculate the liquid number of the pycnometer:

$$n_{liquid} = \frac{M_{liquid}}{\rho_{liquid}} \tag{11.22}$$

- Calculate the volume of liquid in the pycnometer without the rock sample:

$$V_{liquid} = \frac{M_{liquid}}{\rho_{liquid}} \tag{11.23}$$

- Calculate the mass of liquid in the pycnometer with the rock sample:

$$M^1_{liquid} = M_3 - M_{sample} \tag{11.24}$$

- Calculate the volume of liquid in the pycnometer with the rock sample:

$$V^1_{liquid} = \frac{M^1_{liquid}}{\rho_{liquid}} \tag{11.25}$$

- Determine the volume of the rock sample in the pycnometer:

$$V_{sample} = V_{liquid} - V^1_{liquid} \tag{11.26}$$

- Calculate the true density of the minerals composing the rock sample:

$$\rho_{true} = \frac{M_{sample}}{V_{sample}} \tag{11.27}$$

11.4 Determining the Carbonate Content of Rocks

11.4.1 Definition of Carbonate Content

Carbonate content refers to the amount of carbonate salts present in a rock, such as limestone ($CaCO_3$), dolomite ($CaMg(CO_3)_2$), calcined soda (Na_2CO_3), potash (K_2CO_3), siderite ($FeCO_3$), and others (Table 11.3).

Carbonate content is crucial for determining the feasibility of acid treatment for increasing the permeability of productive layers. The carbonate content is measured in percentages or fractions of a unit:

$$K(\%) = \frac{100 \times M_{carb}}{M} \tag{11.28}$$

Table 11.3 Classification by $CaCO_3$ content.

Name	$CaCO_3$ Content, %
Limestone	95–100
Dolomitic Limestone	50–95
Calcareous Dolomite	5–50
Dolomite	0–5

where:

- M is the mass of the rock sample (g);
- M_{carb} is the mass of carbonates in the sample (g).

Most of the carbonates in rocks are represented by **calcite ($CaCO_3$)**. When calcite reacts with 14% hydrochloric acid at 20 °C, carbon dioxide (CO_2) is released according to the following reaction:

$$CaCO_3 + 2HCl = CaCl_2 + H_2CO_3 = CaCl_2 + H_2O + CO_2$$

Where **carbonic acid (H_2CO_3)** immediately breaks down into water and carbon dioxide.

Formula for Calculating Carbonate Content

1. Determine the mass of CO_2 released:

$$m = V \times \rho \tag{11.29}$$

   where:

   - V is the volume of the released CO_2 gas (cm^3);
   - ρ is the density of the CO_2 gas under experimental conditions (g/cm^3).

   The molecular weight of CO_2 is 44 g/mol, and the molecular weight of $CaCO_3$ is 100 g/mol. Therefore, 100 g of $CaCO_3$ corresponds to 44 g of CO_2.

2. Proportion to determine the mass of reacted $CaCO_3$:

$$\frac{100}{x} = \frac{44}{m}$$

   Solving for x, the mass of reacted $CaCO_3$:

$$x = \frac{m \times 100}{44} = \frac{V\rho \times 100}{44}$$

3. Formula for calculating carbonate content:

$$K = 100 \times \frac{100}{44} \times \frac{V\rho}{M} = \frac{V\rho_T}{4.4M} \tag{11.30}$$

   where:

   - M is the mass of the rock sample (g);
   - V is the volume of released CO_2 gas (cm^3);
   - ρ_T is the density of CO_2 under experimental conditions (mg/cm^3).

CO_2 Gas Density at 20 °C and 760 mmHg:

$$\rho_T = 1.878\frac{g}{L} = 1.878 \times 10^{-3}\frac{g}{cm^3}$$

Reaction for Dolomite:

$$CaMg(CO_3)_2 + 4HCl = CaCl_2 + MgCl_2 + 2H_2O + 2CO_2$$

When calculating the carbonate content for dolomite, the volume of gas released over the first 60 seconds is used for calcite, and the remaining gas volume is used for dolomite, which typically reacts over 20–25 minutes.

11.4.2 Determination of Carbonate Content of Rocks by Gasometric Method Using the Clark Apparatus

The objective is to determine the carbonate content of a rock sample by the volumetric gasometric method using the Clark apparatus AK-4 (Figure 11.5). This method is based on the chemical decomposition of carbonates under the action of hydrochloric acid and the measurement of the volume of carbon dioxide (CO_2) released as a result of the reaction:

$$CaCO_3 + 2HCl \rightarrow CaCl_2 + H_2O + CO_2$$

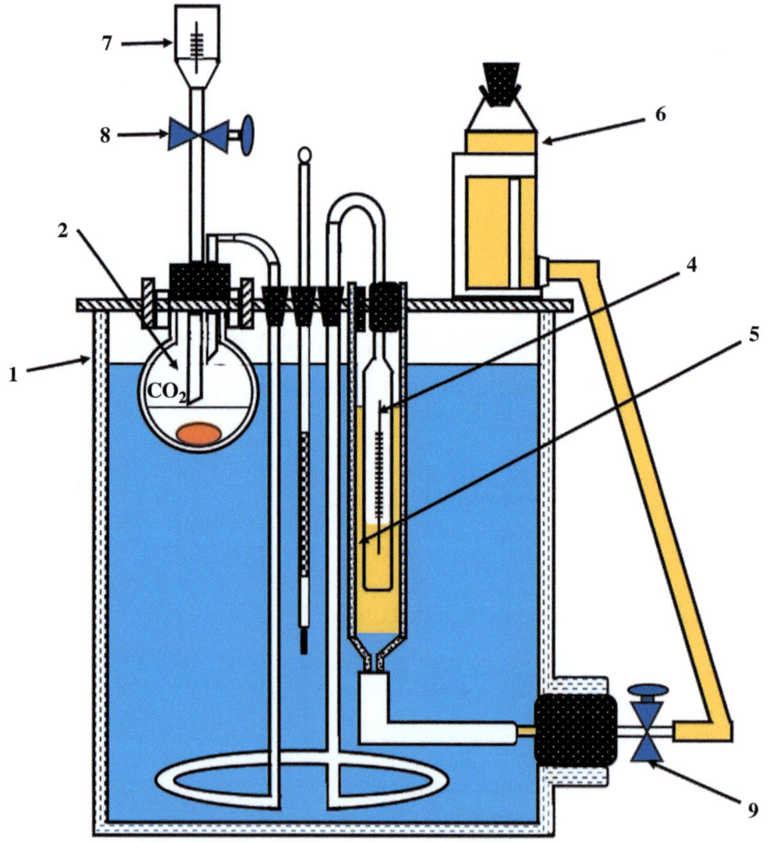

Figure 11.5 Clark apparatus AK-4.

Procedure

1. **Prepare the thermostat:** Fill the thermostat (1) with water.
2. **Set up the gasometer:** Fill the gasometer, consisting of a cylinder (5) and a measuring burette (4), with a 10–15% salt solution from the equalizing flask (6).
3. **Prepare the sample:** Grind the extracted and dried rock sample. Weigh a portion of 0.2–1 g and place it in the reaction flask (2).
4. **Prepare the reaction flask:** Insert a stopper with a measuring burette (7) for hydrochloric acid and a valve (8) into the flask. Ensure the entire system—from the reaction flask (2) to the equalizing flask (6)—is airtight.
5. **Set the initial level:** Adjust the initial level in the gasometer using the equalizing flask (6) with valve (9) open. Then, close valve (9).
6. **Record initial conditions:** Measure the atmospheric pressure using a barometer and the water temperature in the thermostat using a thermometer.
7. **Add hydrochloric acid:** Pour hydrochloric acid into the burette (7) with valve (8) closed and note its level. Quickly open valve (8) and add a portion of the acid to the flask. Close valve (8) and simultaneously open valve (9). The released gas will displace some of the water from the burette into the surrounding space.
8. **Measure the gas volume:** Lower the equalizing flask (6) to bring the salt solution level in the surrounding space down to the level in the burette. Repeat this operation until the reaction stops. Measure the volume of gas released by noting the difference between the initial and final water levels in the burette.

Formula for Carbonate Content Calculation

The carbonate content is calculated using the formula:

$$K = \frac{V \times G}{4.4 \times a} \tag{11.31}$$

where:

- V is the volume of CO_2 released (cm^3);
- G is the weight of 1 cm^3 of CO_2 at the temperature and atmospheric pressure during the experiment (from Table 11.4);
- a is the weight of the rock sample (g); and
- 4.4 is the conversion coefficient from CO_2 to $CaCO_3$.

Chemical Reaction

$$CaCO_3 + 2HCl \rightarrow CaCl_2 + H_2O + CO_2$$

Atomic weights:

- Calcium (Ca) = 40;
- Carbon (C) = 12; and
- Oxygen (O) = 16.

According to the reaction, 100 parts by weight of $CaCO_3$ yield 44 parts by weight of CO_2. To determine the percentage of $CaCO_3$ in the rock, convert the volume of CO_2 into milligrams using the weight of 1 cm^3 CO_2 from Table 11.4.

Table 11.4 Conversion table for CO_2 volume (cm^3) to mass (mg) at various temperatures and pressures.

Temperature (°C)	Pressure (mmHg)													
	742	744.5	747	749	751	753.1	756	758	760	762.5	765	767	769	771
28	1.778	1.784	1.792	1.797	1.804	1.810	1.817	1.823	1.828	1.833	1.837	1.842	1.847	1.852
27	1.784	1.790	1.797	1.803	1.810	1.816	1.823	1.829	1.834	1.839	1.843	1.848	1.853	1.858
26	1.791	1.797	1.803	1.809	1.816	1.822	1.829	1.835	1.840	1.845	1.849	1.854	1.859	1.864
25	1.797	1.803	1.810	1.816	1.823	1.829	1.836	1.842	1.847	1.852	1.856	1.861	1.866	1.871
24	1.803	1.809	1.816	1.822	1.829	1.835	1.842	1.848	1.853	1.858	1.862	1.867	1.872	1.877
23	1.806	1.815	1.822	1.828	1.835	1.841	1.848	1.854	1.859	1.864	1.868	1.873	1.878	1.883
22	1.815	1.821	1.828	1.834	1.841	1.847	1.854	1.860	1.865	1.870	1.875	1.880	1.885	1.890
21	1.822	1.828	1.835	1.841	1.848	1.854	1.861	1.867	1.872	1.877	1.882	1.887	1.892	1.897
20	1.828	1.834	1.841	1.847	1.854	1.860	1.867	1.873	1.878	1.883	1.888	1.893	1.898	1.903
19	1.834	1.840	1.847	1.853	1.859	1.866	1.872	1.878	1.884	1.889	1.894	1.899	1.904	1.909
18	1.840	1.847	1.853	1.859	1.866	1.872	1.879	1.885	1.890	1.895	1.900	1.905	1.910	1.915
17	1.846	1.853	1.860	1.866	1.873	1.879	1.886	1.892	1.897	1.902	1.907	1.912	1.917	1.922
16	1.853	1.860	1.866	1.873	1.879	1.886	1.892	1.898	1.903	1.908	1.913	1.918	1.923	1.928
15	1.859	1.866	1.872	1.879	1.885	1.892	1.898	1.905	1.910	1.915	1.920	1.925	1.930	1.935
14	1.865	1.872	1.878	1.885	1.892	1.898	1.905	1.911	1.916	1.921	1.926	1.931	1.936	1.941
13	1.872	1.878	1.885	1.892	1.899	1.906	1.912	1.919	1.924	1.929	1.934	1.939	1.944	1.950
12	1.878	1.885	1.892	1.899	1.906	1.912	1.919	1.925	1.930	1.935	1.940	1.945	1.950	1.955
11	1.885	1.892	1.899	1.906	1.913	1.919	1.926	1.932	1.937	1.942	1.947	1.952	1.957	1.962
10	1.892	1.899	1.906	1.913	1.920	1.926	1.933	1.939	1.944	1.949	1.954	1.959	1.964	1.969

Table 11.5 Initial experimental data and calculations for determining $CaCO_3$ content in rock samples.

No.	Parameter description	Symbol	Value
1	Rock sample weight (g)	a	
2	Burette water level at atmospheric pressure before reaction (cm^3)	b	
3	Burette water level at atmospheric pressure after reaction (cm^3)	c	
4	Volume of released CO_2 (cm^3)	$V = c - b$	
5	Water temperature (°C)	t	
6	Barometric pressure (mmHg)	P_B	
7	Weight of 1 cm^3 CO_2 at t and P_B (mg)	G	
8	$CaCO_3$ content in the rock (%)	$K = \frac{V \times G}{4.4 \times a}\%$	

Then, divide by 44, multiply by 100, divide by the sample weight a in grams, and multiply by 1,000 for conversion to milligrams. The final formula simplifies to:

$$K = \frac{V \times G \times 100 \times 100}{44 \times a \times 1,000}\%$$

After simplification:

$$K = \frac{V \times G}{4.4 \times a}\% \qquad (11.32)$$

The experimental data are recorded in Table 11.5.

11.5 Collector Properties

11.5.1 Extraction of Oil-saturated Rock Samples

To determine the filtration properties of rocks (such as porosity, permeability, carbonate content, etc.), a sample that is free from oil and water is required. The removal of these substances from the rock's pores is achieved through extraction using a Soxhlet apparatus.

Description and Operation of the Soxhlet Apparatus

The Soxhlet apparatus (Figure 11.6) consists of a flat-bottomed glass flask (1), an extractor vessel (2), and a reflux condenser (4). All parts of the apparatus must be tightly ground to fit each other.

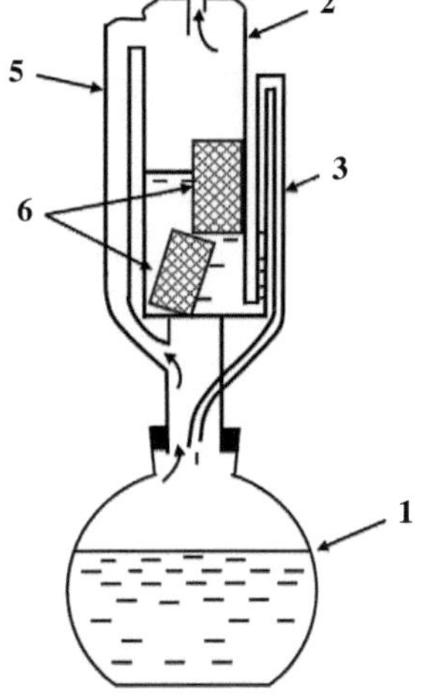

Figure 11.6 Soxhlet apparatus.

The lower outlet of the condenser connects to the water supply, while the upper outlet connects to the drain. When the flask is heated, solvent vapors rise through the tube (5) into the upper part of the extractor and the condenser (4), where they condense.

The condensate drips into the extractor and dissolves the oil in the sample (6). As the condensate accumulates, it drains through the siphon tube (3) from the extractor into the flask. The heating of the flask should be carried out on a closed burner inside a fume hood.

Steps for Performing the Procedure:

1. Clean the dust from the rock samples, weigh them in a cartridge made from filter paper, and place them in the extractor.
2. Fill the flask to about two-thirds of its volume with solvent, assemble the apparatus, turn on the water, and start heating.
3. Solvents such as toluene, carbon tetrachloride (CCl_4, with a density of 1632 kg/m^3, and a boiling point of 80 °C), an alcohol-benzene mixture, or others can be used.
4. Extraction is considered complete when the condensate draining into the flask becomes completely clear.
5. Extraction lasts from 12 to 24 hours, and for dense fine-grained rocks, it may take longer. After extraction is complete, allow the apparatus to cool, disassemble it, and dry the samples in an oven at 102–105 °C for 40 minutes until a constant weight is reached.
6. If the weight of the sample before extraction was P_1 and after extraction is P_2, then the amount of liquid (water, oil) content (P_1) is determined by the formula:

$$P_{liq} = P_1 - P_2$$

In percentage terms relative to the initial sample weight, the liquid content is calculated as:

$$P_{liq}(\%) = \frac{P_1 - P_2}{P_1} \times 100\% \tag{11.33}$$

11.6 Porosity Measurements

11.6.1 Helium Grain Volume and Grain Density

11.6.1.1 Sample Preparation

Before conducting helium grain volume and grain density tests, it is critical to thoroughly clean the rock samples of any fluids and dry them in a manner suited to the purpose of the porosity data being collected. The dry weight of the sample, denoted as W_d, is measured prior to testing, as this value is fundamental for calculating the grain density.

The shape of the sample is flexible, meaning the test can be performed on various types of samples, including cylindrical plugs, irregular rock fragments, or even loose sand. However, when conducting routine core analysis (RCA) or special core analysis (SCAL), cylindrical plugs must be used. Larger plugs, typically 1.5 inches in diameter, are preferred over smaller 1-inch plugs. This is because smaller plugs yield smaller pore volumes, and any error in measurement is magnified when dealing with small volumes. A 1-inch plug has approximately four times less pore volume than a 1.5-inch plug with the same porosity, making it more prone to significant errors during testing. Therefore, in porosity measurements, the potential for errors in core analysis must be carefully considered, especially when using smaller samples.

11.6.1.2 Test Equipment

Grain volume measurements are typically performed using a twin-cell helium porosimeter or pycnometer, which operates based on Boyle's law. This law states that, under constant temperature, the product of pressure and volume in a closed system remains constant:

$$P_1 V_1 = P_2 V_2 \tag{11.34}$$

In these experiments, helium is the preferred gas due to its ideal gas behavior at room conditions, inertness, and its ability to penetrate even the smallest pores because of its tiny molecular size. When handled with care, this method provides accurate and reproducible results.

To understand how Boyle's law applies to grain volume measurements, let's look at a typical porosimeter setup (illustrated in Figure 11.7). Helium is initially allowed into a reference volume V_r, at pressure P_r. Valves V_1 and V_2 keep the reference volume isolated. An empty matrix cup, with volume V_s, is used as the expansion chamber. Valves V_2 and V_3 isolate this chamber.

Once valve V_2 is opened, the helium at pressure P_r expands to fill both the reference volume V_r and the expansion chamber V_s. The final pressure after expansion, P_x, is determined by the relative volumes of V_r and V_s according to the formula:

$$P_t V_t = P_x(V_r + V_s)$$

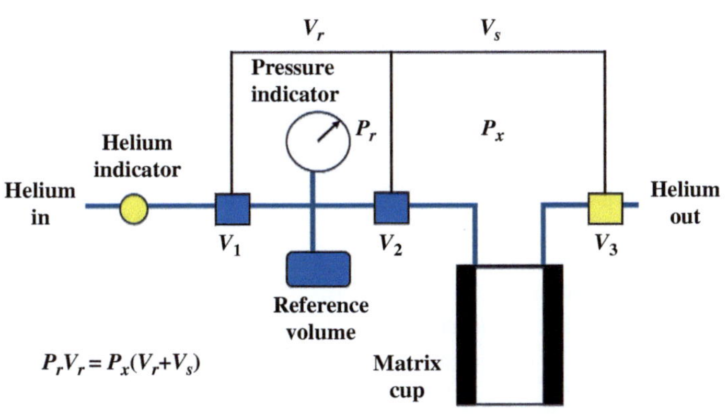

Figure 11.7 Helium grain volume instrument schematic (matrix cup empty).

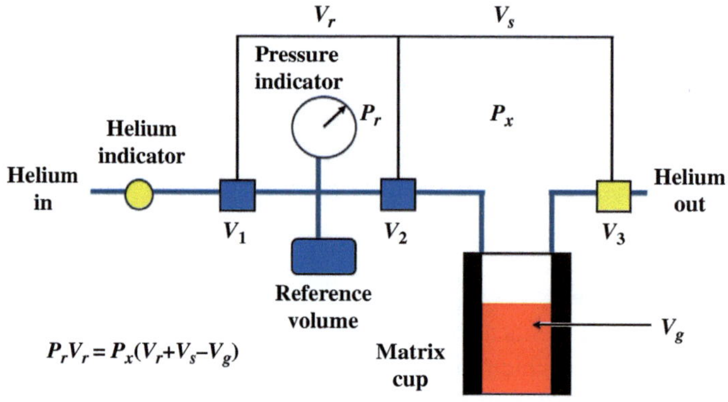

Figure 11.8 Helium grain volume instrument schematic (matrix cup filled with plug).

Now, when a core sample is placed in the matrix cup (illustrated in Figure 11.8), the cup's volume (V_s) is reduced by an amount equal to the grain volume of the sample, (V_g). Any pores in the sample fill with helium as the gas expands. When this expansion process is repeated with the sample in place, Boyle's law is modified as follows:

$$P_tV_t = P_x(V_r + V_s - V_g)$$

By rearranging this equation, we can solve for the grain volume (V_g) as:

$$V_g = \frac{P_x(V_r + V_s) - P_tV_t}{P_x} \tag{11.35}$$

Therefore, with known values of V_r and V_s, and by measuring the initial reference pressure and the expanded pressure, the grain volume (V_g) can be calculated. The values of V_r and V_s are determined during calibration experiments, where solid, nonporous steel, or metal billets of known bulk volume are placed in the matrix cup for reference.

11.6.1.3 Test Procedures

To begin the grain volume test, the core sample is placed into the matrix cup, which is chosen based on the sample's dimensions. If the plug is too short to fill the matrix cup entirely, billets of known grain (or bulk) volume are added to ensure the cup is properly filled. For optimal accuracy and repeatability, any residual air in the system or sample must be removed. This is done by flushing the system with helium either by running the expansion process multiple times or by pulling a vacuum on the matrix cup and then flushing it with helium.

The test proceeds with the matrix cup vented to the atmosphere and valves V_2 and V_3 closed. Next, valve V_1 is opened, allowing helium to fill the reference volume, which is typically regulated to about 100 psi. After closing valves V_1 and V_3 to isolate the reference volume, the pressure in the reference volume is allowed to stabilize. Once stable, valve V_2 is opened, allowing the helium to expand into the matrix cup, and the expanded pressure is measured until it reaches a steady state. This process may take only seconds for high-permeability samples but can take several minutes for low-permeability samples. The operator must ensure the system is completely sealed and leak-free during this process.

Temperature control is critical for accurate measurements. Grain volume measurements should be conducted at the same temperature as the instrument's calibration temperature (within 1 °C). Although temperature variations can theoretically be accounted for using **Charles's law** (which states that $\frac{PV}{T}$ is constant), differences in material expansion or contraction in the instrument's pipework and reference volumes can introduce errors if the temperature fluctuates.

Instrument calibration must be verified using known reference billets or check plugs before, during, and after each set of plug measurements. If any inconsistencies arise during the test, the instrument should be recalibrated to maintain accuracy.

11.6.1.4 Grain Volume and Grain Density Calculation

The grain volume (V_g) is calculated using Formula (11.35), derived from Boyle's law.

Grain density (ρ_g) is then calculated by dividing the dry weight (W_d) of the sample by the measured grain volume:

$$\rho_g = \frac{W_d}{V_g} \tag{11.36}$$

Advantages of the method:

- This method provides relatively fast measurements for most samples.
- It is a relatively inexpensive technique for porosity analysis.
- With proper control, the measurements are both accurate and reproducible.
- The sample remains intact, allowing for further tests if necessary.
- This method accommodates a wide range of sample shapes, including irregular rock chips and sand.

Drawbacks and issues:

- Stabilizing pressure for low-permeability samples can be slow, so sufficient time must be allocated.
- Smaller samples tend to have larger measurement errors compared to larger samples.
- If the sample is not properly dried, residual fluids can suppress the grain density measurements, leading to inaccuracies.

11.6.2 Helium Pore Volume

11.6.2.1 Sample Preparation

For pore volume measurements using helium, sample preparation is critical to ensure accurate and reliable results. Samples must be thoroughly cleaned of any fluids and dried using appropriate methods that align with the intended application of the porosity data. Unlike grain density measurements, the shape of the sample is highly important in pore volume determination. The sample should be as close to a perfect right cylinder as possible.

Irregularly shaped plugs, or those with rounded ends, fractures, or vugs (large cavities in the rock) that are not properly sealed during the measurement, lead to significant inaccuracies. Larger cylindrical plugs are preferred to reduce uncertainty in porosity calculations.

A common practice is to combine helium grain volume and helium pore volume measurements for porosity determination. This method is particularly useful for unconsolidated or weak sands, as well as vuggy carbonates, where mercury intrusion might pose a hazard or could potentially invade the pore system and affect the measurement. This approach provides a safer and noninvasive alternative for fragile or complex samples (Table 11.6).

11.6.2.2 Test Equipment for Helium Pore Volume

Helium pore volume is measured using a modified version of the twin-cell helium porosimeter, which is also used for determining grain volume. The primary difference lies in how the sample is housed during testing. Instead of a matrix cup, the sample is held within a rubber sleeve inside a Hassler-type coreholder, as illustrated in Figure 11.9. To ensure that the sleeve properly conforms to the surface of the plug (surface conformance), a moderate confining pressure is applied. In some cases, the test may be conducted under elevated stress conditions to simulate reservoir stress levels.

Table 11.6 Helium pore volume—advantages and drawbacks.

Advantages	Drawbacks and issues
Relatively low cost	Errors in small sample measurements tend to be larger than those in larger sample measurements.
Fast and convenient: Can be run at elevated stress to study porosity as a function of stress.	Inefficient drying of free fluids may suppress pore volume readings.
Nondestructive: Samples can be reused for further testing (unless stress-sensitive)	The method depends on the sample shape. Irregular shapes, rounded ends, or unsealed fractures, and vugs may yield inaccurately high pore volume values.
	For unconsolidated samples, dead volume between the sample and protective jacket, and the pore volume of the end screens, must be considered in measurements.

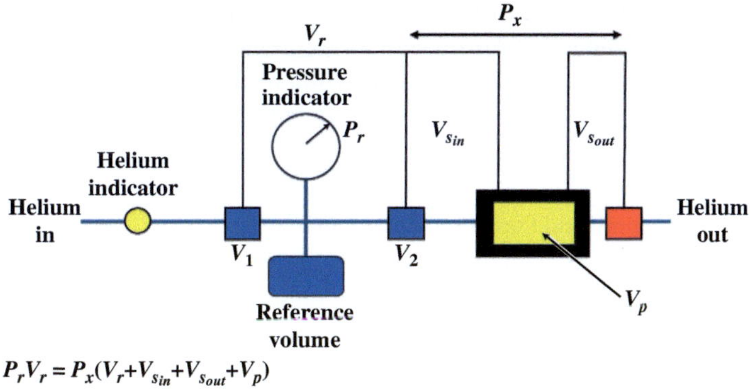

Figure 11.9 Helium pore volume porosimeter schematic.

It's important to note that, unlike helium grain volume or mercury bulk volume porosity measurements, the pore volume depends on the amount of confining stress applied. This variability should be considered when comparing results across different systems or laboratories, as stress differences can impact the measurements.

Automated porosity measurement systems often base their calculations on the combination of pore volume and grain volume.

In this setup (see Figure 11.9), Boyle's law is applied as:

$$P_rV_r = P_x(V_r + V_{S_{in}} + V_{S_{out}} + V_p)$$

Rearranging and solving for V_p (pore volume):

$$V_p = \frac{P_rV_r - P_x(V_r + V_{S_{in}} + V_{S_{out}})}{P_x} \tag{11.37}$$

The reference volume (V_r), inlet volume ($V_{S_{in}}$), and outlet volume ($V_{S_{out}}$) are determined through calibration procedures similar to those used for grain volume measurements. Hollow metal cylinders with known pore volumes are employed for this calibration.

Care should be taken when interpreting pore volume results, as the technique measures all empty spaces between the inlet and outlet platens of the coreholder. This includes the dead volume of the platens and any empty annulus space between the sample and the coreholder's internal sleeve, which may affect the accuracy of the measurement.

11.6.2.3 Test Procedures for Helium Pore Volume

The sample is placed inside the coreholder and subjected to the required confining stress. To eliminate any air, the system and the sample are flushed with helium, either by conducting a "dummy" expansion process or by first drawing a vacuum and then flushing with helium. Once the system is ready, the reference volume is filled with helium to a pressure of approximately 100 psi or 200 psi. After stabilization, helium is allowed to expand into the coreholder, and the pressure is monitored until stable, a process that can take several minutes for lower-permeability samples.

As with grain volume measurements, the pore volume measurements must be conducted at the same temperature as the instrument calibration, with any deviation kept within 1 °C. Calibration of the instrument should be confirmed using hollow billets or reference plugs before, during, and after the measurements. If needed, the instrument is recalibrated to ensure accuracy.

11.6.2.4 Pore Volume and Porosity Calculation

The pore volume (V_p) is calculated using Eq. (11.37).

To determine porosity (φ) when combining helium pore volume with grain volume, the following formula is used:

$$\varphi = \frac{V_p}{(V_p + V_g)} \tag{11.38}$$

where:

- V_p is the pore volume, and
- V_g is the grain volume.

11.6.3 Bulk Volume

In RCA, three primary methods are commonly used to determine the bulk volume (V_b) of core samples:

- Caliper system
- Mercury immersion system
- Mercury pycnometer

1. **Caliper System**
   The caliper method measures the length and diameter of the core sample to calculate its bulk volume. This approach typically overestimates bulk volume because real core plugs are rarely perfect right cylinders. This method is only recommended for ambient condition measurements on samples where **mercury** may enter the pore space, such as in vuggy carbonates. In these cases, it is advised to use the largest possible sample, such as whole core samples, to reduce the impact of measurement inaccuracies.
2. **Mercury Immersion System**
   The mercury immersion method leverages the fact that mercury is a **high-density, nonwetting liquid** that does not spontaneously enter the pores of the core sample unless pressure is applied. This allows the bulk volume to be calculated based on the displacement of mercury by the sample.
   Key factors:
   - The **air in the pore volume and the grain volume** displaces an equivalent volume (or weight) of mercury.
   - Mercury does **not enter the pore space** without applied pressure, meaning that the bulk volume can be accurately measured as long as mercury penetration into the pore system is avoided.

- If mercury penetrates the sample's pores, it can result in an underestimation of bulk volume. Moreover, some mercury may get trapped during withdrawal, rendering the samples **unsuitable** and **hazardous** for further testing.

Precautions:

- Mercury must be free from **rock dust** or other contaminants, which could lead to **surface wetting** and penetration into the pores.
- The **largest pore throat radius** of the sample's true pore system must be sufficiently large to prevent mercury penetration.
- Samples with **surface vugs, fractures, or coarse grain fabrics**, or those sleeved in **metal/Teflon** (e.g., unconsolidated samples), should not be measured with this method, as mercury will penetrate, resulting in a lower bulk volume.

For these types of samples, the **caliper method** or using a fluid of known density to **saturate the pore space** is recommended to determine bulk volume.

3. **Mercury Pycnometer**

The mercury pycnometer uses the **displacement principle** of mercury without applying significant pressure to the sample. This method is advantageous when the sample is small or has an irregular shape, but caution must be exercised to avoid mercury penetration into the pore system. This method is widely used due to its **accuracy and repeatability** under the right conditions.

Calculation of Pore Throat Radius During Mercury Immersion

The depth of mercury immersion affects the **pore throat radius** that mercury can penetrate. The following formula can be used to calculate the pore throat radius (r) based on the depth of immersion:

$$r = \frac{c 2\sigma \cos\theta}{\left(\rho_{Hg} - \rho_{air}\right) g H} \tag{11.39}$$

where:

- r = Pore throat radius (cm)
- ρ_{Hg} = Density of mercury (g/cc)
- ρ_{air} = Density of air
- θ = Mercury contact angle (°)
- σ = Air–mercury IFT (dyne/cm)
- g = Gravitational constant
- H = Depth of immersion (cm)
- c = Units constant

For a mercury density of 13.6 g/cc, the pressure at the base of a 1.500 cm diameter and 300 cm long plug immersed vertically in mercury is about 1.5 psi. This pressure would allow mercury to fill **pore bodies connected to pore throats** larger than a **70 μm radius**. Therefore, if the pore throat radii are smaller than this, mercury will **not invade** the pore system.

However, if the core plug is held horizontally, the **minimum pore throat radius** that mercury can invade increases to **140 μm**.

11.6.3.1 Sample Preparation

When using mercury in core analysis, certain precautions must be strictly followed due to the highly **toxic** nature of mercury vapor. Many labs have banned its use and even those that allow it enforce stringent safety measures. Proper management includes:

- Mercury vapor monitoring systems
- Suitable extraction systems
- Employee health monitoring and surveillance protocols

Under these conditions, mercury immersion methods can be used regularly. Ideally, all mercury-related equipment should be contained within an isolated and chilled room or a fume cupboard to minimize the risk of exposure.

Sample Shape and Size

Unlike grain volume measurement, the shape of the test sample is not critical for mercury bulk volume measurements. However, regular right cylindrical plugs are still preferred for consistency and accuracy. Larger plugs are also preferred to reduce the uncertainty in porosity data.

Unsuitable Samples

The method is unsuitable for samples with:

- Surface vugs
- Fractures
- Very coarse-grained rock fabrics

These samples can result in mercury intrusion into the pore volume, leading to an **underestimation of bulk volume**. For accurate measurements, it is crucial that mercury does not enter the sample's pores.

Sleeved Samples

For samples sleeved in metal or PTFE, it can be challenging to differentiate between false volume and pore volume, making this method unreliable for such samples. As a result, this method is not recommended when samples are sleeved in materials that obscure clear volume measurements.

Weight Check for Contamination

The sample dry weight should be measured both before (W_{di}) and after (W_{df}) the mercury immersion test. This helps ensure that no mercury has contaminated the sample. Any increase in weight after testing indicates possible mercury contamination, which would invalidate the results and require sample disposal or further testing.

11.6.3.2 Test Equipment—Mercury Pycnometer

A mercury pycnometer (illustrated in Figure 11.10) is a device used to measure the bulk volume of core samples by determining the volume of mercury required to fill the space around the sample. All pycnometers share two main components:

1. **Pressure chamber**: This accommodates the sample during the test.
2. **Mercury pump**: Equipped with a known bore and piston diameter, this pump injects mercury into the chamber. The amount of mercury injected is measured either with a digital readout or an analog readout using a scale and Vernier.

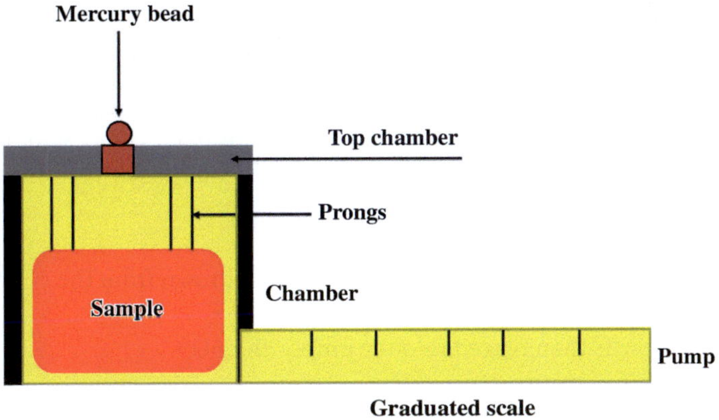

Figure 11.10 Typical mercury pycnometer schematic.

The sample chamber has a small hole at the top to ensure that, once the chamber is filled, a small bead of mercury emerges and touches a needle datum. To ensure the sample remains submerged in mercury and does not block the exit hole, steel forks or prongs are often used to counteract the buoyancy force of the mercury (Table 11.7).

Table 11.7 Mercury bulk volume: advantages and drawbacks.

Advantages	Drawbacks and issues
Relatively cheap method	Unsuitable for samples with surface vugs, fractures, coarse-grained fabric, or samples sleeved in metal/Teflon (unconsolidated samples). Mercury can enter the pores, leading to an underestimation of bulk volume.
Fairly rapid for most samples	Using incorrect mercury density for the measurement temperature can introduce errors. A temperature variation of 5 °C can cause a 0.02% systematic error in bulk volume.
Minimizes immersion depth if horizontal immersion is possible: This reduces the chance of mercury entering sample pores since the pressure acting on the sample base is minimized.	Air bubbles can adhere to the sample, causing an overestimation of bulk volume.
Most accurate method for suitable samples	Small sample measurements may have larger errors than those for larger samples.
	Mercury must be kept scrupulously clean; contamination can reduce surface tension and allow mercury to enter pores. Cleaning with acetic acid is common practice.
	Potential mercury contamination can invalidate further testing, especially for electrical property measurements.

11.6.3.3 Test Procedures—Mercury Pycnometer

The operation of the mercury pycnometer follows these steps:

1. Initial Filling of the Empty Chamber:
   - The mercury pump is used to inject mercury into the empty sample chamber.
   - Mercury is pumped until it just appears at the hole at the top of the chamber.
   - The pump is then zeroed.
2. Withdrawal of Mercury:
   - Once the chamber is full, the mercury is withdrawn from the chamber.
3. Sample Placement:
   - The sample is then placed into the empty chamber.
4. Refilling the Chamber:
   - Mercury is again injected into the chamber until it just appears at the top hole, reaching the same level as before.
5. Volume Calculation:
   - The amount of mercury required to reintroduce to the same level as the initial step represents the bulk volume (V_b) of the sample.

11.6.3.4 Test Equipment—Mercury Immersion System

The mercury immersion system (illustrated in Figure 11.11) is another method used to determine the bulk volume of a sample by immersing it in mercury. The system consists of the following key components:

1. Mercury bath: A plastic or metal container filled with mercury, designed to hold the sample during the immersion process.
2. Electronic balance: A balance equipped with a tare (zero) facility that allows precise measurements of the mass. The mercury bath is placed on this balance to measure the changes in weight as the sample is immersed.
3. Sample cradle and immersion mechanism: A cradle holds the sample, and a screw-type immersion mechanism is used to lower the sample into the mercury bath gradually and in a controlled manner.

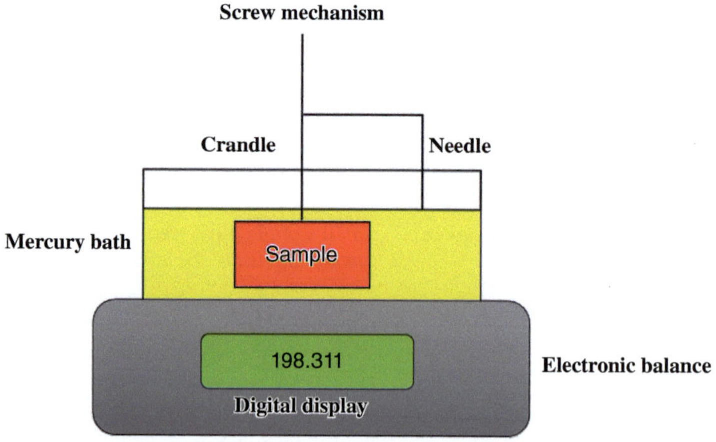

Figure 11.11 Typical mercury immersion system schematic.

4. Indicator system: A needle connected to a DC battery and bulb serves as an indicator. The system is designed to signal when the needle touches the mercury's surface, ensuring proper immersion.
5. Temperature probe: Used to monitor the temperature of the mercury as temperature fluctuations can affect the density of mercury and the accuracy of the measurement.

11.6.3.5 Test Procedures—Mercury Immersion System

The procedure for determining bulk volume using a mercury immersion system involves the following steps:

1. Initial setup: The sample cradle is lowered into the mercury bath up to a predefined reference mark, and the balance is tared (zeroed).
2. Immersion: The cradle is withdrawn, and the sample is placed in the cradle. It is then carefully re-immersed to the same reference mark. The balance measures the increase in weight, referred to as the immersed weight (W_{imm}).
3. Bulk volume calculation: The bulk volume of the sample is calculated using the formula:

$$V_b = \frac{W_{imm}}{\rho_{Hg}} \tag{11.40}$$

   where:

   - V_b is the bulk volume of the sample,
   - W_{imm} is the immersed weight, and
   - ρ_{Hg} is the density of mercury at the test temperature.

Key Considerations

- The possibility of mercury entering the sample must be carefully monitored, as it can result in underestimating bulk volume and contaminating the sample.
- For samples with large surface pores or unconsolidated sands, this method is unsuitable due to the risk of mercury intrusion.
- To ensure accurate results and avoid contamination, permeability is often measured first, with mercury bulk volume being the final measurement in the RCA porosity and permeability test sequence.
- Some laboratories now prefer helium pore volume and helium grain volume methods, as these eliminate mercury-related health and safety risks. However, helium-based measurements can increase uncertainty in porosity calculations.
- Laboratory selection should include a review of mercury handling protocols, ensuring all necessary HSE procedures are in place.

11.6.4 Liquid Saturation Porosity

The liquid saturation porosity method determines the porosity of a sample that has been saturated and immersed in a liquid of known density. Although this method is not commonly used in RCA programs due to its relative expense and time consumption, it remains valuable in certain contexts, particularly for SCAL tests.

Historically, toluene was used to saturate samples, but modern practices typically employ helium porosity methods. However, SCAL tests, such as electrical resistivity, capillary pressure, wettability, and relative permeability measurements, often require samples to be saturated with formation water (FW). In these instances, re-saturation porosity is a more accurate and reliable method compared to helium grain volume or mercury bulk volume measurements. In particular, for electrical property measurements, where mercury contamination could interfere, re-saturation porosity is the preferred method (Table 11.8).

This method is based on the sample's pore space being completely saturated with brine of known density. The total volume of brine (occupying both the pore space and grain volume) displaced when the sample is immersed in a brine of the same density gives the measurement of the sample's porosity.

11.6.4.1 Sample Preparation

Prior to the liquid saturation porosity test, samples must undergo thorough cleaning and drying. The extraction of fluids must be complete, and proper drying methods must be followed to ensure accurate results. It is advisable to measure the porosity of the sample using one of the methods outlined earlier (such as helium porosity) to cross-check the accuracy of the re-saturation porosity measurement. Special care

Table 11.8 Re-saturation porosity: advantages and drawbacks.

Advantages	Drawbacks and Issues
High measurement accuracy: Since the test is based on weight measurements rather than volume, it can achieve higher accuracy, provided there is no grain loss.	**Brine density sensitivity**: Using the correct brine density at the test temperature is crucial. A temperature variation of 5 °C can introduce a **0.02% error** in bulk volume.
Nondestructive method: The method is nondestructive, allowing samples to be used for further tests.	**Expensive and time-consuming**: This method is more expensive and time-intensive than other porosity methods.
Integration with SCAL experiments: Saturating the sample in FW is the first stage of many SCAL tests, making the method useful.	**Full saturation requirement**: The sample must be fully saturated with brine, which can be challenging.
	Air bubbles: Air bubbles adhering to the sample surface may result in an overestimation of bulk volume.
	Sample size sensitivity: Smaller samples are more prone to measurement errors.
	Friable samples: Errors can occur in more friable samples due to grain loss during testing.
	Unconsolidated/sleeved samples: Difficult to use on unconsolidated samples unless the sleeve material's weight and volume are fully accounted for.

should be taken when using mercury bulk volume methods due to the potential risk of contamination, which could affect subsequent testing.

11.6.4.2 Test Equipment

The pore volume of the sample is measured by calculating the weight difference between the saturated weight and the dry weight of the plug using an electronic balance. The bulk volume is determined using the Archimedes immersion principle, where the sample is suspended in a liquid and the volume is calculated based on the displacement of the liquid. The experimental setup is similar to that shown in Figure 11.12b. The electronic balance should feature a weighing hook underneath, allowing the core cradle to be immersed in the same brine (synthetic formation water, SFW) used for saturation.

11.6.4.3 Test Procedures

1. Measuring dry weight: The dry weight of the sample (W_d) is measured using the electronic balance.
2. Sample saturation: The sample (plug) is placed in a vacuum desiccator and saturated in deaired SFW. For low-permeability samples, the plugs are loaded into a pressure vessel and pressure-saturated in SFW at approximately 2,000 psi for 2–3 days to fully saturate the pore space.
3. Saturated weight: Once saturation is complete, the saturated weight (W_s) of the sample is measured by placing it on top of the balance (Figure 11.12a).
4. Bulk volume measurement: The bulk volume is determined by immersing the plug in the brine (Figure 11.12b):
   - The plug is suspended under the weight balance using a cradle attached to the underside of the balance.
   - The saturated weight of the plug is first recorded on top of the balance.
   - The core cradle is then lowered (or the brine container raised) until a reference mark on the wire is reached, ensuring consistent immersion.

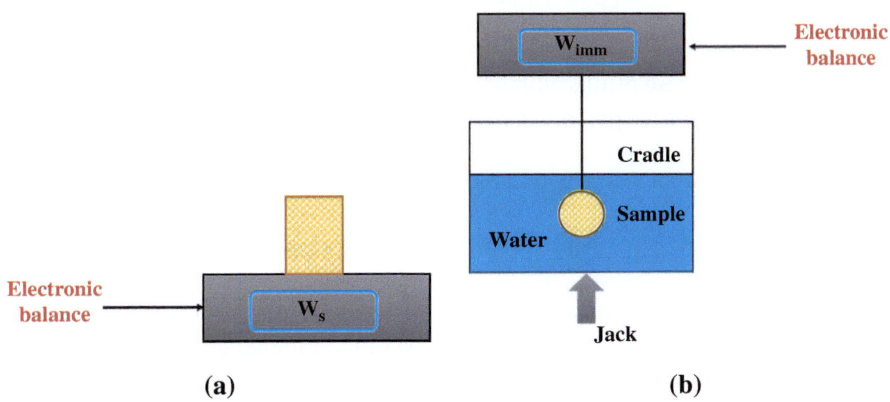

Figure 11.12 Archimedes brine immersion system: (a) saturated weight (pore volume); (b) immerse weight (bulk volume).

- The balance is tared to account for the displacement caused by the cradle.
- The cradle is removed, and the plug is placed inside. The cradle is then re-immersed in the SFW to the same reference mark on the wire, and the immersed weight (W_{imm}) is recorded.

11.6.4.4 Porosity Calculation

To determine the re-saturation pore volume (V_{ps}) and the immersed bulk volume (V_{bimm}), the following equations are used:

1. Re-saturation pore volume (V_{ps}):

$$Vp_s = \frac{W_s - W_d}{\rho_{FW}} \tag{11.41}$$

where:

- W_s = Saturated weight of the sample
- W_d = Dry weight of the sample
- ρ_{FW} = Density of the FW (brine)

2. Immersed bulk volume (V_{bimm}):

$$Vb_{imm} = \frac{W_s - W_{imm}}{\rho_{FW}} \tag{11.42}$$

where:

- W_{imm} = Immersed weight of the sample in brine
- ρ_{FW} = Density of the FW (brine)

3. Re-saturation porosity (φ):

$$\varphi = \frac{Vp_s}{Vb_{imm}} \tag{11.43}$$

where:

- V_{ps} = Re-saturation pore volume
- V_{bimm} = Immersed bulk volume

Cross-checking these results with helium porosity measurements ensures consistency and accuracy in the porosity determination process.

11.6.4.5 Re-saturation Porosity Quality Control Issues, Checks, and Diagnostics

- The re-saturation porosity should be within 0.2 porosity units (p.u.) of the helium porosity.
- If the re-saturation porosity is too low compared to the helium porosity, it may indicate that not all of the pore space was fully saturated with SFW.
- If the re-saturation porosity is too high, it may suggest that the sample has experienced grain loss during the procedure, potentially affecting the accuracy of the porosity measurement.

11.6.5 Accuracy and Repeatability of Porosity Measurements

The accuracy and reproducibility of porosity measurements are crucial, especially for comparing results between laboratories and ensuring consistency across methods. API RP 40, though withdrawn in 2021, established important guidelines for helium porosity measurements with a reproducibility error of 0.5 porosity units (p.u.) between laboratories. Most reputable laboratories still aim for a stricter repeatability of 0.25 p.u. This standard remains a useful reference for setting expectations in core analysis.

Factors Affecting Accuracy:

1. Sample size: Smaller plugs tend to have larger errors in porosity measurements compared to larger plugs.
2. Helium pore volume/caliper bulk volume: If the plugs are not perfect cylindrical shapes, errors in helium pore volume or bulk volume measured by calipers can increase.
3. Grain density/matrix density (ρ_m): This is essential for calibrating wireline density logs, which calculate porosity from bulk density (ρ_b) and fluid density (ρ_f). Errors in grain density will result in errors in log porosity calculations.

The formula used for log porosity is:

$$\varphi = \frac{\rho_m - \rho_b}{\rho_m - \rho_f} \tag{11.44}$$

1. Grain density error (Figure 11.13): For low porosity formations, an error of 0.03 g/cc in grain density corresponds to an error of 2 p.u. in porosity. Lower porosity samples are more sensitive to grain density errors, while higher porosity samples experience less error.

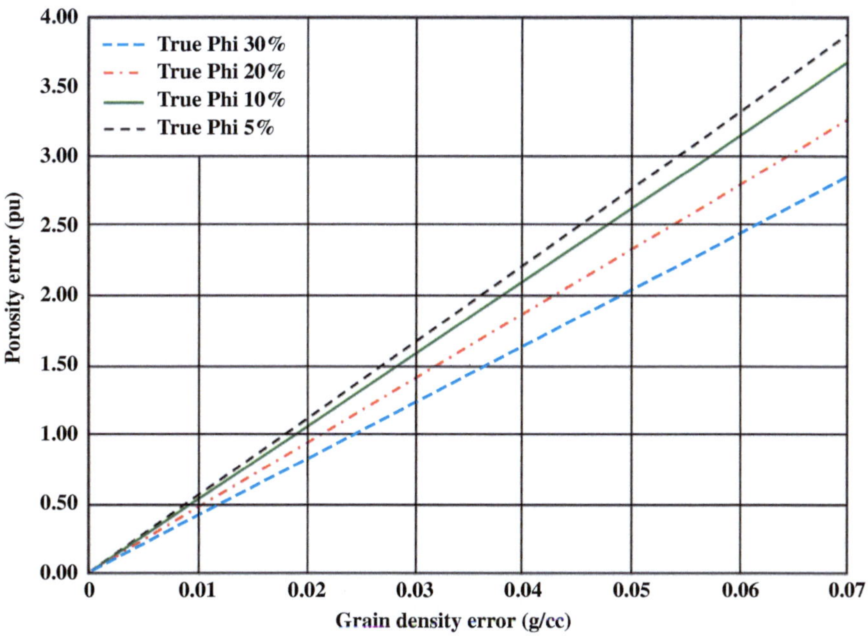

Figure 11.13 Grain density errors in density porosity calculations.

2. API standard for grain density: Errors around 0.01 g/cc (the API standard) typically result in a porosity error of less than 0.5 p.u. for most reservoir rocks. Adopting a stricter standard of 0.25 p.u. requires a grain density standard error of 0.005 g/cc.

Core and Log Measurement Discrepancies:

Core porosity measurements made under ambient conditions in RCA programs often cannot be directly reconciled with log measurements, due to:

1. Stress effects: Logs respond to bulk densities at *in situ* stress conditions, while core measurements are made at ambient conditions.
2. SCAL testing: Stress effects on core porosity must be accounted for in SCAL programs.

11.7 Wettability and Wettability Tests

Wettability is a key property that determines how reservoir rock components interact with pore-filling fluids. It influences the location and behavior of fluids within the porous rock structure and affects how fluids flow during the production process. In laboratory experiments that aim to simulate reservoir conditions, wettability plays a crucial role, especially in capillary pressure and relative permeability tests.

Ensuring that core samples maintain reservoir-representative wettability during lab analyses is critical, as inaccurate wettability data could render experimental results unreliable. However, measuring wettability downhole remains a challenge, and current laboratory techniques can only assess the wettability of a sample at the time of measurement. Various factors during the extraction and handling process can alter the natural wettability, including:

- contact with drilling mud,
- pressure and temperature changes during core recovery,
- oil oxidation due to inadequate core preservation,
- core cleaning processes,
- aging of the core sample, and
- core testing.

These factors can modify the original wettability, leading to potential deviations from the true reservoir conditions.

Wettability States

In a typical rock/oil/brine system, different wetting states are observed, including:

1. Water wet: Water coats the rock grains and occupies the smaller pores, while oil fills the larger pore centers.
2. Oil wet: Oil coats the grains and occupies the smaller pores, with water confined to the larger pore spaces.
3. Neutrally wet: The rock exhibits no preference for either oil or water. This state can be further divided into:
   - Fractional wettability: Different areas of the rock show distinct preferences. Some minerals prefer water, while others favor oil (e.g., calcite-cemented quartz sandstone or clay-lined pores).

- Mixed wettability: A specialized form of fractional wettability where oil coats the larger pore surfaces, while water occupies the smaller pores. Though not definitively proven, this state allows for very low residual oil saturations and can explain unique relative permeability behaviors.

Both fractional and mixed wettability suggest that wettability varies at the pore level, potentially at the micron scale.

Wettability Measurement Methods

Several methods exist to quantitatively determine wettability in the rock/fluid system, with the most common being:

1. Contact angle method: This technique measures the wettability of a specific mineral surface. However, its relevance to the reservoir as a whole may be limited, as it focuses on individual surfaces rather than the entire system.
2. Amott method (or Amott–Harvey): Measures the average wettability of a core sample, providing a bulk assessment.
3. USBM method: Another method for measuring the bulk wettability of a core sample, similar to the Amott method in its approach.

The contact angle method is limited in its application to the broader reservoir formation, while the Amott and USBM methods offer insights into the overall wettability of the sample.

11.7.1 Contact Angle Method

The contact angle method is advantageous in wettability testing because it eliminates the potential for mud filtrate, surfactants, or other compounds to alter the wettability of the mineral surface. It is particularly useful in determining if crude oil can change wettability and in studying the effects of temperature, pressure, and brine chemistry on wettability. However, there are challenges in applying this method directly to reservoir conditions due to the complexities of replicating reservoir environments in the laboratory.

Methods Used

1. **Sessile drop method**: A drop of liquid (oil or brine) is placed on a flat, solid surface, and the angle formed at the contact point between the liquid, solid, and air (contact angle) is measured. This angle indicates the wetting preference of the solid.
2. **Modified sessile drop method**: A variation of the sessile drop method tailored to better simulate reservoir conditions, adjusting for factors such as temperature and pressure.

Purpose of the Test

The contact angle method measures the wetting preference of a solid by a liquid, offering insight into whether a surface is water-wet, oil-wet, or neutrally wet.

Sample Requirements

Two flat polished mineral surfaces* are needed, either made of:

- Quartz (for sandstone)
- Calcite (for carbonate formations)

Required Input Data for Laboratory Testing

1. SFW composition to replicate brine chemistry
2. A sample of reservoir oil to examine how it interacts with the mineral surface

11.7.1.1 Sample Preparation for Sessile Drop Method

In the sessile drop method, a single flat, polished mineral surface is used (Figure 11.14), typically made of quartz (for sandstones) or calcite (for carbonates). The modified sessile drop method involves two such polished mineral surfaces mounted parallel on adjustable posts, simulating reservoir conditions more closely by incorporating dual-surface interactions.

11.7.1.2 Equipment Setup

1. Mineral crystal surfaces: Prepared quartz or calcite crystals simulate the behavior of sandstone and carbonate reservoirs, respectively.
2. Mounting on adjustable posts: In the modified sessile drop method, these surfaces are mounted on posts to ensure accurate parallel alignment (Figure 11.15).
3. Contact angle goniometer: This instrument measures the contact angle by capturing the profile of the liquid droplet on the solid surface using an optical system. Modern systems use high-resolution cameras and advanced software to analyze contact angles (Figure 11.16).

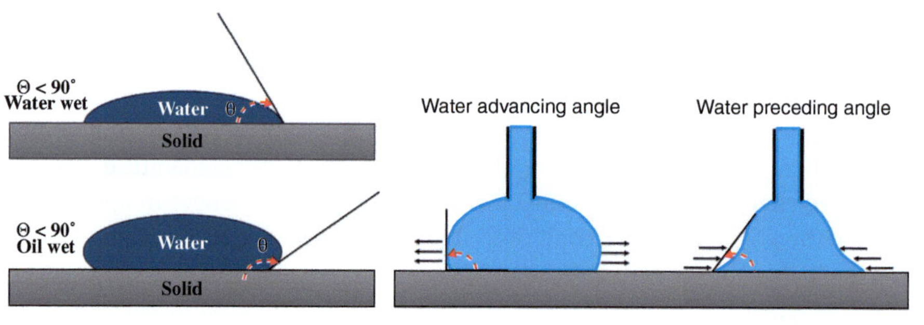

Figure 11.14 Sessile drop method.

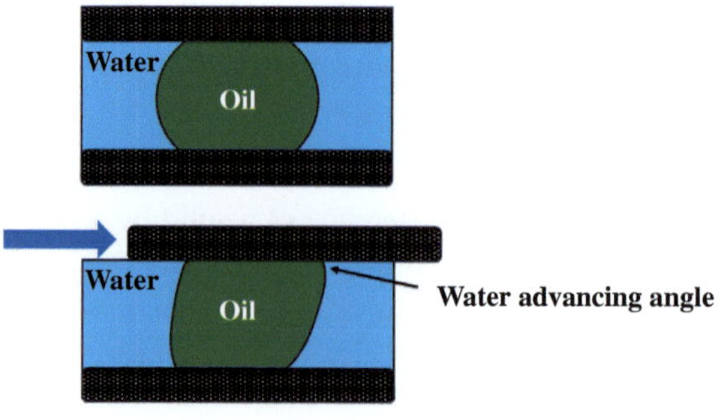

Figure 11.15 Modified sessile drop method.

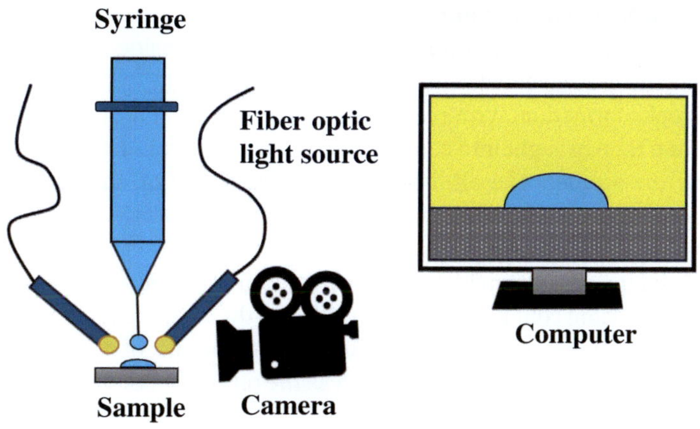

Figure 11.16 Goniometer experimental setup.

Advancing and Receding Contact Angles

1. Advancing contact angle (θ_a): This angle is measured during imbibition when water displaces oil. It tends to be larger due to the presence of hydrophobic domains on the crystal surface, which can obstruct the movement of the contact line as the liquid advances.
2. Receding contact angle (θ_r): This angle is observed during drainage when oil displaces water. Hydrophilic domains resist the motion of the receding liquid, resulting in smaller contact angles.

Heterogeneity Effects

Surface heterogeneity, such as chemical domains on the crystal, plays a role in the difference between advancing and receding angles. For instance:

1. Hydrophobic domains: These areas resist the advancing liquid, increasing the contact angle.
2. Hydrophilic domains: These areas slow down the receding motion, decreasing the contact angle.

11.7.1.3 Test Procedures

The procedure for measuring contact angles using the sessile and modified sessile drop methods involves careful preparation, ensuring that both the mineral surfaces and liquids used are as clean and uncontaminated as possible. Here are the key steps involved:

Preparation and Contamination Control

1. Cleaning: All equipment, including the cell and capillary tubes, must be thoroughly cleaned before testing. Even trace amounts of contaminants or the presence of an oxide layer on the solid surface can alter the contact angle significantly.
2. Purified liquids: Purified, deoxygenated synthetic formation brine and crude oil are used to ensure reproducibility of results, typically within a few degrees.

Sessile Drop Method Procedure

1. Setting up: The cleaned mineral surface is placed inside a test cell, which is filled with deoxygenated synthetic formation brine.
2. Oil droplet formation: A droplet of crude oil is formed at the tip of a fine capillary tube and then brought into contact with the flat mineral surface (Figure 11.14).
3. Aging the droplet: The oil droplet is allowed to age on the surface to ensure proper interaction between the oil and the mineral surface.
4. Measuring contact angles:
   - Advancing contact angle: By increasing the droplet size with the capillary tube, the advancing contact angle is measured as the oil spreads across the surface.
   - Receding contact angle: By reducing the volume of the oil droplet, the receding contact angle is measured as the oil contracts.

Modified Sessile Drop Method Procedure

1. Dual surface setup: Two flat mineral surfaces (e.g., quartz or calcite) are mounted parallel to one another on adjustable posts inside a test cell filled with synthetic formation brine (Figure 11.15).
2. Placing the oil droplet: An oil droplet is placed between the two surfaces, ensuring that it contacts a large area of each crystal.
3. Aging process: The oil/crystal interface is allowed to age for 2–3 days to stabilize the contact.
4. Crystal displacement: The two crystals are displaced parallel to each other to shift the oil droplet, allowing the brine to move over the previously oil-covered surface.
5. Measuring water-advancing contact angles: The water-advancing contact angles are measured after moving the oil droplet. Initially, a nonequilibrium angle is observed, which decreases over time until it reaches a constant value.

Equilibrium Time and Surfactant Effects

1. Aging time: When crude oil contains natural surfactants, the water-advancing contact angle tends to increase as the oil/crystal interface ages. This process continues until equilibrium is reached, which can take hundreds or thousands of hours of interface aging time.
2. Surfactant influence: The presence of surfactants influences the wettability as they adsorb onto the mineral surface, increasing the contact angle.

11.7.1.4 Results

The results of contact angle experiments are reported as an angle (θ), which correlates to the wettability conditions of the system. As shown in Table 11.9, certain contact angle ranges are indicative of water-wet, neutrally wet, or oil-wet

Table 11.9 Contact angle wettability classification.

	Water wet	Neutrally wet	Oil wet
Minimum	0°	60–75°	105–120°
Maximum	60–75°	105–120°	180°

conditions. The contact angle values help in classifying the wettability of the sample and can be compared to cut-off values similar to those identified by Treiber et al., 1972.

11.7.1.5 Data Reporting Requirements

For accurate and comprehensive data reporting from contact angle experiments, the following information should be provided:

1. Procedures and apparatus: A brief description of the experimental methods and the equipment used.
2. Mineral type: Identification of the type of mineral tested (e.g., quartz for sandstone, calcite for carbonate, or other minerals).
3. Test conditions: The temperature and pressure under which the tests were conducted.
4. Oil properties: API gravity and oil density at test conditions.
5. Water properties: Composition and density of the SFW at test conditions.
6. IFT: Oil/water IFT at test conditions, if available.
7. Measured contact angles: Both advancing and receding contact angles (for the modified sessile drop method).
8. Contact angle over time: The change in contact angle as a function of time to establish equilibrium.

11.7.1.6 Contact Angle Summary

Table 11.10 summarizes key points regarding contact angle measurements, such as the fluid systems and lithologies these measurements apply to, and the pros and cons of the method.

QC Checks

To ensure the validity and reliability of the results, the following quality control (QC) checks should be conducted:

1. Review laboratory report: A thorough examination of the lab report to verify that the procedures followed are acceptable. The proper methodology should be adhered to at each step.
2. Limitations: Recognize that this method only tests the interaction of fluids with mineral surfaces, without replicating the full complexity of formation pore systems. The results are useful for aging requirements but should not be directly extrapolated to estimate formation wettability.

Suitable Laboratories

Contact angle tests are typically conducted in the following settings:

1. PVT laboratories: These labs specialize in pressure, volume, and temperature (PVT) analyses and often perform contact angle measurements.
2. University petroleum engineering departments: Many academic departments with petroleum engineering programs also have the capability to carry out these tests.

Table 11.10 Summary of contact angle measurements.

Advantages	Drawbacks and issues
• **Low-cost technique**: It is a relatively inexpensive method for determining wettability.	• **Equilibrium delay with crude oil**: Contact angles can take a significant time to reach equilibrium, particularly when crude oil is used due to the presence of surfactants. Early measurements may indicate water-wet conditions, while equilibrium might show oil-wet conditions.
• **Pure fluids and rock crystals**: Best method for measuring wettability with pure fluids and pure rock crystals as it avoids interference from surfactants and other compounds. Ideal for research and development studies.	• **Contact angle hysteresis**: The difference between advancing and receding contact angles (hysteresis) can be significant, sometimes exceeding 60°. Hysteresis results from surface roughness, composition heterogeneity, and immobility at the macromolecular level.
• **Wettability restoration**: Useful for determining aging times and conditions for restoring wettability on clean, water-wet core samples.	• **Idealized conditions only**: The test is done on pure mineral surfaces, and it doesn't account for real reservoir rock complexities such as surface roughness, heterogeneity, and pore-scale geometry.
	• **Surface roughness effects**: Surface roughness can skew the observed contact angle. For surfaces with true contact angles less than 90°, roughness decreases the observed angle, while for angles greater than 90°, roughness increases it (Figure 11.17). A contact angle of 50° on a flat surface can appear as 0° on a rough surface (Anderson 1987).
	• **Nonrepresentative results**: Despite careful, reproducible measurements, contact angle tests on pure mineral surfaces are not fully representative of wettability in real reservoir cores.

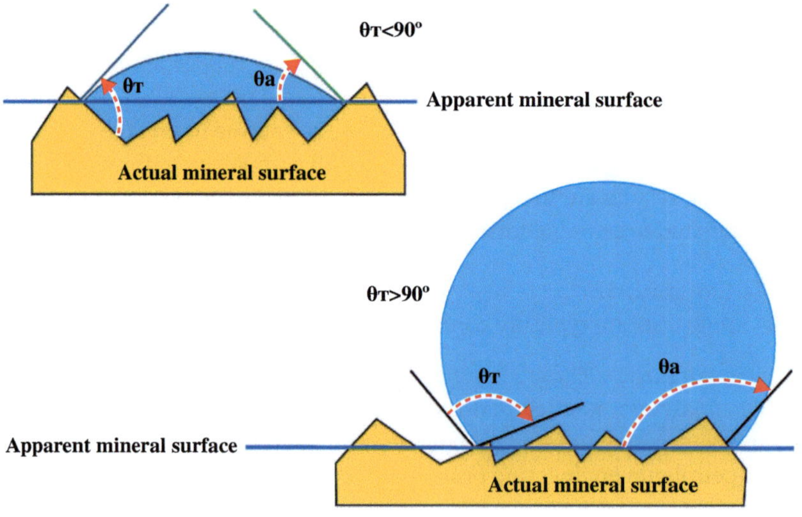

Figure 11.17 Surface roughness effects on measured contact angle (Popiel, 1978 / Exposition Press).

11.7.2 Amott (Amott–Harvey) Method

The **Amott method** (1959) combines imbibition and forced displacement to measure the "average" or complex wettability of a core. This test can use both reservoir samples and fluids (Table 11.11). The principle is based on the behavior that the wetting fluid will spontaneously imbibe into the core, displacing the nonwetting fluid (Figure 11.18).

Table 11.11 Overview of Amott wettability methods.

Purpose of Test	To determine wettability preference of core plugs
Sample Requirements	• Cylindrical core plug samples of 1″ or 1½″ diameter. A larger sample provides more pore volume, reducing the impact of saturation measurement errors. • Cleaned (water-wet) sample, a preserved core sample, or a sample after wettability restoration. • Alternatively, a plug from a RCA, provided the plug has not been pre-stressed or damaged in ways that alter wettability.
Required Input Data	• **Formation of water salinity** and **ionic composition** (as close as possible to reservoir conditions). Lab will prepare SFW accordingly. • **Reservoir crude oil sample** (e.g., STO or recombined crude oil) for aging, or **mineral oil** for testing. • **Maximum capillary pressure** (if forced displacement by centrifuge) or **overburden pressure** (if in coreholder). • **Equilibration time criteria** to complete spontaneous phases. • Time required to spin sample (normally 24 or 48 h per pressure, depending on sample permeability) if using a centrifuge for forced displacements.

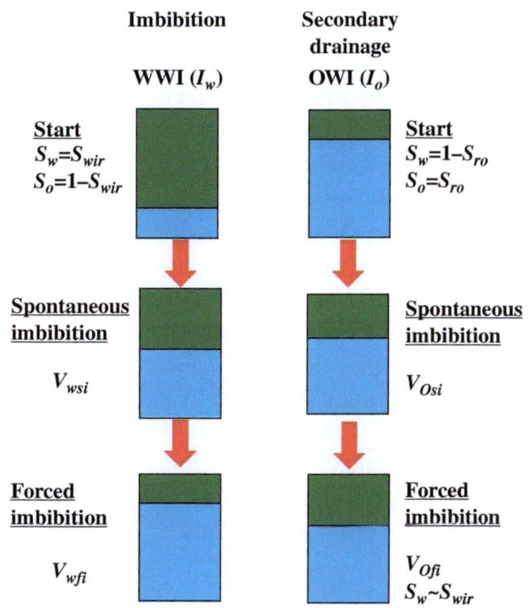

Figure 11.18 Schematic of Amott test.

Key Definitions:

1. **Drainage**: A reduction in the wetting phase saturation.
2. **Imbibition**: An increase in the wetting phase saturation.

For water-wet rocks, drainage means the water saturation decreases, while for oil-wet rocks, drainage refers to the oil saturation decreasing. To avoid confusion, the convention is:

- Water saturation decreasing is **drainage**.
- Water saturation increasing is imbibition—regardless of wettability.

Amott and Amott–Harvey Methods:

1. Original Amott method: The core is saturated with oil and centrifuged under brine until residual oil saturation (S_{ro}) is achieved.
2. Modified Amott–Harvey method: Before running the test, the core is centrifuged under brine followed by oil to reach the initial water saturation (S_{wir}).

In both methods, forced imbibition is typically done using a centrifuge, although in some tests, it can be achieved by dynamic displacement in a coreholder.

11.7.2.1 Sample Preparation

Wettability tests such as Amott and USBM can be carried out on core plugs under various initial conditions, each with specific procedures and characteristics:

1. Fresh state:
   - Definition: Core samples tested "as received" in the laboratory.
   - Not to be confused with: "Native state," which refers to cores obtained and stored to preserve the reservoir's wettability.
   - Water-based mud (WBM): For cores drilled with WBM, oil saturation may be near residual saturation due to filtrate invasion. These samples are flushed or centrifuged with SFW to reach residual oil saturation (S_{ro}), and the test begins with the spontaneous imbibition of oil (secondary drainage).
   - Oil-based mud: For cores drilled with oil-based mud, water saturation is close to irreducible saturation (S_{wir}). Samples are flushed or centrifuged with oil to reach S_{wir}, and the test begins with the spontaneous imbibition of water (primary imbibition).
   - Precautions: Before injecting oil, two to three pore volumes of decalin are flushed to prevent asphaltene deposition when in contact with stock tank oil (STO).
2. Cleaned state:
   - Definition: Cores where all fluids and adsorbed organic material are removed by flushing or solvent refluxing.
   - Objective: Return the cores to a water-wet state, hypothesized to be the reservoir condition before hydrocarbon accumulation.
   - Initial Condition: The cleaned cores are saturated in water and must be desaturated to irreducible water saturation (S_{wir}).
   - Desaturation techniques:
     i. Porous plate method: Preferred because it ensures evenly distributed water saturation. It produces the most reliable S_{wir}.

ii. Dynamic displacement: Uses oil to replace brine but may result in higher and less uniform S_{wir}.
iii. Capillary desaturation: Done via a porous plate or centrifuge, with capillary pressure tailored to reservoir height and rock quality.
iv. Desaturation phases: Either air or oil. If air is used for desaturation, it must be displaced by oil afterward under back pressure.

3. Restored state:
   - Definition: Cores treated to reestablish native wettability.
   - Procedure:
     i. The core is first cleaned to return it to a water-wet state.
     ii. The core is then saturated in brine and desaturated to S_{wir} using STO or recombined crude oil to simulate hydrocarbon accumulation.
     iii. It is then aged at reservoir temperature and pressure for up to 40 days, the industry standard for wettability restoration.
   - Starting condition: Tests on restored-state samples typically begin from S_{wir}, the condition after restoration.
   - Live oil vs. dead crude oil:
     i. Live oil (containing dissolved gases) is preferred, especially for oils prone to asphaltene precipitation.
     ii. Dead crude oil can be used for aging if the gas-oil ratio is less than 400–500 Mcf/bbl.
     iii. Challenges: Restoring wettability with live oil can suffer from leaks, leading to premature termination of the aging process. There are also HSE issues with crudes containing H_2S, and few labs can handle these.

Core cleaning as part of wettability restoration is critical for inducing a strong water-wet state in test samples, as this simulates the reservoir formation prior to oil accumulation. The primary goal of cleaning is to ensure the complete removal of pore contaminants such as mud components, altered hydrocarbons, FW, and precipitated salts, all of which could alter the wettability of the core. Proper cleaning is crucial, particularly when dealing with cores that contain significant asphaltene content or have been drilled with oil-based muds, as any residual contaminants can make the core excessively oil-wet after restoration.

Challenges in Core Cleaning:

1. Contaminants in pores: Asphaltene-rich oils or oil-based muds can leave behind oil-wetting contaminants in the pore spaces, making subsequent wettability restoration less effective.
2. Soxhlet extraction limitations:
   - Traditional hot Soxhlet extraction using toluene and methanol is the industry standard but has limitations.
   - While toluene can effectively remove some crude oil fractions and water around grains, it may not remove all components, particularly from clay interlayers, leading to incomplete cleaning and insufficient water-wet conditions.
   - Hot Soxhlet extraction may also damage clay minerals by removing water from clay interlayers, affecting their structure.

3. Residual oil components: Remnants of oil can remain in direct contact with mineral surfaces after cleaning, altering the core's wettability. This issue is exacerbated when cores have low initial water saturation, are stored long-term, or are exposed to air, leading to further deterioration of wettability.

Improved Cleaning Methods:

1. Lower temperature immersion: Total immersion cleaning at lower temperatures, or using mild miscible flush cleaning, is generally considered more effective than hot Soxhlet extraction. Solvent mixtures such as toluene–isopropanol–methanol can be better at removing oil-wetting contaminants without damaging the core material.
2. Verification: In any wettability study, it is crucial to verify the cleaning method by performing wettability tests on cleaned-state cores to ensure that the procedure successfully renders the core water-wet.

Special Considerations for Forced Imbibition and Drainage:

1. Centrifuge testing: If forced imbibition or drainage is conducted in a centrifuge, care must be taken to ensure that the selected core plugs can endure elevated spin speeds.
   - The lab may need to evaluate the integrity of the plugs and reduce the maximum spin speed (Pc) if necessary to prevent damage.

11.7.2.2 Test Conditions

The test conditions under which wettability is evaluated—such as temperature, pressure, and the type of oil used—can significantly influence the measurement results. However, studies in the literature present contradictory conclusions on these effects:

1. Temperature:
   - Cuiec (1995) observed a tendency for wettability to shift toward water-wet behavior as temperature increases.
   - Contrarily, other studies have shown that water wettability diminishes at higher temperatures, indicating that temperature may affect wettability differently depending on the rock and fluid system being studied.
   - This contradictory nature suggests that temperature's effect on wettability is complex and reservoir specific.
2. Pressure:
   - There is limited research on the effect of pressure on wettability, and most studies indicate that pressure has little to no influence on wettability.
   - This suggests that pressure may not be a major factor in altering the wettability of a reservoir system.
3. Challenges in testing at reservoir conditions:
   - In commercial labs, it is often difficult to evaluate wettability at full reservoir conditions (temperature and pressure). The primary issue is the lack of adequate equipment to maintain reservoir conditions throughout the entire testing process (Amott cell, Figure 11.19).

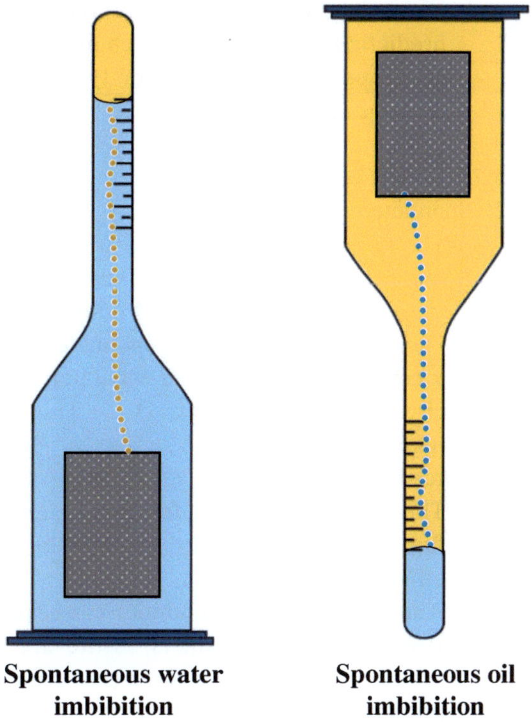

Figure 11.19 Amott cell schematic for spontaneous imbibition of water and oil.

- For example, while spontaneous imbibition may be carried out at room temperature or slightly elevated temperatures, forced displacements using centrifuges or coreholders typically occur at limited reservoir conditions.
- It is not possible to conduct all four Amott stages (spontaneous and forced imbibition and drainage) while maintaining full reservoir conditions due to equipment limitations, as pressure is often released and temperature is reduced when transferring samples between different apparatuses.

4. Use of mineral oil vs. crude oil:
   - Commercial wettability tests often rely on mineral oil rather than crude oil, even though crude oil may be used for aging the samples. Mineral oil is preferred for testing because it has relatively constant properties and is easier to work with compared to crude oil.
   - It is particularly challenging to conduct tests using live oil or maintaining full reservoir conditions (temperature, pressure, and crude oil properties) with standard commercial equipment.
5. Cuiec's conclusions on reservoir types:
   - For carbonate reservoirs, Cuiec concluded that changes in thermodynamic conditions (such as temperature and pressure) have no significant effect on wettability.
   - For sandstone reservoirs, the trend observed was that water wettability increased with temperature, though this was not always consistent, with some cases showing the opposite trend.

- Ultimately, Cuiec's results highlight that each reservoir is unique, and predicting the influence of changes in thermodynamic conditions (such as temperature, pressure, or type of oil) on wettability is difficult and must be approached on a case-by-case basis.

11.7.2.3 Test Equipment

1. Spontaneous imbibition:
   - The Amott cell (Figure 11.19) is the standard test equipment used for measuring spontaneous imbibition. It features a graduated section to determine the volume of imbibed fluid.
   - For unconsolidated cores, spontaneous imbibition can be conducted using a coreholder, a graduated separator system, and a pump within a closed circulating system (Figure 11.20). This setup allows better control but incurs higher costs.
2. Forced imbibition:
   - The preferred method for forced imbibition is via a high-speed centrifuge. This method helps achieve the required displacement forces more efficiently.
   - Alternatively, forced imbibition can be performed by flooding in a coreholder. However, it's important to ensure that both imbibition and drainage use the same displacement force (i.e., maintain the same constant differential pressure).
   - One limitation of the flooding method is that true residual saturations may not be achieved, potentially skewing the test results.

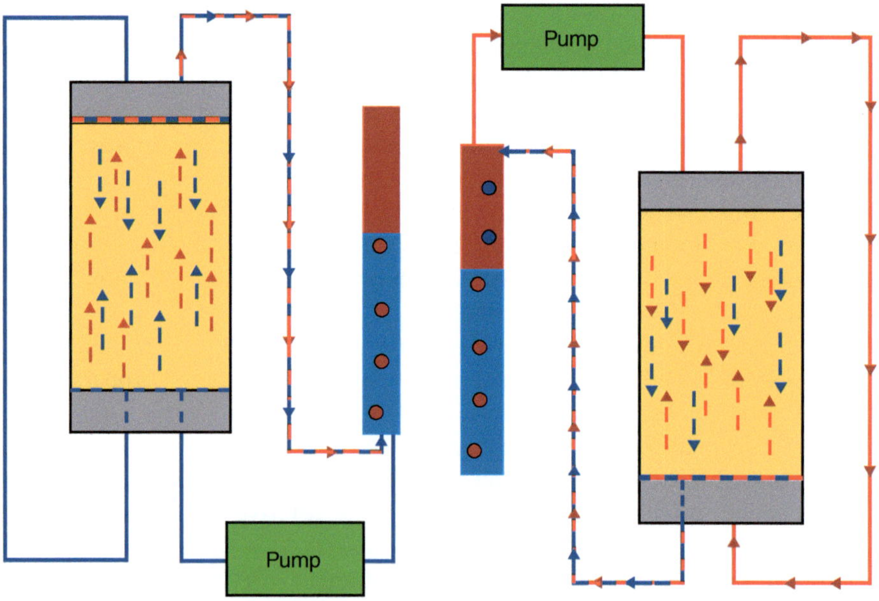

Figure 11.20 Spontaneous imbibition schematic for unconsolidated material (using coreholder, separator, and pump).

11.7.2.4 Test Procedures

The Amott and Amott–Harvey methods consist of four stages, with the order depending on the initial saturation of the test sample. If the core sample is reduced to initial water saturation (S_{wir}), the process proceeds as follows, illustrated in Figure 11.18:

1. Spontaneous brine imbibition (Amott A):
   - Brine (SFW) is allowed to spontaneously imbibe into the core, which is initially at S_{wir}, using an Amott cell (Figure 11.19).
   - As brine imbibes, oil is displaced from the core and rises to the top of the graduated tube, where volumes are monitored.
   - The volume of oil displaced (equal to the volume of brine imbibed) is tracked over time. In water-wet cores, a significant volume of brine is imbibed, while in non-water-wet cores, spontaneous brine imbibition may be small or zero.
   - The original Amott technique involved a 20-hour "cut-off" for spontaneous imbibition, but this is now considered too short, particularly in neutral-wet rocks where imbibition can be slow. Periods of up to 480 hours are now common.
   - Fluid production is monitored as a function of log time or square root of time to assess when to stop the cycle. For unconsolidated core material, a coreholder (Figure 11.21) is used to perform spontaneous imbibition. Fluid production is tracked similarly and plotted as a function of square root or logarithmic time.
2. Forced brine imbibition (Amott B):
   - Brine is forcibly imbibed into the core by applying pressure to displace oil toward residual oil saturation (S_{ro}). This is done using either a centrifuge (preferred method) or a coreholder.

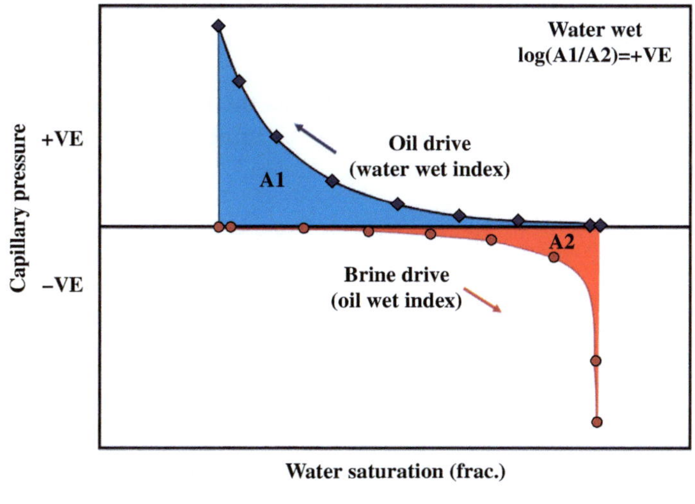

Figure 11.21 USBM example—water wet sample.

- Viscous instability and capillary end effects during dynamic flooding can affect the measured volumes of forced imbibition. Displacement by flooding may not achieve true S_{ro}, and thus the total volumes and wetting index could be inaccurate.
- To minimize errors, it is recommended to use the constant pressure method rather than the constant rate. The volume of fluid displaced (brine imbibed) is determined by tracking the displaced oil volume.

3. Spontaneous oil imbibition (Amott C):
   - At residual oil saturation (S_{ro}), oil is spontaneously imbibed into the core using the Amott cell, which is inverted for this stage.
   - The volume of brine displaced (equal to the volume of oil imbibed) is monitored over time. Oil-wet cores will show significant oil imbibition, while non-oil-wet cores will display minimal or no spontaneous oil imbibition.
   - Sufficient time should be allowed to achieve fluid production equilibrium.
4. Forced oil imbibition (Amott D):
   - Oil is forcibly imbibed into the core by applying an equivalent force (similar to that used during forced brine imbibition) to displace brine toward irreducible water saturation (S_{wir}). This is done via dynamic displacement using either a centrifuge (preferred) or coreholder flooding.
   - It is essential that the displacement force used in forced oil imbibition is the same as that used during forced water imbibition, ensuring comparability between the two processes.
   - The volume of fluid displaced (brine imbibed) is determined by monitoring the displaced oil volume.

11.7.2.5 Amott–Harvey Wettability Index Calculation

The Amott–Harvey wettability index is a widely used method to quantify the wettability of rock samples in a core test, specifically evaluating the relative preference of a rock for water or oil. This index combines the displacement ratios of both brine (water phase) and oil, determined through spontaneous and forced imbibition.

The water-wetting index (δ_w) is calculated as the ratio of the volume of brine spontaneously imbibed to the total volume of brine imbibed (spontaneous and forced). The formula is expressed as:

$$\delta_w = \frac{V_{wsi}}{V_{wsi} + V_{wfi}} \tag{11.45}$$

where:

- V_{wsi} is the volume of spontaneously imbibed brine (cc) and
- V_{wfi} is the volume of forcibly imbibed brine (cc).

For a water-wet rock, δ_w will approach 1, as most of the brine will enter the rock spontaneously. In contrast, for a less water-wet or oil-wet rock, spontaneous brine imbibition is minimal or even 0, resulting in a lower δ_w value.

Similarly, the oil-wetting index (δ_o) measures the ratio of oil spontaneously imbibed to the total volume of oil imbibed (both spontaneous and forced). It is expressed as:

$$\delta_o = \frac{V_{osi}}{V_{osi} + V_{ofi}} \tag{11.46}$$

where:

- V_{osi} is the volume of spontaneously imbibed oil (cc).
- V_{ofi} is the volume of forcibly imbibed oil (cc).

For an oil-wet rock, δ_o approaches 1, indicating that the rock strongly prefers oil and most of the oil imbibed enters spontaneously. In water-wet or neutral-wet rocks, the δ_o value will be closer to 0, indicating minimal or no spontaneous oil imbibition.

The Amott–Harvey wettability index (AHWI) is calculated by subtracting the oil-wetting index from the water-wetting index. This provides a single value indicating the rock's wettability preference:

$$AHWI = \delta_w - \delta_o \tag{11.47}$$

- A value of +1 indicates complete water wetness.
- A value of −1 indicates complete oil wetness.
- A value near 0 indicates neutral wettability.

The absolute value of the AHWI provides insight into the strength of the rock's wetting preference—larger values (positive or negative) indicate a stronger preference for either water or oil.

The wettability of the rock can be classified based on the values of δ_w, δ_o, and the AHWI. Table 11.12 summarizes the interpretation of these indices.

Once the Amott test is complete, a Dean–Stark extraction is often performed to verify the final water saturation in the core sample. Following this, the core is leached with methanol and dried. Measurements of the pore volume, bulk volume, and air permeability are then taken for further analysis.

In addition to calculating the wettability index, plotting the spontaneous fluid production against dimensionless time can provide valuable qualitative insights. According to Zhang et al. (1996), the shape of these production curves offers a further understanding of the rock's wettability behavior. This analysis is often used to complement the numerical index with a visual representation of fluid behavior over time (Table 11.13).

Table 11.12 Amott wetting indices.

Index	Water wet	Neutral wet	Oil wet
δ_w	Positive	0	0
δ_o	0	0	Positive
$AHWI$	+0.3 to +1.0	−0.3 to +0.3	−0.3 to −1.0

Table 11.13 Amott–Harvey summary.

Advantages	Drawbacks and issues
More sensitive than the USBM method for neutral wettability, showing positive displacement ratios for both water (δ_w) and oil (δ_o).	Results are dependent on the duration of the spontaneous imbibition phase, potentially leading to inconsistencies.
Suitable for unconsolidated samples if forced displacement is conducted in a coreholder. Centrifuge use may damage friable/unconsolidated plugs.	Maximum pressure applied during forced stages must be the same for both drainage (+Pc) and imbibition (−Pc).
Measures the average wettability of the core, compared to the localized nature of contact angle measurements.	Spontaneous brine imbibition depends on the distribution of interstitial water after primary drainage, with wettability changes increasing as S_{wir} decreases.
	Adhesion of expelled phase droplets to the core surface during spontaneous imbibition can lower the measured volume of the imbibed phase.
	Dynamic flooding may lead to viscous instability and capillary end effects, which impact forced imbibition volumes.
	Centrifuge-based forced imbibition can cause fracturing, especially at high velocities, and is unsuitable for poorly consolidated cores. Stopping and restarting the centrifuge may further damage the samples.
	Not valid as an absolute measurement but is widely accepted as an industry standard for comparing core wettability.

11.7.3 USBM Method

The USBM method is designed to measure the average wettability of a core sample by comparing the work needed for one fluid to displace another. Due to favorable free-energy changes, the work required to remove the nonwetting fluid is less than the opposite process. This work is proportional to the area beneath the drainage and imbibition capillary pressure curves.

Unlike the Amott test, the USBM method only involves forced primary imbibition and forced secondary drainage, without specifically measuring spontaneous imbibition. However, during the initial low-pressure centrifugation, some spontaneous imbibition naturally occurs. In most cases, the combined Amott/USBM method has replaced the standalone USBM test. The combined method allows for spontaneous imbibition measurements, and both Amott and USBM wetting indices can be calculated (see Table 11.14 for an overview of the USBM method).

Table 11.14 Overview of the USBM wettability method.

Category	Details
Purpose of Test	Determine the wettability preference of core plugs.
Sample Requirements	Cylindrical plug (100 or 1½00 diameter), with a preference for cleaned water-wet samples. Can use preserved plugs or those restored for wettability.
Required Input Data	FW salinity and ionic composition, crude oil sample from the reservoir (STO or recombined). Mineral oil can also be used for testing.
Capillary Pressure Data	Information from mercury injection capillary pressure tests can help design the pressure levels for each test. Use fluid density difference to determine maximum Pc pressure.
Time for Each Step	Spin times for each pressure step are typically 24–48 h but might extend further depending on sample permeability. Production must cease before proceeding to the next step.

Key Features of the USBM Method:

- Capillary pressure considerations: The original USBM criteria set a maximum capillary pressure of 10 psi. However, for samples with poorer reservoir quality, this might not be enough to define the irreducible water saturation (S_{wir}) or residual oil saturation (S_{ro}), so higher pressures may be needed. Both drainage and imbibition capillary pressures should remain equal.
- Sensitivity to core quality: The test's effectiveness can depend on the permeability of the core sample. For samples with lower permeability, especially common in regions like the Caspian Basin, higher pressure is required to overcome capillary forces.

11.7.3.1 Sample Preparation

The wettability test typically begins with samples at irreducible water saturation (S_{wir}). These samples can be in a fresh state (drilled with an oil-based mud), cleaned state, or restored state. For samples drilled with WBM, the test may start at residual oil saturation (S_{ro}).

11.7.3.2 Test Equipment

The key equipment used in this test includes a high-speed centrifuge. The centrifuge facilitates forced primary imbibition and secondary drainage.

11.7.3.3 Key Processes

1. Sample preparation:
   - Prepare the sample according to its initial state (fresh, cleaned, or restored).
   - For cleaned and restored plugs, measure the base permeability, pore volume, and bulk volume (porosity).

2. Primary imbibition process:
   - Spin the samples under brine in the centrifuge at predetermined pressure steps (controlled by rotational speeds).
   - Monitor the volume of oil produced, which is equivalent to the volume of water imbibed.
   - Calculate saturation as a function of pressure, producing a capillary pressure curve.
3. Secondary drainage process:
   - Spin the samples under oil in the centrifuge at predefined pressure steps.
   - The maximum capillary pressure can be the standard 10 psi or a user-defined value, depending on the rock quality, to ensure S_{wir} or S_{ro} is properly defined.
4. Water saturation measurement:
   - Measure the average (not end-face) water saturation and plot the relationship between the average water saturation (S_w) and capillary pressure (P_c) for both drainage ($+P_c$) and imbibition ($-P_c$) curves.
5. Wettability index calculation:
   - Calculate the wettability index based on the area beneath the primary imbibition and secondary drainage curves.
6. Dean–Stark extraction:
   - Perform a Dean–Stark extraction as an independent method to verify the endpoint water saturation.

11.7.3.4 Test Procedures

The USBM method consists of two primary stages: **primary forced imbibition** and **secondary forced drainage**. The procedure is typically carried out on samples previously prepared to irreducible water saturation (S_{wir}). Below are the detailed steps for each stage.

1. **Primary Forced Imbibition**
   - *Initial setup*: The sample, prepared to initial water saturation (S_{wir}), is placed in brine and centrifuged. The speeds are incrementally increased to simulate rising capillary pressures. The first pressure is kept as low as possible (~$\pm$0.5 psi or $\pm$1 psi) to replicate spontaneous imbibition and minimize the forced imbibition effect.
   - *Imbibition process*: As the centrifuge speed increases, oil is displaced from the sample and replaced by brine. The expelled oil collects in the centrifuge's receiving cup. The volume of expelled oil is recorded at each step, and the average saturation of the core (excluding end-face effects) is calculated.
   - *Data plotting*: Oil production is plotted as a function of logarithmic time or square root time to assess when production has ceased. The resulting curve is identified as Curve A2 in the figures (Figures 11.21–11.23). By the end of this stage, the sample is considered to be at or near irreducible oil saturation (S_{ro}).

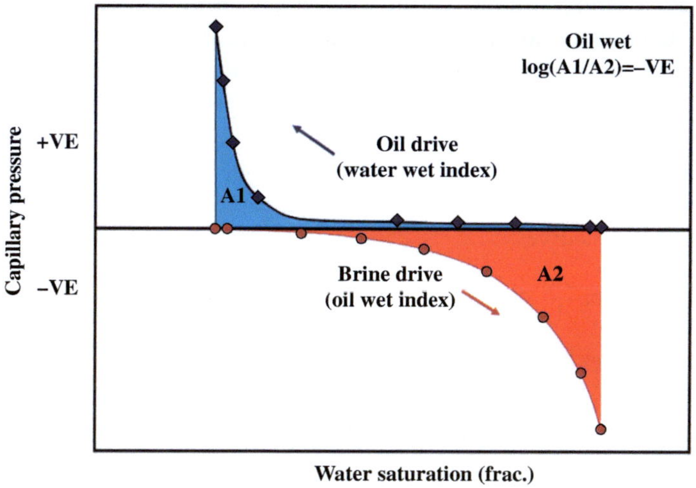

Figure 11.22 USBM example—oil wet sample.

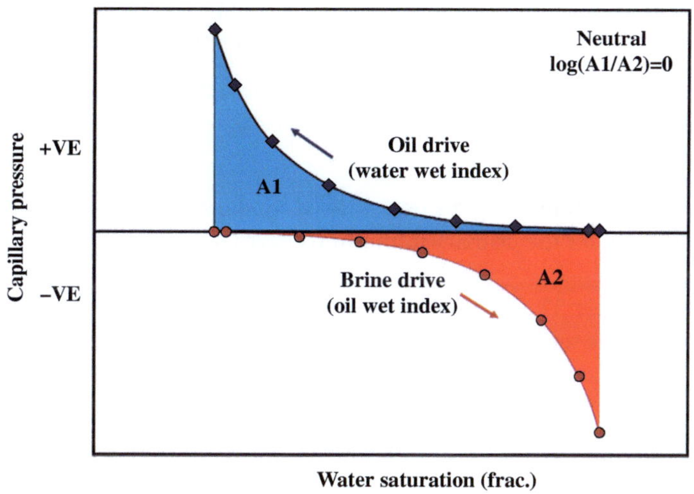

Figure 11.23 USBM example—neutral wet sample.

2. **Secondary Forced Drainage**
   - *Initial setup*: The sample, now at residual oil saturation (S_{ro}), is placed under oil and centrifuged at increasing speeds, similar to the first stage. The speeds are selected to match the pressure increments used in the primary forced imbibition process. The initial pressure remains low (~±0.5 psi or ±1 psi) to mimic spontaneous secondary drainage as closely as possible.
   - *Drainage process*: Brine is displaced from the sample and replaced by oil. The expelled brine is collected in the centrifuge's receiving cup, and the volume of the produced brine is recorded at each pressure step.

- *Data plotting*: Similar to the first stage, brine production is plotted as a function of logarithmic time or square root time. The results from the secondary drainage stage are represented as Curve A_1 in the figures (Figures 11.21–11.23).

11.7.3.5 USBM Index Calculation

The USBM wettability index is determined by comparing the areas under the two capillary pressure curves:

$$USBM = \log\frac{A_1}{A_2} \tag{11.48}$$

- A_1 is the area under the **secondary drainage** (oil drive) capillary pressure curve.
- A_2 is the area under the **primary imbibition** (brine drive) capillary pressure curve.

Interpretation of USBM Index:

- A **positive** USBM value indicates that the sample is **water-wet**.
- A **negative** USBM value indicates that the sample is **oil-wet**.
- A value close to **zero** suggests **neutral wettability**.

The greater the absolute value of the USBM index, the stronger the preference for either water or oil as the wetting phase.

Example Curves:

1. Water-wet sample: The area under the secondary drainage curve (A_1) is larger than that under the primary imbibition curve (A_2), leading to a positive USBM index.
2. Oil-wet sample: The area under the primary imbibition curve (A_2) is larger than that under the secondary drainage curve (A_1), leading to a negative USBM index.
3. Neutral-wet sample: The areas under both curves are equal, resulting in a USBM index of 0 (Table 11.15).

Table 11.16 summarizes advantages and disadvantages of USBM method.

11.7.4 Combined Amott–USBM (Combination) Method

This method combines both the Amott and USBM wettability tests, allowing for the determination of Amott and USBM wettability indices in a single procedure. It incorporates spontaneous imbibition (Amott) in an Amott cell, followed by forced imbibition in a high-speed centrifuge using multiple speeds, similar to the USBM test (Table 11.17).

11.7.4.1 Sample Preparation

The test begins with samples at irreducible water saturation (S_{wir}) or residual oil saturation (S_{ro}). These samples can be either in fresh, cleaned, or restored states. Starting conditions depend on whether the samples were drilled with WBM.

Table 11.15 USBM wetting indices.

Wettability state	USBM value
Water-wet	Positive
Neutral-wet	0
Oil-wet	Negative

Table 11.16 Summary of USBM method.

Advantages	Drawbacks and issues
• Relatively rapid method, typically taking only a few days since no spontaneous imbibition is involved.	• Amott method is more sensitive to neutral wettabilities.
• Measures the **average wettability** of the core, offering broader insight than contact angle measurement, which focuses on a localized scale.	• Drainage and imbibition **maximum Pc must be identical** to ensure consistent USBM index calculation.
• Provides additional information on **capillary pressure curves**, especially during imbibition. Pressure versus saturation plots use the average saturation, whereas Pc curves are more accurate when based on inlet face saturation.	• Centrifuge forced imbibition can create **unusual stress regimes**, with risks of fracturing at high spin velocities, making it unsuitable for fragile samples. Fracturing can also occur when the centrifuge is stopped and restarted.
	• The method is not valid as an absolute measure of wettability, but it is widely accepted as an **industry standard** for comparing wettability among different core plugs.

Table 11.17 Overview of the combined Amott–USBM Method.

Purpose of Test	• To determine wettability preference of core plugs using two analytical methods (Amott and USBM indices).
Sample Requirements	• Cylindrical-shaped plug sample of 1″ or 1½″ diameter. • Cleaned (water-wet) samples, preserved core, or restored-state plugs. • Alternatively, a plug previously used in RCA programs, provided there is no damage to clay minerals or altered wettability.
Required Input Data for Lab	• FW salinity and ionic composition, as close to reservoir conditions as possible. • Reservoir crude oil or mineral oil for testing. • Specific equilibration time for completing spontaneous phases. • Estimated capillary pressure for each test pressure station. • Time for spinning samples (24–48 h per pressure step).

11.7.4.2 Test Equipment

1. Spontaneous stages: Carried out in an Amott cell (Section 11.7.2.3 [Test Equipment]).
2. Forced stages: Performed using a high-speed centrifuge.

11.7.4.3 Test Procedures

The combined Amott–USBM method involves four stages (Figure 11.24) and is typically initiated from irreducible water saturation (S_{wir}), although it can also begin from residual oil saturation (S_{ro}) for samples drilled with WBM. The procedure is as follows, assuming the initial water saturation (S_{wir}):

1. Spontaneous brine imbibition (Amott):
   - Brine is allowed to spontaneously imbibe into the sample (starting at S_{wir}) using either an Amott cell for consolidated material or a coreholder for unconsolidated material. This stage is carried out similarly to the Amott method described in Section 11.7.3.4 (Test Procedures).
2. Forced brine imbibition (Amott + USBM):
   - Brine displaces oil under centrifugal force to reach residual oil saturation (S_{ro}). The plug sample is centrifuged at incrementally increasing speeds, and fluid production is recorded at each step, as detailed in Section 11.7.3.4 (Test Procedures).
   - The total fluid production at the highest speed is recorded as the Amott forced fluid volume (V_{wfi}). The Amott water-wet index (δ_w) is calculated (Section 11.7.2.5 [Amott–Harvey Wettability Index Calculation]).
   - The production at each speed is plotted, and the area under the curve (A_2) is used to calculate the USBM wettability index (Section 11.7.3.5 [USBM Index Calculation]).

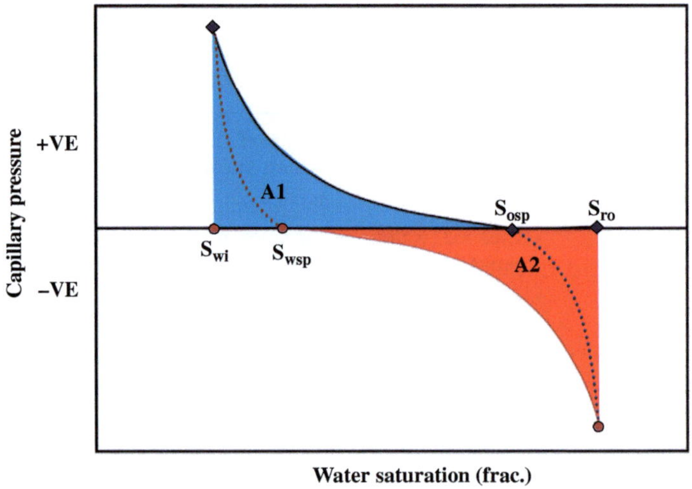

Figure 11.24 Combined Amott-USBM curve example.

3. Spontaneous oil imbibition (Amott):
   - Oil is allowed to spontaneously imbibe into the sample, starting at residual oil saturation (S_{ro}). This is performed in an inverted Amott cell or a coreholder. Details are similar to those in Section 11.7.2.4 (Test Procedures).
4. Forced oil imbibition (Amott + USBM):
   - Oil displaces brine under centrifugal force, driving the sample toward irreducible water saturation (S_{wir}). The sample is centrifuged at increasing speeds, and fluid production is measured at each step.
   - The total fluid production at the final speed is recorded as the Amott forced fluid volume (V_{ofi}). The Amott oil-wet index (δ_o) is calculated (Section 11.7.2.5 [Amott–Harvey Wettability Index Calculation]).
   - The production at each speed is plotted, and the area under the curve (A1) is used to calculate the USBM wettability index (Section 11.7.3.5 [USBM Index Calculation]).

11.7.4.4 USBM and Amott–Harvey Index Calculation

The Amott–Harvey (AHWI) and USBM indices are computed independently, using the same formulas detailed in their respective sections (Sections 11.7.2.5 [Amott–Harvey Wettability Index Calculation] and 11.1.6.3 [USBM Index Calculation]).

- For the Amott–Harvey index, the total cumulative production volume at the highest centrifuge speed serves as the forced fluid volume for the index equation, denoted as V_{wfi} (for brine) and V_{ofi} (for oil).
- Both indices should be calculated and compared for each sample to ensure consistency and to provide a clearer understanding of the wettability.

Table 11.18 summarizes advantages and disadvantages of combined Amott-USBM method.

Table 11.18 Combined Amott–USBM summary.

Advantages	Drawbacks and Issues
• Combines the benefits of the USBM method with the spontaneous wetting resolution provided by the Amott method.	• The maximum pressure for both **drainage** and **imbibition** phases must be identical to ensure consistency in USBM index calculations.
• Both **Amott** and **USBM indices** can be determined in a single experiment using the same core plug. Results can be cross-plotted, making it easier to detect outliers, as the methods should yield similar results, especially for samples with strong wetting conditions.	• The method is not suitable for samples that cannot endure the high-speed rotation in the centrifuge.
• Preferred method for assessing core sample wettability.	• Like the USBM method, it is not an absolute measurement of wettability, but it is an industry-standard for comparative analysis across various core samples.

11.8 Interfacial Tension

IFT plays a crucial role in translating laboratory data to conditions representative of a reservoir. Laboratory analyses, whether static or dynamic, are often carried out using synthetic fluids such as SFW, mineral oil, or nitrogen. However, the IFT between these fluids at lab conditions can significantly differ from the IFT in the reservoir, leading to potential inaccuracies if uncorrected.

For instance, most laboratories report an IFT value for standard laboratory fluids (SFW and mineral oil) around 48 mN/m (48 dyne/cm). However, as illustrated in Figure 11.25, IFT can vary based on salt concentration, ranging from slightly over 30 to 45 mN/m. In dynamic testing, where *in situ* saturation monitoring involves doped mineral oil, the use of iododecane to dope the oil can lower the laboratory IFT to approximately 16 mN/m at elevated temperatures. This discrepancy necessitates corrections to both capillary pressure and relative permeability to account for IFT differences.

IFT Correction Factor

To adjust laboratory data to more accurately reflect reservoir conditions, an IFT correction factor is used. This factor is calculated using the following formula:

$$IFT\ factor = \frac{\sigma \cos\theta_{res}}{\sigma \cos\theta_{lab}} \tag{11.49}$$

where:

- σ_{res} is the IFT under reservoir conditions,
- σ_{lab} is the IFT under laboratory conditions, and
- θ_{res} and θ_{lab} are the contact angles in the reservoir and lab, respectively.

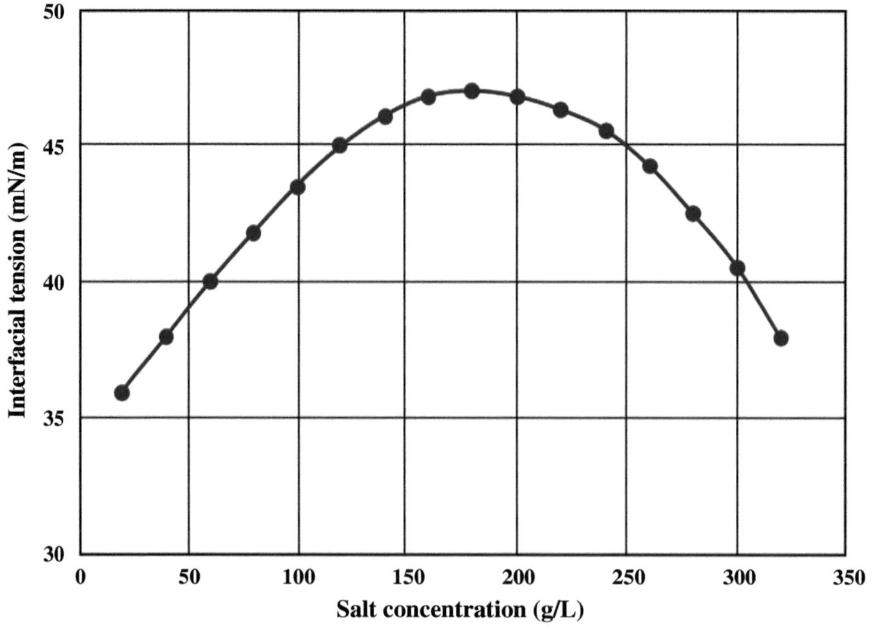

Figure 11.25 SFW-mineral oil IFT versus salt concentration.

For example:

- If the IFT in the reservoir is 24 mN/m and the standard lab IFT is 48 mN/m, the IFT factor would be 0.5.
- If the IFT is measured to be 35 mN/m, the IFT factor would increase to 0.686.
- For iododecane-doped oil at elevated temperatures (where IFT drops to 16 mN/m), the IFT factor could reach 1.5.

This variance can result in an error exceeding 200% in the conversion factor if laboratories assume a standard IFT without measuring the actual IFT of the fluids used. To minimize uncertainty, it is strongly recommended that IFT be measured for the specific test fluids at all relevant test conditions, as well as for the reservoir fluids under reservoir conditions.

11.8.1 Methods to Determine IFT

IFT values can vary significantly with time as fluids interact, even if they have been equilibrated outside the testing equipment. For example, in an oil–water system, IFT might initially register between 35 and 40 mN/m within seconds, but it may decrease over time to 35–30 mN/m as the system equilibrates. This emphasizes the need for laboratories to provide detailed descriptions of their procedures for measuring IFT (Table 11.19).

11.8.1.1 From Compositional Data

IFT can be estimated using fluid compositional data through a mathematical approach. The following equation is commonly used:

$$\sigma = \left[6.95 \frac{\Delta\rho}{T_r^{0.3125}}\right]^4 \tag{11.50}$$

where:

- $T_r = \dfrac{T}{T_c}$
- T is the reservoir temperature (°R [Rankine scale of temperature]),
- T_c is the critical temperature of the fluid mixture (°R), and
- $\Delta\rho$ is the water–oil density difference at reservoir temperature (lb/ft^3).

11.8.1.2 The Pendant Drop Method

The pendant drop technique is the most widely used experimental method for measuring IFT in the petroleum industry. It relies on analyzing the shape of a drop suspended in a denser fluid. The shape of the drop provides a means to calculate IFT, using the Young–Laplace equation, which relates the drop's curvature to the forces acting on it—surface tension and gravity. This process is typically automated by software.

Key Assumptions of the Method:

- The drop is symmetric around a central vertical axis, making the direction of observation irrelevant.
- The drop is static, meaning that viscosity and inertia do not affect its shape, and only surface tension and gravity influence the shape.

Table 11.19 Summary of IFT measurement methods.

Method	Advantages	Disadvantages
Pendant Drop	• Simple calibration, only requiring optical magnification. • Wettability of the apparatus is irrelevant, so cleanliness is not a concern. • Can operate under high pressure (17,000 psi) and high temperature (200 °C).	• Recent method, limiting its use for comparison with legacy data. • Not as widely available as traditional methods.
Du Noüy Ring	• Extensive data exists for comparison purposes. • Higher accuracy due to the larger wetted length of the ring (3 times more than a plate). • Useful for small IFT measurements, especially for substances with poor wetting properties.	• Requires mathematical corrections for lifted liquid volume. • Density of fluids must be known. • Possible phase mixing during measurement. • Ring movement may cause erroneous results, especially for high-viscosity solutions. • Limited to ambient or slightly elevated temperatures. • Measures static tension, which may not suit reactive systems (e.g., microbial oil recovery).
Wilhelmy Plate	• No correction factors needed. • Liquid densities do not need to be known. • Prevents phase mixing as it only touches the surface. • More stable since the plate does not move after detection. • Suitable for long-term surface tension monitoring.	• Poor wetting properties for some substances, such as cationic surfactants. • Less accurate for small IFT due to smaller wetted length compared to the ring method. • Limited to ambient pressure and slightly elevated temperatures. • Measures static tension, not ideal for reactive systems.

The dimensions of the drop are measured from the image and combined with the density data of the surrounding fluid to calculate IFT (Figure 11.26).

Advantages of the Pendant Drop Technique:

1. *Simple calibration*: The primary calibration requirement is optical magnification, which can be measured accurately as long as the density is known.
2. *No special cleanliness*: Unlike other methods such as the Wilhelmy plate, the apparatus' surface cleanliness does not affect results since wettability is not a factor.
3. *High pressure and temperature operation*: The technique can be conducted under extreme conditions, up to 17,000 psi and 200 °C.
4. *Sensitivity*: Digitized image processing allows the method to measure IFTs as low as 10^{-2} mN/m, which is important for EOR and gas condensate processes.

Figure 11.26 Illustration of pendant drop method.

$$\gamma = \frac{g\rho(de)^2}{H}$$

where:
Υ: surface (interfacial) tension
g: acceleration due to gravity
ρ: droplet density
de: maximum droplet diameter
$1/H$: correction factor dependent upon de/ds

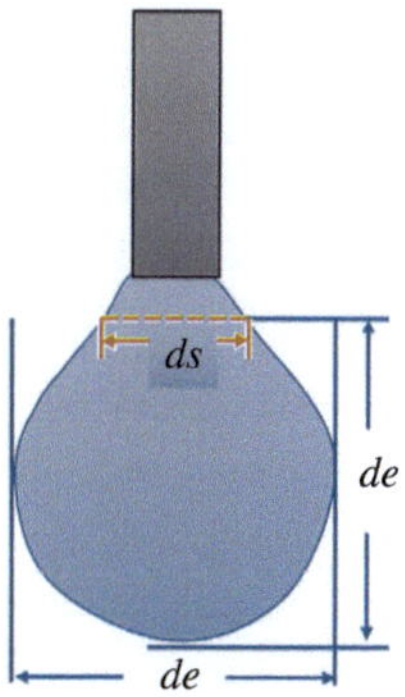

Some laboratories provide only **first-contact IFT** (averaged from 5 to 10 drops), while others report only **equilibrated IFT** (also averaged over 5–10 drops). It is recommended that every drop be recorded over time to offer both first-contact and equilibrated values. IFT vs. time curves can serve as a quality control tool, helping detect potential interference from impurities and assessing variance.

11.8.1.3 Force Tensiometry

Force tensiometry is a method for measuring surface and IFT through force measurements of a probe interacting with the interface between two fluids. This technique is generally performed at ambient pressure and slightly elevated temperatures. In this method, a probe is submerged into the lighter phase of the two fluids to begin the experiment. The probe is suspended from a balance, and the forces acting on it as it comes into contact with the liquid interface are used to calculate the IFT.

Factors Influencing Force Tensiometry Measurements:

1. *Size and shape of the probe*: Controlled precisely.
2. *Contact angle*: Fixed to zero for complete wetting by using high-energy probes, typically made of a platinum/iridium alloy. This ensures complete wetting and easy cleaning.
3. *Surface tension* of the liquid: Influences the measured force.

Two common probes are used for IFT measurements in force tensiometry: the **Du Noüy ring** and the **Wilhelmy plate**.

11.8.1.4 Du Noüy Ring

The Du Noüy ring method involves a platinum ring interacting with the surface of the liquid. The process is as follows:

- The ring is submerged below the interface between the two fluids.
- It is then raised upward, creating a meniscus of liquid around the ring.
- As the ring continues to rise, the meniscus stretches until it reaches a point where it breaks off and returns to its original position.

During this process, the force exerted by the meniscus increases, reaching a maximum before it begins to decrease, and the meniscus finally breaks (Figure 11.27). The maximum force is the key to calculating the surface tension or IFT.

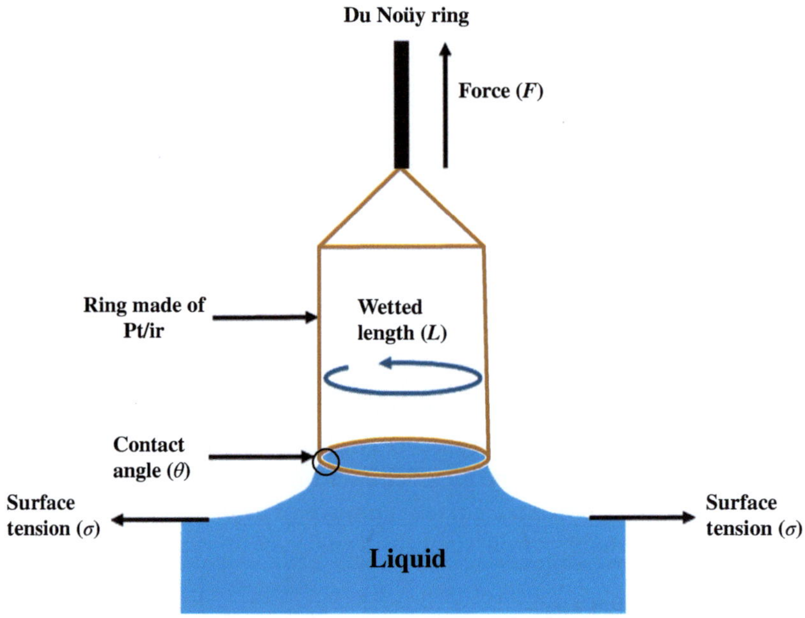

Figure 11.27 Schematic of Du Noüy ring method.

The formula for calculating surface or IFT using the Du Noüy ring method is:

$$\sigma = \frac{F_{max} - F_V}{L \cos\theta} \tag{11.51}$$

where:

- σ is the surface (interfacial) tension,
- F_{max} is the maximum force exerted by the meniscus,
- F_V is the weight of the volume of the liquid lifted,
- L is the wetted length of the ring, and
- θ is the contact angle.

Contact Angle Considerations:

- The contact angle (θ) decreases as the meniscus extends.
- At the point of maximum force, the contact angle reaches 0°, which gives $\cos\theta = 1$.
- Due to the volume of liquid hanging from the ring when it is pulled above the surface, a correction factor F_V must be applied to account for the weight of the liquid. Several correction methods are available for this purpose.

11.8.1.5 Wilhelmy Plate

The Wilhelmy Plate method is a technique used to measure surface tension by examining how a platinum plate interacts with the surface of a liquid. The key to this method is the behavior of the plate as it comes into contact with the liquid surface. When the plate touches the surface, the tensiometer (a device that measures force)

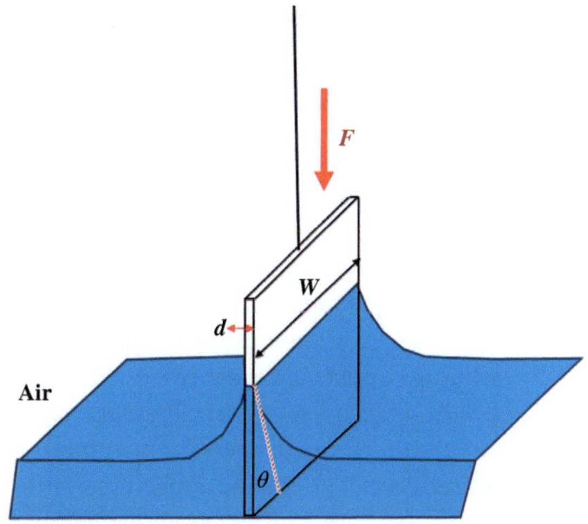

Figure 11.28 Illustration of Wilhelmy plate method.

detects a change in the forces it experiences. This moment is noted as the "zero depth of immersion."

To ensure that the plate is fully wetted, it is lowered further into the liquid until the entire plate is coated, which ensures the contact angle (the angle between the liquid and the plate) is 0°. This step guarantees complete wetting of the plate.

Afterward, when the plate is returned to the "zero depth of immersion," the force measured by the tensiometer is used to calculate surface tension using the following equation:

$$\sigma = \frac{F}{L\ \cos\theta} \tag{11.52}$$

In this formula:

- σ is the surface or IFT,
- F is the force measured by the tensiometer,
- L is the length of the wetted part of the plate, and
- θ is the contact angle between the liquid and the plate.

Since the contact angle (θ) is approximately 0° due to the complete wetting of the plate, the term $\cos\theta$ becomes roughly equal to 1. As a result, the calculation of surface tension depends mainly on the measured force (F) and the wetted length (L) (Figure 11.28).

An important advantage of the Wilhelmy Plate method is that, unlike other techniques such as the ring method, no additional correction factors are needed.

11.9 Steady-State Permeability Measurements

11.9.1 Sample Preparation

Before testing, samples must be thoroughly extracted from fluids and dried according to the standard cleaning and drying procedures.

11.9.2 Test Equipment

Steady-state gas permeability measurements require a specific setup, as shown in Figure 11.29. All gas permeameters share key features:

- *Coreholder (Hassler cell)*: The sample is held in a coreholder, which includes a rubber sleeve to confine the test sample. The end platens allow gas to enter and leave the sample, promoting even gas distribution. The end platens can move against the plug when pressure is applied to the rubber sleeve, ensuring isostatic loading conditions (equal axial and radial stress). This seals the sample, preventing gas from bypassing it along the sleeve. Testing can be done under confining stress to simulate reservoir conditions.
- *Pressure application system*: Pressure is applied to the rubber sleeve either hydraulically (for high stress) or pneumatically (for low stress). A pressure gauge or transducer monitors the applied pressure.
- *Gas pressure control*: The gas entering the sample is regulated, usually limited to a maximum injection pressure of 100 psi. This is controlled using a gas regulator upstream of the sample.
- *Pressure measurement*: Pressure gauges or transducers measure upstream (P_1) and downstream (P_2) pressures. In some designs, a differential pressure transducer measures the pressure drop ($P_1 - P_2$) across the sample. For high-permeability samples, higher flow rates are used to improve accuracy. However, high flow rates can lead to nonlaminar flow, resulting in significant permeability underestimation, especially for samples with permeabilities greater than 5,000 mD. To avoid this, pressure should be measured directly at the core faces (P_1 and P_2) using separate pressure taps.
- *Outlet pressure control*: A backpressure regulator controls the outlet pressure, particularly important for high-permeability samples and Klinkenberg measurements.
- *Flow rate measurement:* The gas flow rate is measured to accommodate various sample permeabilities, typically up to 2,000 mL/min. Modern permeameters often use a mass flow meter upstream of the sample to measure gas flow at the

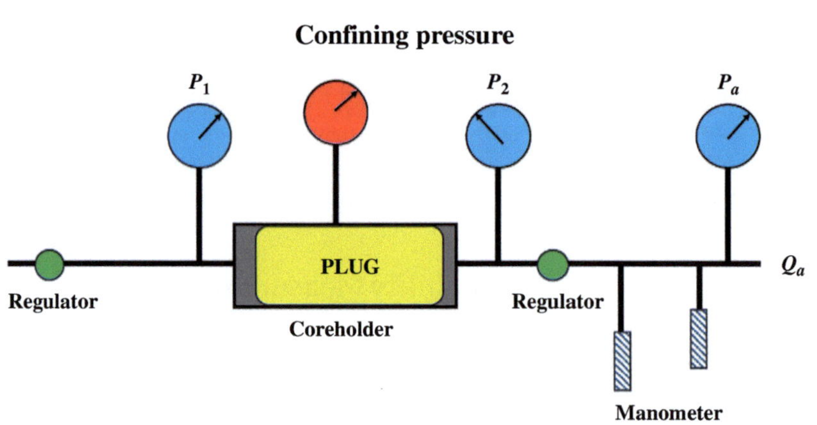

Figure 11.29 Gas permeameter schematic.

inlet pressure (Q_1) instead of atmospheric conditions (Q_a). Flow is corrected to mean pressure conditions using the equation:

$$Q_m = \frac{Q_1 P_1}{P_m} \tag{11.53}$$

Alternatively, flow can be directed through a capillary tube downstream of a backpressure regulator at atmospheric pressure (P_a), where the pressure drop correlates to the flow rate and is measured by a manometer.

- *Temperature and atmospheric pressure measurement*: A thermocouple measures gas temperature, aiding in gas viscosity predictions, while a barometer measures atmospheric pressure (P_a) if the flow rate is recorded at atmospheric conditions. The flow rate is then corrected by:

$$Q_m = \frac{Q_a P_a}{P_m} \tag{11.54}$$

11.9.3 Test Procedures

1. *Sample measurements*: Measure the sample's length and diameter at least three times. If the bulk volume (V_b) has been determined during porosity measurements, the cross-sectional area (A) is calculated using the formula:

$$A = \frac{V_b}{L} \tag{11.55}$$

   where L is the sample length.
2. *Loading the sample*: Load the sample into the coreholder and apply a confining stress. The confining stress must be sufficient to prevent nitrogen from flowing between the sample and the confining sleeve.
3. *Flowing nitrogen*: Oxygen-free nitrogen is injected through the sample at a constant flow rate or pressure drop. Low injection pressures are preferred for accuracy. For high permeability samples, using a nominal backpressure helps stabilize the flow rate. In low permeability samples, higher injection pressures may be required, and the confining stress should be adjusted to maintain a constant net confining stress (NCS). The NCS is defined as:

$$\text{NCS} = \text{TCS} - p_p \tag{11.56}$$

   where TCS is the total applied stress and p_p is the pore pressure. For example, if the NCS is 300 psi and the mean pore pressure is 50 psi, the TCS must be 350 psi.
4. *Gas flow rate selection*: While there is no definitive guide for selecting gas flow rates, a maximum flow rate of 2,000 mL/min is typical for high permeability samples.
5. *Recording data*: Once steady-state conditions are reached (when pressure drops and flow rates stabilize), record the following:
   - Flow rate
   - Differential pressure
   - Temperature
   - Atmospheric pressure

If no backpressure is applied, the outlet pressure (P_2) is typically equivalent to atmospheric pressure (P_a). When backpressure is used, P_2 will be higher than P_a. The use of backpressure is recommended because it provides better control of the gas flow rate and differential pressure, ensuring Darcy's viscous flow under higher pressures.

6. *Klinkenberg permeability (if required)*: To measure Klinkenberg permeability, increase the mean pressure in the sample. This can be done by increasing the inlet pressure (P_1) or flow rate, or by using the backpressure regulator to increase the downstream pressure (P_2). After the system reaches stability, record the same parameters as before. Repeat this procedure two or three times at increasing mean pressures, while maintaining the same NCS, to obtain multiple gas permeability measurements at different pressures.

11.9.4 Gas Permeability and Klinkenberg Permeability

The gas permeability (k_g) of nitrogen is determined using the following equations, depending on where the gas flow rate (Q) is measured:

- If the flow rate (Q_a) is measured at atmospheric conditions:

$$k_g = \frac{2Q_a P_a \mu_g L}{A(P_1^2 - P_2^2)} \tag{11.57}$$

- If the flow rate (Q_1) is measured at inlet (upstream) pressure (P_1):

$$k_g = \frac{2Q_1 P_1 \mu_g L}{A(P_1^2 - P_2^2)} \tag{11.58}$$

where:

- P_1 and P_2 are the inlet and outlet pressures, respectively,
- μ_g is the gas viscosity,
- L is the sample length, and
- A is the cross-sectional area of the sample.

Klinkenberg permeability (k_1) is calculated using linear regression from at least three pairs of apparent gas permeability (k_g) values and the inverse of the mean pressure ($1/P_m$).

11.9.5 Evaluation of Klinkenberg and Non-Darcy Effects in Steady-State Flow

To evaluate non-Darcy flow and Klinkenberg effects, three graphical analysis techniques can be used:

1. *Klinkenberg plot* (Figure 11.30): This plot analyzes gas slippage by graphing inverse mean pressure ($1/P_m$) against apparent gas permeability (k_g). Klinkenberg-corrected permeability can be estimated from the linear trend.

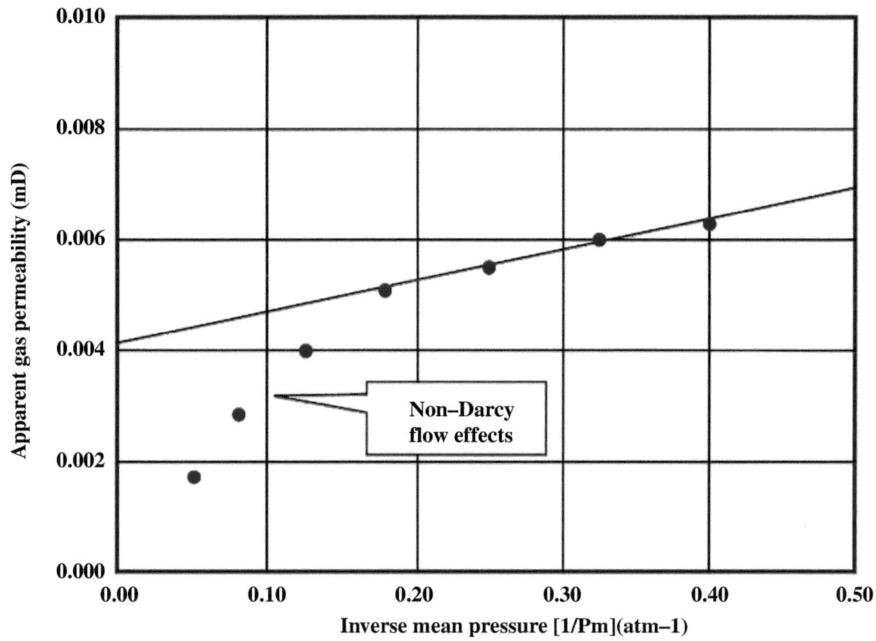

Figure 11.30 Hypothetical Klinkenberg plot showing non-Darcy flow behavior.

Any deviation from the straight line at lower ($1/P_m$) values (higher pressures) suggest the onset of non-Darcy inertial effects.

2. *Slippage-corrected pressure plot* (Figure 11.31): This plot corrects for slippage effects by plotting the Darcy pressure drop against the mass flow rate. A straight line through the origin, with a slope equal to the inverse of Klinkenberg permeability (k_1), should be observed. The equation used is (Rushing et al., 2004):

$$\frac{M\left(1+\frac{b}{P_m}\right)\left(P_1^2-P_2^2\right)}{2\mu zRTL}=\frac{\dot{m}}{k_1} \tag{11.59}$$

where:

- M is the molecular weight of the gas,
- b is the gas slippage factor,
- P_m is the mean pressure,
- P_1 and P_2 are the inlet and outlet pressures,
- μ is the gas viscosity,
- z is the gas deviation factor,
- R is the universal gas constant,
- T is the absolute temperature, and
- $\dot{m}$ is the mass flow rate.

Non-Darcy effects appear as an upward deviation from the straight line.

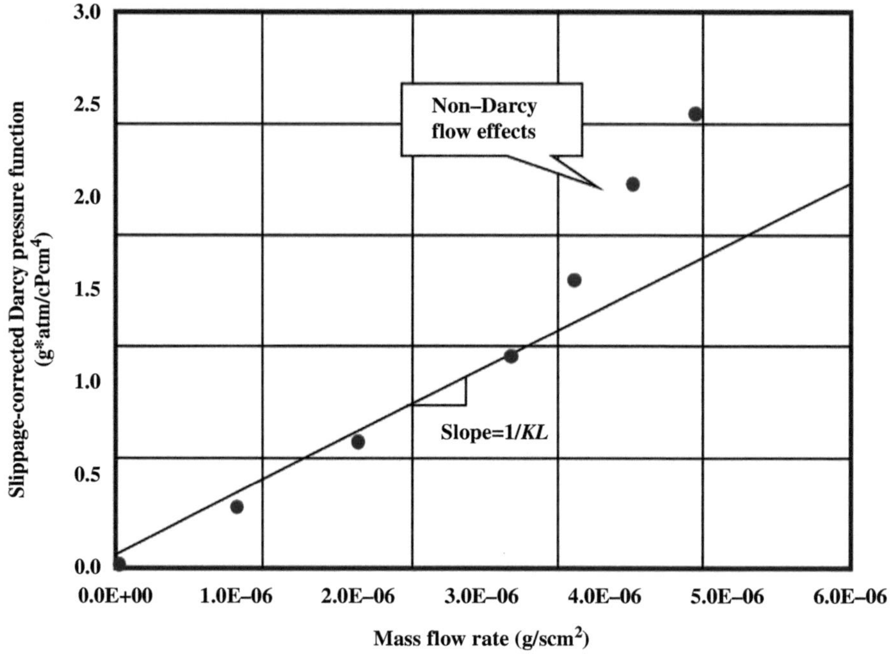

Figure 11.31 Hypothetical slippage—corrected Darcy plot showing non-Darcy flow behavior.

3. *Inertial flow-corrected pressure plot* (Figure 11.32): This plot corrects for inertial effects by plotting the slippage-corrected Forchheimer pressure change against the slippage-corrected mass flow rate (Rushing et al., 2004). The slope of the line represents the inertial resistance coefficient (b), while the intercept gives the inverse of Klinkenberg permeability (k_1). The equation used is:

$$\frac{M\left(1+\frac{b}{P_m}\right)\left(P_1^2-P_2^2\right)}{2\mu zRTL\dot{m}}=\frac{m\left(1+\frac{b}{P_m}\right)\beta}{\mu}+\frac{1}{k_1} \tag{11.60}$$

Where β is the Forchheimer inertial resistance coefficient.

11.10 Unsteady-state Permeability Measurements

The **pressure decay method** is an unsteady-state technique used to measure permeability, particularly for rocks with very low permeability (<0.1 mD), where reaching steady-state conditions can take a long time. The time required to achieve steady-state conditions generally increases with the square of the sample length and decreases with higher intrinsic permeability.

In this method, the core sample is first saturated with an inert gas, then a pressure transient is induced across the sample. The change in differential pressure over time is recorded and analyzed to determine permeability (Table 11.20).

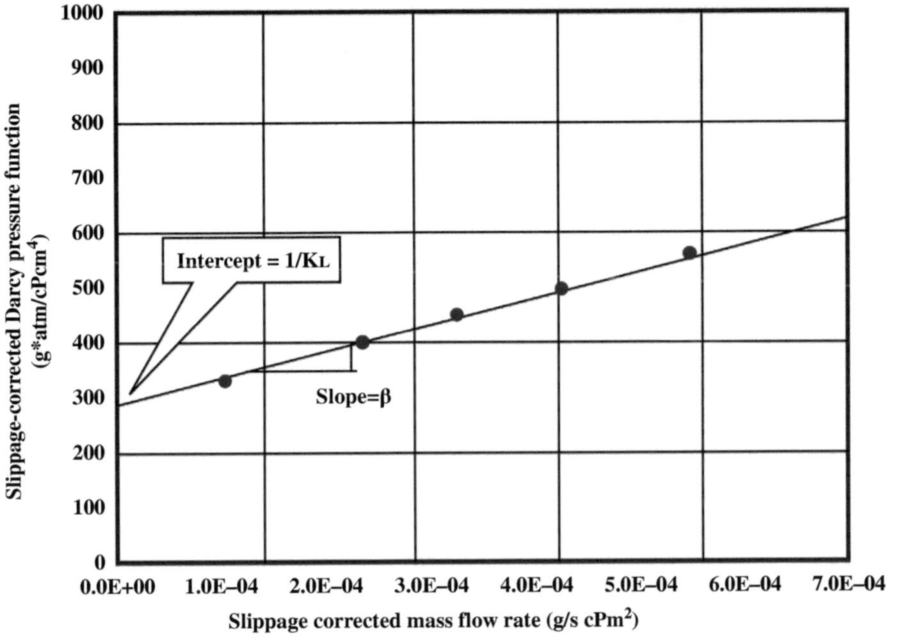

Figure 11.32 Hypothetical inertial flow and slippage-corrected Forchheimer plot.

Table 11.20 Unsteady-state gas permeability summary.

Advantages	Drawbacks and issues
Much faster than steady-state methods, especially for low-permeability samples and Klinkenberg permeability.	Samples need to have a cylindrical shape.
No flow meter calibration required.	Results are highly dependent on sample preparation (preserving, cleaning, and drying).
Klinkenberg permeability (*kl*), slippage coefficient (*b*), and Forchheimer inertial resistance factor (β) calculated in one experiment.	Gas permeability is back-calculated from Klinkenberg permeability and slippage factor, not measured directly.
Measurement time and accuracy can be optimized by selecting the appropriate tank volume.	Minimum confining stress in CMS-300 is 800 psi, which may lead to inconsistent data with older, lower-stress systems.
Gas (nonreactive) used instead of brine, crucial for clay-rich samples, and less prone to mobilize fines.	Klinkenberg coefficient measurement is difficult for samples above 1,000 mD; vacuum or steady-state methods are recommended.
Nondestructive method, allowing for further tests on the same sample.	Adequate thermal insulation is required to minimize temperature changes from ambient fluctuations.

To ensure accurate results, the analysis must account for Klinkenberg effects (gas slippage) and Forchheimer effects (non-Darcy flow), as the induced pressure pulse can affect the mean pressure and lead to nonlaminar flow.

11.10.1 Test Equipment

There are two main types of equipment for unsteady-state permeability measurements: the **pressure drawdown method** and the **pulse decay system**.

1. *Pressure drawdown method*: In this setup (Figure 11.33), a calibrated gas tank is connected to a coreholder via a quick-opening valve. The downstream end of the coreholder is open to the atmosphere, allowing gas to flow directly through the sample. This setup is a simplified version of the pulse decay system, where the downstream reservoir is considered infinite, maintaining constant pressure at the sample outlet.
2. *Pulse decay method*: In this system (Figure 11.34), a tank with known volume is connected to the downstream side of the coreholder. This setup helps determine permeability by measuring pressure changes over time as gas flows through the sample. Estimation accuracy improves when the upstream and downstream tank volumes are similar and close to the sample's pore volume.

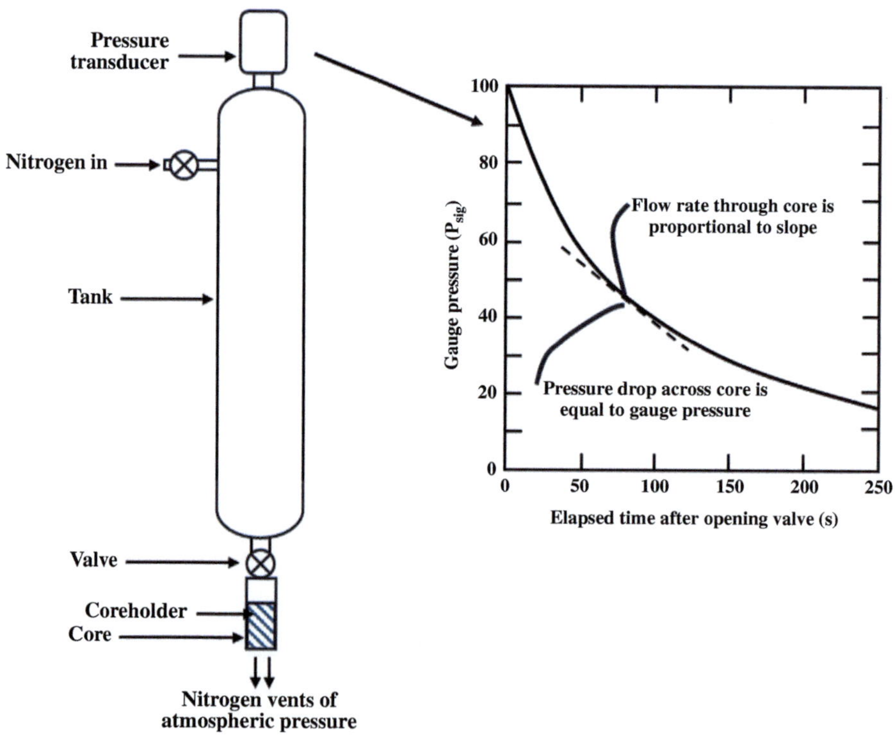

Figure 11.33 Example of unsteady state pressure drawdown permeameter.

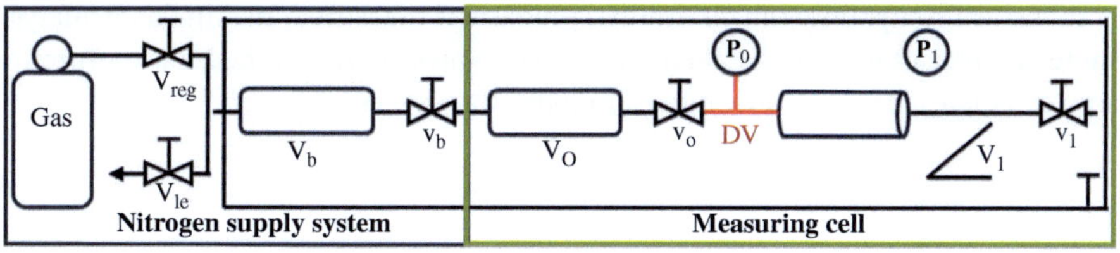

Figure 11.34 Example pulse decay apparatus.

A common commercially available system typically measures permeability using helium, although nitrogen can also be used. Helium is contained in one of several reservoirs with precisely known volumes. A valve isolates the coreholder from the helium tank for porosity measurement, but it opens for permeability measurement, allowing gas to flow through the sample.

The core plug is automatically loaded into a coreholder designed to withstand confining stresses up to high pressures (≥10,000 psi). As gas flows through the sample, a microcomputer tracks the pressure decay in the reference tank over time. Instantaneous flow rates and pressure differences across the core are then calculated to determine permeability.

11.10.2 Test Procedures and Permeability Calculation

In permeability testing, the core plug is placed inside a coreholder, where it is subjected to specific confining stress. The core is connected to upstream and, if applicable, downstream reference volumes, which are initially sealed off by valves. Gas, such as helium or nitrogen, is introduced into a source chamber. When the valve is opened at time zero, the gas expands through the core either into the atmosphere, vacuum, or downstream tank, depending on the setup. The pressure across the core is continuously recorded over time until it reaches a predetermined value or stability criterion.

Since the pressure in the core declines as the gas flows, and initial flow rates are fast, the calculation must account for both viscous (Klinkenberg) and inertial (Forchheimer) effects. The general equation for pressure drop across the core is:

$$-\frac{dP}{dL} = \frac{\mu v}{k_1\left(1 + \frac{b}{P_m}\right)} + 3.238E^{-8}\beta\rho v^2 \tag{11.61}$$

where:

- μ is the gas viscosity,
- v is the velocity,
- k_1 is the Klinkenberg permeability,
- b is the slip factor, and
- β is the Forchheimer factor.

In the test setup, only the differential pressure is measured, making data analysis more complex than steady-state methods. The Jones (1972) method for interpreting data calculates the mass flow rate exiting the core:

$$V_t = \frac{d\rho}{dt} = Q_{out} \tag{11.62}$$

where:

- V_t is the source tank volume,
- ρ is the gas density, and
- t is time.

The volumetric flow rate of helium leaving the chamber is approximated by:

$$Q_{out} = Vtf_1\left(\frac{-d \ln\ p_o}{dt}\right) \approx \frac{Vtp_g f_1}{p_g + p_a}\left[\frac{\ln\frac{p_1}{p_2}}{t_2 - t_1}\right] \tag{11.63}$$

where:

- f_1 is a correction factor for gas deviation, $f_1 = \left[1 - \frac{p_0}{z_0}\left(\frac{dz_0}{dp_0}\right)\right]$
- p_0 is the pressure,
- z_0 is the gas deviation factor,
- p_1 and p_2 are pressures at selected times t_1, t_2,
- p_a is atmospheric pressure, and
- p_g is the geometric mean pressure between p_1 and p_2 (Figure 11.35).

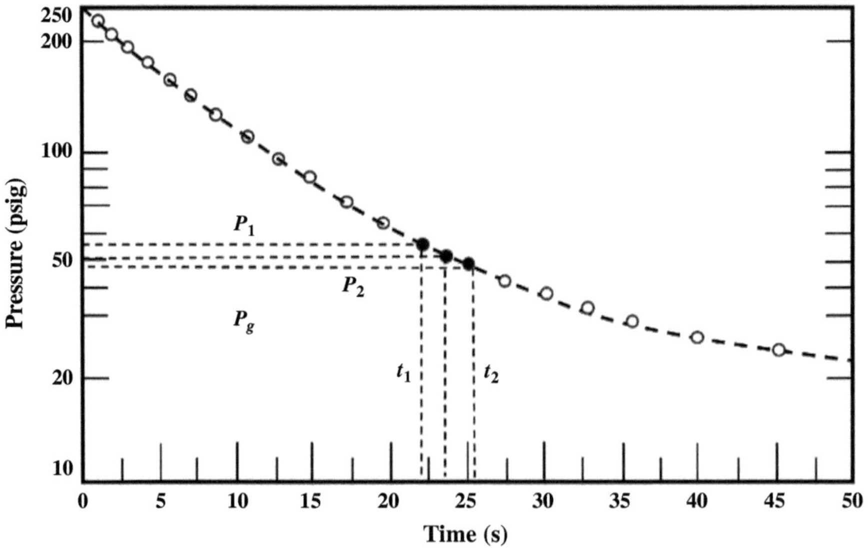

Figure 11.35 Example pressure decay curve.

The Darcy velocity at any point is calculated from the instantaneous flow. This analysis accommodates the nonconstant flow rate through the core, influenced by pore volume, tank volume, and pressure changes. The solution yields an equation of the form:

$$Y(n) = A_0 + A_1X_1(n) + A_2X_2(n) \tag{11.64}$$

where:

- A_0 relates to the slip factor b,
- A_1 relates to the Klinkenberg permeability k_1, and
- A_2 relates to the Forchheimer factor β.

A least squares method is applied to the pressure-time data to solve for the unknowns: b, k_1, and β. However, this method is approximate and may introduce errors, particularly above 1,000 mD permeability. Ruth and Kenny (1989) provided a numerical solution using a finite difference method to address these issues more accurately.

Maintaining isothermal conditions during testing is crucial, as rapid gas expansion (Joule-Thomson effect) can cause temperature drops that invalidate assumptions. Therefore, labs often impose a permeability cut-off of around 1,000 mD for unsteady-state tests to ensure accurate results.

11.11 Steady-State Liquid (Absolute) Permeability Measurements

Although liquid permeability measurements are not usually part of a RCA program, many laboratories now offer these measurements as a standard service. The equipment used is relatively simple, but handling a large number of core plugs and preparing samples can be time-consuming, making these tests more involved, slower, and more expensive than gas permeability measurements. For these reasons, liquid permeability tests are typically performed as part of a SCAL program.

Compared to gas permeability tests, liquid permeability measurements provide a better approximation of reservoir fluid flow. Oil, usually mineral oil, is preferred over brine because it is less reactive and helps prevent rehydration or reflocculation of clays, which could damage the core structure (Table 11.21).

The following sections detail the preparation and test procedures for conducting absolute, single-phase liquid permeability measurements using the steady-state method.

11.11.1 Sample Preparation

Core samples for liquid permeability testing are usually prepared following established procedures. In sensitive rock types, permeability to FW can be measured after cleaning the sample with a miscible flush. Brine is used instead of drying the plugs to prevent damage to clays and protect unconsolidated core material from alteration.

Table 11.21 Steady-state absolute liquid permeability summary.

Advantages	Drawbacks and issues
Permeability measurements are more representative of reservoir fluids than air or Klinkenberg permeability.	Samples must have a cylindrical shape.
Brine permeability can be conveniently added to SCAL protocols, such as permeability before capillary desaturation, relative permeability, and formation factor measurements.	Results are dependent on sample preparation (preserving, cleaning, and drying).
Oil is nonreactive with clays, preventing rehydration or other adverse reactions.	Slower and more expensive than gas permeability measurements (steady-state or unsteady-state).
Multirate measurements help assess and correct pressure transducer errors.	Samples must be fully saturated with liquid; any air can suppress permeability due to relative permeability effects.
	Re-saturation with the formation of water can cause adverse clay reactions in sensitive lithologies.
	Higher NCS is required in low-permeability samples to prevent differential compaction.
	High flow rates may lead to fines migration if the critical velocity is exceeded.
	For tests on chalk or certain carbonates, FW must match *in situ* concentrations, requiring equilibrium with crushed material.

11.11.2 Saturation Procedures

Whenever possible, core plugs should be saturated in deaired brine or mineral oil under a pressure of 2,000 psi for at least 3 days. Vacuum saturation can help speed up the process, but for brine, the time under vacuum should not exceed 20 minutes to prevent excessive water evaporation, which could alter the brine's density and conductivity.

The viscosity of the brine or oil should be measured under test conditions to ensure accurate results. If helium porosity data is available, the core's saturated weight can be used to verify that the core is fully saturated. If a sample appears to be undersaturated, it should be returned to the saturator, and pressure saturation should be continued until full saturation is achieved.

11.11.3 Test Procedures and Permeability Calculation

In this procedure, the core plugs are loaded into a coreholder under brine or oil to avoid air contamination. The samples are then confined at the desired NCS. Liquid injection is controlled either through a constant pressure system (e.g., nitrogen pressure regulation in a transfer cylinder) or constant rate pumps (e.g., HPLC pumps), with the latter typically preferred. A backpressure regulator is recommended, as applying a nominal backpressure (e.g., 200 psi) helps compress and flush trapped air and reduces pressure and flow fluctuations.

To avoid fines migration, the initial flow rate should be limited to 0.5–1 mL/min. Once the flow rate (Q) and pressure drop ($P_1 - P_2$) stabilize, after at least two pore volumes (PV) of brine or oil have been injected, liquid permeability (k) is calculated using Darcy's law:

$$k = \frac{Q\mu L}{A(P_1 - P_2)} \tag{11.65}$$

where:

- Q = Flow rate,
- μ = Liquid viscosity (measured at lab temperature),
- L = Length of the core,
- A = Cross-sectional area of the core, and
- $(P_1 - P_2)$ = Pressure drop across the core.

Since liquids like oil and brine have much higher viscosities than gas, pressure differentials during the test can be significant. It's essential to keep the NCS constant throughout the experiment. In low-permeability samples, large pressure differentials can cause the NCS at the outlet to be lower than at the inlet, potentially leading to differential pore volume compaction, which affects permeability measurements. In such cases, the test should be conducted at higher confining stresses to minimize differential NCS.

Measurement Accuracy

The accuracy of permeability measurements, especially in high-permeability samples, depends on the accuracy and resolution of the pressure measurement system. Modern digital pressure transducers have varying accuracy depending on their pressure range and calibration. These transducers require regular calibration with a deadweight tester at zero and maximum pressure. The error, termed full-scale deflection (FSD), is expressed as a percentage of the transducer's full range. Gauge transducers are usually less accurate than differential transducers, depending on type, calibration, and FSD error.

Small transducer errors, such as nonlinearity or zero shift, can significantly affect permeability calculations if a single flow rate is used. To improve accuracy, a multipoint technique can be applied, measuring the pressure differential ($\delta P = P_1 - P_2$) at three or four incremental flow rates, typically 0.5, 1.0, 1.5, and 2.0 mL/min, while ensuring that the flow velocity does not exceed the critical value for fines mobilization.

Darcy Equation and Linear Regression

The Darcy equation is a linear relationship:

$$\frac{Q}{A} = \frac{k}{\mu}\frac{\delta p}{L} \tag{11.66}$$

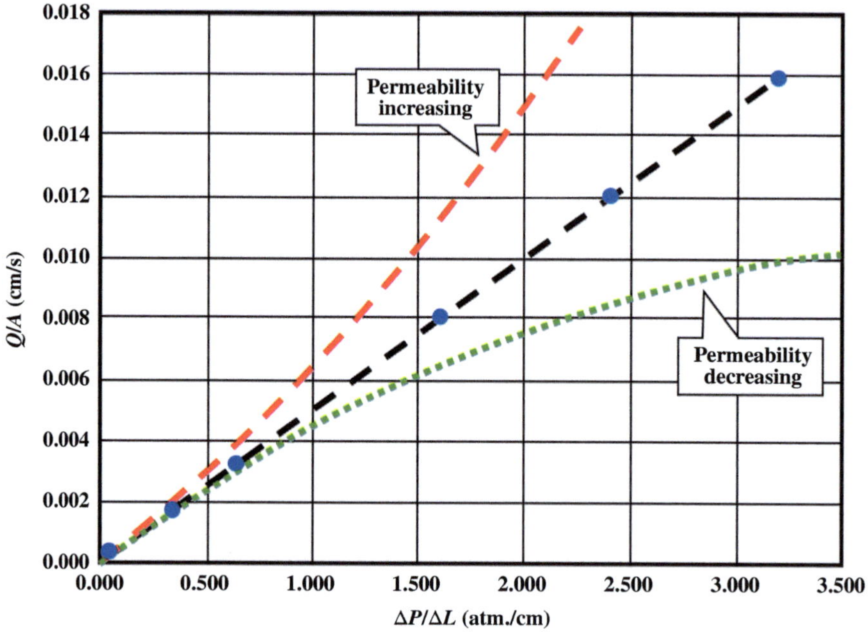

Figure 11.36 Example of Darcy plot for liquid flow.

A plot of $\frac{Q}{A}$ versus $\frac{\delta p}{L}$ should form a straight line:

$$y = mx + c \tag{11.67}$$

The slope (m) equals $\frac{k}{\mu}$, and permeability can be determined by linear regression of the data points (Figure 11.36). A nonzero intercept (for $Q = 0$) on the $\frac{\delta p}{L}$ axis indicates a transducer gauge error (zero shift). Deviation from linearity may suggest a transducer FSD error but is often associated with fines migration. An increasing slope indicates permeability is increasing due to fines being flushed from the core, while a decreasing slope signals pore throat blockage.

To detect fines movement, reverse the flow after reaching stability at each flow rate and monitor the pressure response. A short-lived pressure transient indicates dynamic fines movement. After measuring the highest flow rate, return the rate to the initial low (or base rate) and check for pressure repeatability. If fines are blocking pores, the differential pressure at the base rate will increase. If fines are flushed from the plug, the differential pressure will decrease.

Bibliography

American Petroleum Institute. (1988). *Recommended Practices for Core Analysis* (API RP 40) [Withdrawn]. American Petroleum Institute.

Amott, E. (1959). Observations relating to the wettability of porous rock. *Transactions of the AIME*, *216*(01), 156–162.

Anderson, W.G. (1987). Wettability literature survey part 5: the effects of wettability on relative permeability. *Journal of Petroleum Technology*, *39*(11), 1453–1468.

Boneau, D.F. & Clampitt, R.L. (1977). A surfactant system for the oil-wet sandstone of the North Burbank Unit. *Journal of Petroleum Technology*, *29*(05), 501–506.

Cuiec, L. (1995). Wettability laboratory evaluation under reservoir conditions: a new apparatus. *SCA Symposium, Paper* (Vol. 9529).

Jones, S.C. (1972). A rapid accurate unsteady-state Klinkenberg permeameter. *Society of Petroleum Engineers Journal*, *12*(05), 383–397.

Khanin, A.A. (1969). *Породы-коллекторы нефти и газа и их изучение* [Oil and gas reservoir rocks and their study]. Nedra. (In Russian)

Ph, F. (1901). Wasserbewegung durch boden. *Zeitschrift des Vereines Deutscher Ingenieure*, *45*(50), 1781–1788.

Popiel, W.J. (1978). *Introduction to Colloid Science*. Exposition Press.

Rushing, J.A., Newsham, K.E., Lasswell, P.M., et al. (2004, September). Klinkenberg-corrected permeability measurements in tight gas sands: steady-state versus unsteady-state techniques. *SPE Annual Technical Conference and Exhibition* (Paper No. SPE-89867). Society of Petroleum Engineers. https://doi.org/10.2118/89867-MS

Ruth, D.W. & Kenny, J. (1989). The unsteady-state gas permeameter. *Journal of Canadian Petroleum Technology*, *28*(03), PETSOC-89-03-06.

Treiber, L.E. & Owens, W.W. (1972). A laboratory evaluation of the wettability of fifty oil-producing reservoirs. *Society of Petroleum Engineers Journal*, *12*(06), 531–540.

Zhang, X., Morrow, N.R., & Ma, S. (1996). Experimental verification of a modified scaling group for spontaneous imbibition. *SPE Reservoir Engineering*, *11*(04), 280–285.

12

Economic Assessment of Enhanced Oil Recovery (EOR)

12.1 Introduction

When exploiting oil fields, the issue of increasing the efficiency of oil production becomes of great economic importance. Effective evaluation of EOR methods primarily consists of determining technological efficiency. The technological efficiency of EOR is the basis for predicting economic evaluation and determining the feasibility of industrial application of these technologies. This direction provides for an increase in resources, production, and economic potential and is determined by improving the mechanisms for implementing organizational and economic reserves for increasing the efficiency of oil production.

12.2 Determining the Optimal Time to Start EOR

To improve the accuracy of assessment and forecasting of the technological and economic efficiency of measures to increase oil recovery from productive formations, it is necessary to study the possibilities and conditions for the use of their econometric models.

As the main independent factors, the models include not only labor costs and material resources but also production dependencies influenced by natural factors (physicochemical and hydrodynamic processes, geological structure of the oil reservoir).

To determine the optimal time for starting measures to increase oil recovery (t), we consider a model of the stages of field development (I–IV) with two variables—the volume of oil production (Q) and the net profit of the oil and gas producing enterprise (P).

Such a model allows you to visually identify the influence of geological, technical, technological, and economic processes on the further direction of field development (Figure 12.1).

Graphic analysis of the position of the curves (Q and P) shows that in order to increase the economic efficiency of field development, reduce direct capital investments, and maximize the use of reinvestments, the entire period of field development can be divided into four main stages.

Methods for Enhanced Oil Recovery: Fundamentals and Practice, First Edition.
Baghir A. Suleimanov and Elchin F. Veliyev.

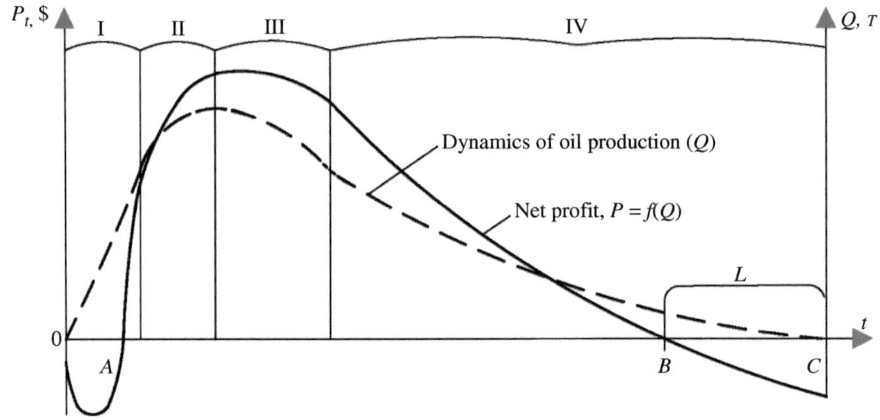

Figure 12.1 Typical chart of the changes in the oil production and net profit of the enterprise by stages of field development.

Oil production at stages I and II of field development is gradually increasing due to the possibility of maximum use of the natural energy of the reservoir (elastic energy, energy of dissolved gas, energy of edge waters, gas cap, and potential energy of gravitational forces). During this time, the initial costs spent on the development and construction of the well stock usually pay off, and the company receives a stable profit.

At stage III, due to the natural depletion of the field, oil production decreases, but production costs increase quite slowly, and the company also makes a profit.

The most difficult is the longest stage of field development. The impact of the main natural geological factor—depletion of the field—leads to a constant increase in the cost of oil production; the cost of one tonne of oil increases, which leads to a decrease in the profit of the oil-producing enterprise. At the moment of time indicated in the figure by point *B*, the cost of oil production may exceed the selling price of oil, and field development becomes unprofitable. In this situation, the enterprise can stop developing the field, although it is technically possible to extract oil for the entire period *L*, limited by points *B* and *S*.

That is why, when the level of oil production decreases and when profitability falls, it becomes necessary to resort to measures to increase oil recovery and maintain the level of income of the enterprise.

Detailed analysis of *P* curves and *Q* allows you to identify the optimal time to start activities to increase oil recovery, which will change the position of point *B* on line *t*, i.e., it will make it possible to extend the time for an oil-producing company to make a profit.

12.3 Technological Efficiency of EOR

In all cases, the introduction of EOR methods raises the need to assess their technological effectiveness based on field data. At the stage of experimental work, this is necessary in order to make a decision on the feasibility of industrial application of the method, and at the implementation stage—to determine the effectiveness of the money spent. At the same time, an objective, reliable assessment of the effectiveness of the method is required so as not to overestimate or underestimate its potential capabilities.

At the present stage of development, oil-producing enterprises, based on the results of an assessment of technological efficiency, are faced with the task of choosing priority methods for increasing oil recovery, since the efficiency of the enterprise as a whole depends on the effectiveness of these measures. When choosing methods for enhancing oil recovery, the main issue is predicting the increase in oil production from these activities. It is also necessary to take into account that carrying out one EOR at different field development sites can lead to different results since the effectiveness of the measures is influenced by many geological and technological parameters.

The volume of its implementation depends on the correct and objective assessment of the effectiveness of any measure, which is one of the main means of implementing processes to stabilize oil production during the exploitation of fields at a late stage of development.

An indicator of the technological efficiency of EOR methods is the effect for the period under review. It can be determined based on a comparison of the basic and implemented field development options using the oil production volume method. When determining technological efficiency, the base option is taken to be the method that was used at the stage of the current development of the facility before the start of the use of EOR. The difference in oil production represents the additional production due to the method. This is where we can see the direct relationship between technological and economic efficiency (Figure 12.2).

In the graph, curves *1* and *1′* are, respectively, the production and cost of oil of the base development option; curves *2* and *2′*, *3* and *3′*, *4* and *4′*—oil production and

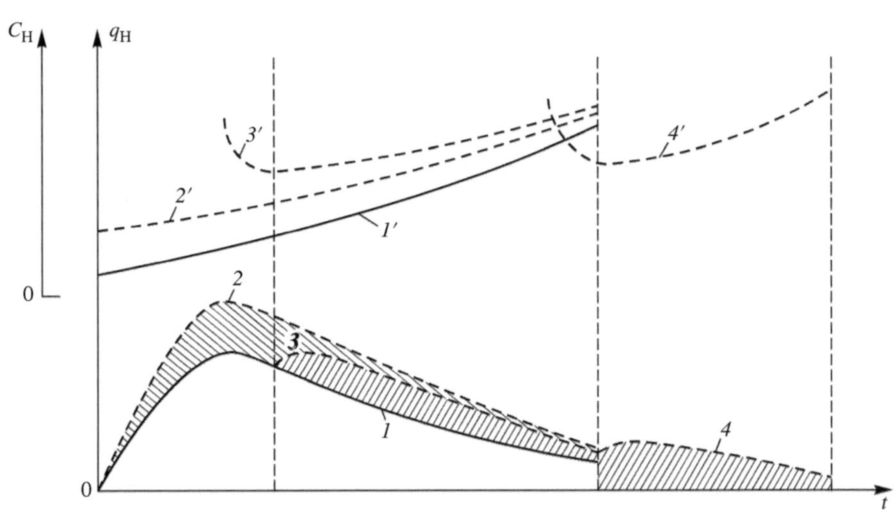

Figure 12.2 The structure of the dependence of technological (q_H) and economic (C_H) effects on the time of implementation (t).

cost when using EOR methods implemented at different stages; shaded areas are additional oil production from the methods used.

If field development using the basic method or depletion mode is technologically impossible, or the use of the method begins after reaching the profitability limit with conventional technology, then all oil production can be considered additional through the use of EOR—this is the so-called "zero option" (curve *4*).

Justification of the effectiveness of the proposed measures to increase oil recovery is carried out on the basis of a comparison of the proposed option with the base option, which takes the situation before the implementation of the recommended measure.

The technological effectiveness of different methods for increasing oil recovery, used in different geological and physical conditions, needs to be determined in different ways, depending on the nature of the effect and the most representative indicators.

However, to justify the use of the EOR method, the presence of the expected technological effect, which is expressed in additional oil production, a decrease in the water cut of wells, and an increase in the oil recovery factor, is not enough. It is necessary to calculate the economic effect of the event, since the costs of the event and additional oil production may exceed the revenue from the sale of additionally produced oil and the event will be unprofitable.

12.4 Economic Efficiency of EOR

Carrying out measures to increase oil recovery can be considered an investment project; their economic efficiency can be defined as the ratio of the results to the costs that caused this result. In other words, the smaller the cost and the larger the value in which the result is embodied, the higher the efficiency.

The effectiveness of investment projects is characterized by a system of economic indicators that make it possible to judge the economic attractiveness of the project and the economic advantages of some projects over others.

It is recommended to compare various investment projects and select the best one using various indicators, which include:

- net present value (*NPV*), or integral income,
- discounted index of return (*DPI*), or profitability index (*PI*),
- internal rate of return (*IRR*),
- payback period.

Discounting is the determination of the cash flow of an enterprise by bringing the cost of all payments to a certain point in time, i.e., discounting is the basis for calculating the cost by taking into account the time factor.

The most important indicator of the effectiveness of an investment project is *NPV*—the accumulated discounted effect for the calculation period (*NPV*), which is defined as the difference between integral income and expenses reduced to this period.

NPV is closely related to an indicator such as the project *PI*. The latter is an important criterion for whether the project will be profitable. To determine it, the amount of discounted income should be divided by the amount of all planned costs, otherwise, we can say that this is the ratio of discounted inflows to discounted outflows.

If the *PI* of discounted investments $PI > 1$ ($NPV > 0$), then the investment will pay off, and the higher it is, the better the project in economic terms. If $PI < 1$ ($NPV < 0$), then such a project will not ensure the receipt of a normative (standard) rate of profit and, therefore, will lead thc cnterprise to potential losses. If *PI* is equal to 1, then there will be no result from the investment ($NPV = 0$).

The payback period is the duration of the period from the initial moment of investment to the "moment of payback taking into account discounting." The payback point, taking into account discounting, is that point in time in the calculation period after which the current *NPV* becomes and remains positive in the future. An investment project is accepted if the calculated payback period does not exceed the duration of the project itself.

Next, we will consider how the above-listed economic indicators are used when calculating the effectiveness of measures aimed at increasing oil recovery.

The economic effect of implementing measures to enhance oil recovery (*E*) is determined by the difference between the cost estimate of the results and the cost estimate of the costs of their implementation for the calculation period using the formula

$$E = P_t - Z_t, \tag{12.1}$$

where P_t is the cost estimate of the results of implementing EOR measures for the calculation period, n.u. (notional unit); Z_t—cost estimate of costs when carrying out EOR measures for the calculation period, n.u.

The valuation of the results (P_t), as a result of the implementation of EOR measures, represents the increase in revenue due to an increase in the volume of oil production, taking into account its sale on the market, and is determined by the formula

$$P_t = \Delta Q_t \times C_i, \tag{12.2}$$

where ΔQ_t is the increase in oil production for the calculation period, tonnes; C_i is the wholesale price of oil sold by the enterprise, n.u./tonnes.

The increase in oil production (ΔQ_t) is one of the most important factors used to determine the economic feasibility of measures taken to increase oil recovery. It is caused by an increase in the average daily flow rate and is determined by the formula

$$\Delta Q_t = \Delta q \times t \times N, \tag{12.3}$$

where Δq is the increase in the average daily flow rate, tonnes/day; *t* is the date of operation of wells, where the effect of well stimulation is observed, days; *N* is the number of wells producing additional oil, number of wells.

The valuation of operating costs (Zt) is the sum of current production costs resulting from the implementation of EOR measures and the costs of their implementation. In this case, operating costs are determined by the formula

$$Z_t = \Delta Z_{inc} + Z_i, \tag{12.4}$$

where ΔZ_{inc} is the incremental variable costs of additional oil production, n.u.; Z_i—costs of carrying out activities for the calculation period, n.u.

The increase in conditionally variable costs (ΔZ_{inc}) of additional oil production is determined by the formula

$$\Delta Z_{inc} = \Delta Q_t \times Z_{inc}, \tag{12.5}$$

where ΔQ_t is the increase in oil production due to the stimulation, tonnes; Z_{inc}—variable costs of additional production of 1 tonne of oil, n.u./tonnes.

Substituting the value of formulas (12.2), (12.4), and (12.5) into Formula (12.1), we obtain

$$E = \Delta Q_t \times (C_i - Z_{inc}) - Z_i. \tag{12.6}$$

The cost of additional production of 1 tonne of oil (Z_{inc}) is the sum of conditionally variable costs according to the items of calculation of the cost of production of 1 tonne of oil. The change in these costs due to the event is considered in the context of the following costing items:

- energy costs for mechanized fluid extraction (Z_{ec}),
- costs of maintaining reservoir pressure (Z_{mrp}),
- costs of collecting and transporting oil and gas (Z_{ct}),
- costs for technological preparation of oil (Z_{tp}).

1. Energy costs (Z_{ec}) are calculated depending on the volume of mechanized fluid extraction. When calculating these costs, they are based on the average cost of electricity and its specific consumption. Energy costs for liquid extraction are calculated using the formula

$$Z_{ec} = R_{ec} \times C_{ec} \times q_{lp}, \tag{12.7}$$

where R_{ec}—specific energy consumption when extracting liquid mechanically, kWh/tonne; C_{ec}—cost of 1 kWh of electricity, n.u.; q_{lp} is the increase in liquid production by mechanical means in the calculation period, tonnes.

2. The costs of maintaining reservoir pressure (Z_{mrp}) consist of the costs of servicing injection wells (Z_{inj}) and the costs of water injection (Z_{wi}):

$$Z_{mrp} = Z_{inj.} + Z_{wi} \tag{12.8}$$

Calculation of costs for servicing injection wells (Z_{inj}) is determined by the formula

$$Z_{inj} = R_{inj.} \times N_{inj} \tag{12.9}$$

where R_{inj}—expenses for servicing the existing stock of injection wells, n.u./well; N_{inj}—number of injection wells in the calculation period, well.

When determining the cost of water injection (Z_{wi}), they proceed from the volume of water injected into the reservoir, its cost, and energy costs. The standard for determining energy costs when pumping water into a reservoir is established based on the specific energy consumption and the cost of 1 kWh of electricity. The cost of water injection is determined by the formula

$$Z_{wi} = (R_{ec} \times C_{ec} + C_w) \times q_w, \tag{12.10}$$

where R_{ec}—specific energy consumption when pumping water, kWh/m^3; C_{ec}—cost of 1 kWh of electricity, n.u.; C_w—cost of water, n.u./m^3; q_w—increase in the volume of pumped water during the calculation period, m^3.

3. Costs for collection, transportation, and technological preparation of oil ($Z_{ct.}$) are calculated depending on the volume of extracted liquid according to the formula

$$Z_{ct} = R_{ct.} \times q_f, \tag{12.11}$$

where $R_{ct.}$—expenses for the collection and transportation of oil and gas, n.u./tonne; q_f is the increase in liquid production in the calculation period, tonnes

4. The costs of technological preparation of oil (Z_{tp}) are calculated using the formula

$$Z_{tp} = R_{tp.} \times q_f, \tag{12.12}$$

where $R_{tp.}$—expenses for technological preparation of oil, n.u./tonne; q_f is the increase in the volume of liquid production used for technological preparation of oil in the calculation period, tonne.

The total conditionally variable costs for additional oil production (Z_{inc}) during EOR are determined by Formula (12.13).

$$Z_{inc} = Z_{ec} + Z_{mrp} + Z_{ct} + Z_{tp}. \tag{12.13}$$

It should be noted that all components of the formula must be calculated for 1 tonne of additional oil. The remaining expenses for the items calculating the cost of production of 1 tonne of oil remain unchanged. Expenses for depreciation of wells remain at the same level, since new wells are not commissioned during the activities. Expenses for preparation and development of production also remain unchanged. Expenses for basic and additional salaries and contributions for social needs also do not change, since the implementation of EOR does not lead to an increase or decrease in the number of employees.

In addition to the increase in the conditionally variable costs discussed above in the total operating costs (Formula 12.8), carrying out measures to increase oil recovery requires additional costs for their implementation (Z_i), which are carried out either by the enterprise itself or through the involvement of contractors.

In the case where measures to enhance oil recovery are carried out by a contracting organization, the amount of the bilateral contract is the cost of carrying out these measures. If EOR activities are carried out by the enterprise itself, then the costs of carrying them out during the calculation period consist of the following costs:

- costs of materials (Z_m),
- costs of workers' wages (Z_{ww}),
- costs of transport services (Z_t).

1. The costs of reagents (Z_m) (acids, various reagents, process fluids, organic solvents, surfactants, etc.) necessary for the behavior of EOR are the sum of the costs for each of the materials and are determined by the formula

$$Z_m = \sum V_m \times C_m, \tag{12.14}$$

where V_m is the volume of the injected reagent, m^3; C_m—price of the reagent, n.u./m^3.

2. The cost of wages for workers (Z_{ww}) employed in carrying out EOR activities, taking into account social insurance contributions, is determined by the formula

$$Z_{ww} = C \times T_i \times s, \tag{12.15}$$

where C is the total hourly rate of workers involved in the event, taking into account bonuses and additional wages, n.u./h; T_i—duration of work on the implementation, h; s—contributions to social insurance and pensions, fraction of one.

3. Costs for transport services (Z_t) are the sum of the costs for each type of equipment used in carrying out EOR and are determined by the formula

$$Z_t = \sum C_t \times T_i,$$

where C_t is the cost of 1 hour of operation of this type of technical equipment, n.u./h; T_i—hours worked for this type of technical equipment, hours.

Total costs for carrying out activities to enhance oil recovery (Z_i) are determined by the formula

$$Z_i = Z_m + Z_{ww} + Z_t.$$

The economic effect of measures to increase oil recovery (Formula 12.6) is the increase in the enterprise's balance sheet profit from implementation, which in turn is regulated by the Tax Code. Then, the profit remaining at the disposition of the enterprise, i.e., the net income of the calculation period, is determined by the formula

$$NI = [\Delta Q \times (C_i - Z_{inc}) - Z_i] \times (1 - n) \tag{12.16}$$

where n is the profit tax rate, the fraction of 1.

The economic efficiency of investments in EOR activities can be determined on the basis of discount calculations for bringing costs and income to a point in time (calculation year), since the effect of their implementation can be observed for quite a long time. Bringing different time costs and results to a single point in time (discounting) reflects the value of future income from modern positions.

Discount calculations are based on the fact that the value of a monetary unit (ruble, euro, and dollar) spent and received in different years will be different, and therefore their simple algebraic summation will not have a real effect.

When making an economic justification for investment projects, the first (or previous to the first, i.e., zero) year is taken as the calculation year to which the reduction is carried out. The reduction is carried out using the discount factor (K_d), which is determined separately for each year of the billing period.

Discounted balance (discounted cash flow, *DCF*), i.e., the difference between receipts and expenses over a certain period of time, is determined by the formula

$$DCF = (P_t - Z_t) \times (1 - n) \times K_d \tag{12.17}$$

or (for EOR activities):

$$DCF = [\Delta Q \times (C_i - Z_{inc}) - Z_i] \times (1 - n) \times K_d, \tag{12.18}$$

where K_d—discount coefficient.

Discount coefficient is a coefficient used for discounting, that is, bringing the value of cash flow at the *n*th step of a multistep calculation of the effectiveness of an investment project to a moment called the reduction moment.

The discount factor shows how much cash we will receive taking into account the time factor, how much the cash flow will decrease in the *n*th year, based on the given discount rate. When setting the value of the discount factor, they usually focus on the average level of the discount rate. The discount factor is determined by the formula

$$K_d = \frac{1}{(1 + R)^t}, \tag{12.19}$$

where *R*—established value of the discount rate, a fraction of one; *t*—number of years (months) from the future to the current moment of introduction, $t = 1, \ldots, n$.

The discount factor is always less than one. The level of the discount factor may also take into account the risk of the investment being made.

The discount rate (*R*) characterizes the efficiency of investments. Under conditions of a stable economy, it is taken at the level of 0.1, that is, with a return on capital of 10% in a year.

The discounted cash flow accumulated over the entire calculation period (accumulated discounted balance) represents a *NPV*, which characterizes real (net) income taking into account the time factor.

In other words, *NPV* is defined as the sum of current effects for the entire calculation period, reduced to the initial year:

$$NPV = \sum_{t=0}^{T} \frac{(P_t - Z_t) \times (1 - n)}{(1 + R)^t} \tag{12.20}$$

or, for EOR activities,

$$NPV = \sum_{t=0}^{T} \frac{[\Delta Q \times (C_i - Z_{inc}) - Z_i] \times (1 - n)}{(1 + R)^t}. \tag{12.21}$$

To find the *PI* of discounted costs first, discounted inflows and discounted outflows are determined for the entire period of the effect, then the ratio of discounted inflows to the absolute value of discounted outflows is found.

$$PI = \frac{CI}{CO}, \tag{12.22}$$

where *CI*—cash inflow for a certain period of time, *CO*—cash outflow over a certain period of time.

The discounted payback period is the period of time over which the accumulated income becomes equal to the cost of the activity. The moment at which the *NPV* changes sign from "−" to "+" is the payback period of the project.

Carrying out an EOR event becomes profitable at the moment when the *IRR* meets the following condition for the equality of the *NPV*:

$$\sum_{t=0}^{T} \frac{[\Delta Q \times (C_i - Z_{inc}) - Z_i] \times (1 - n)}{(1 + R)^t} = 0. \tag{12.23}$$

The *IRR* answers the question about the economic feasibility of carrying out implementation: if $IRR > R$ (discount rate), then they are appropriate; if $IRR \leq R$, then the implementation is impractical.

12.4.1 Sensitivity to Risks

Since calculations are carried out for the future, and actual indicators are used as initial data, which in the future may change both up and down, which may be a certain risk in achieving economic results, it is necessary to analyze the sensitivity of the project to risk. For sensitivity analysis of an investment project, the main task is to assess the degree of influence of changes in each of the initial parameters in order to foresee the worst-case scenario.

When analyzing the sensitivity of EOR activities, the impact on *NPV* of changes in production growth, oil prices, semi-variable costs, costs of implementation, discount rates, and income tax rates are assessed. In addition to the project's *NPV*, the *IRR* or discounted payback period can be used as an output parameter. Carrying out this type of analysis allows you to assess the project's resistance to risk factors.

When conducting a sensitivity analysis of an investment project, the following algorithm should be followed.

1. The value of the output parameter (for example, *NPV*) is calculated for the base scenario—NPV_{base}.
2. The value of the output parameter is calculated when one of the input parameters changes (for example, oil prices, *C*) provided that the remaining input parameters remain unchanged, *NPV*(*C*).
3. The percentage change (ΔNPV) of the input and output parameters relative to the base scenario is calculated.

$$\Delta NPV = \frac{NPV(C) - NPV_{base}}{NPV_{base}} \times 100 \tag{12.24}$$

4. The degree of sensitivity to the risk factor (F_r) of the output parameter to a change in the input parameter is calculated *by* dividing the percentage change in the output parameter (ΔNPV) by the percentage change in the input parameter (ΔC).

$$F_R = \frac{\Delta NPV}{\Delta C}. \tag{12.25}$$

In general, the formula for assessing sensitivity can be written as follows:

$$\text{Risk Factor} = \frac{\text{percentage change in the output parameter}}{\text{percentage change in the input parameter}}.$$

If the sensitivity is negative, this means that with an increase in the input parameter by 1%, the *NPV* of the project will decrease by % of the risk factor (F_r), and, conversely, with a decrease in the input parameter by 1%, the *NPV* of the project will increase by % of the risk factor.

If the sensitivity is positive, then with an increase in the input parameter by 1%, the *NPV* of the project will increase by % of the risk factor, and, conversely, with a decrease in the input parameter by 1%, the *NPV* of the project will decrease by % of the risk factor.

Such calculations are carried out for all input parameters. This allows us to identify risk factors that represent the greatest vulnerability for an investment project, in this case, for carrying out EOR activities.

12.4.2 Calculation example

Let's consider an abstract example of calculating the economic efficiency of an event involving the injection of slow-acting acid compositions into formations in order to increase oil recovery.

The input data for calculating the economic efficiency of this activity are presented in Table 12.1.

The planned period of effect from carrying out implementation to increase oil recovery is 1 year. The calculation period is divided into steps, one step is equal to one month, and then the time in the calculation period (t) is measured in days.

Variable costs for oil extraction are equal to 247.78 notional unit per tonne, constant throughout the entire calculation period.

Income tax is currently 20% of the book profit remaining on revenue after compensation for operating costs and payment of all taxes.

The discount rate is 10%. The calculation of economic efficiency indicators is carried out at constant prices. The assessment of economic efficiency is generally calculated based on the determination of cash flows from operating activities and investing activities.

Table 12.1 Input data for calculation of economic efficiency.

No.	Parameters	Unit	Value
1	Number of wells, N	Pcs.	4
2	Costs of implementing the event, Z_i	n.u.	50,425
3	Increase in average daily flow rate per well, Δq	tonne/day	10.9
4	Variable costs of oil extraction, Z_{inc}	n.u./tonne	247.78
5	Selling price for 1 tonne of oil, C_i	n.u./tonne	316.0
6	Discount rate, R	%	10.0
7	Income tax, n	%	20.0

Table 12.2 Calculation results for economic efficiency of the EOR implementation.

Calendar time (t), days	Increase in oil production (ΔQ_t), tonnes (Formula 12.3)	Sales revenue (P_t), n.u. (Formula 12.2)	Increase in conditionally variable costs (ΔZ_{inc}), n.u. (Formula 12.5)	Economic effect (E), n.u. (Formula 12.6)	Net income (NI), n.u. (Formula 12.16)
31	1,351.6	427,105.6	334,899.45	41,781.15	33,424.92
28	1,220.8	385,772.8	302,489.82	83,282.98	66,626.38
31	1,351.6	427,105.6	334,899.45	92,206.15	73,764.92
30	1,308.0	413,328.0	324,096.24	89,231.76	71,385.41
31	1,351.6	427,105.6	334,899.45	92,206.15	73,764.92
30	1,308.0	413,328.0	324,096.24	89,231.76	71,385.41
31	1,351.6	427,105.6	334,899.45	92,206.15	73,764.92
31	1,351.6	427,105.6	334,899.45	92,206.15	73,764.92
30	1,308.0	413,328.0	324,096.24	89,231.76	71,385.41
31	1,351.6	427,105.6	334,899.45	92,206.15	73,764.92
30	1,308.0	413,328.0	324,096.24	89,231.76	71,385.41
31	1,351.60	427,105.6	334,899.45	92,206.15	73,764.92
Total	**15,914.0**	**5,028,824.0**	**3,943,170.92**	**1,035,228.08**	**828,182.46**

The results of calculating the economic efficiency of the implementation of EOR are summarized in Table 12.2.

The average daily production of wells where the event is being carried out increases by an average of 10.9 tonnes/day, which will allow the enterprise to obtain an additional 1,351.6 tonnes of oil in the first month of applying the technology, and at the end of the first year of the event, additional oil production will amount to 15,914 tonnes.

The amount of revenue from the sale of extracted oil in the first year of operation of the technology will be 5,028.8 thousand. n.u. The amount of profit after tax and minus all expenses for the event will be 828.2 thousand. n.u.

The second stage of assessing the effectiveness of the EOR implementation measure is carried out using discount calculations, the results of which are summarized in Table 12.3.

With an initial investment of 50,425 n.u., the technology will pay for itself in 1.3 months, i.e., a discounted payback period *of DPP* = 1.3. This indicates that the event is becoming profitable and *the NPV* corresponds to the equality condition of Formula (12.23). At the end of the implementation period, the event will bring additional profit to the enterprise in the form of *NPV* in the amount of 500.7 thousand. n.u. The dynamics of the *NPV* for the implementation of EOR are shown in Figure 12.3.

Table 12.3 Results of second stage calculations assessing effectiveness of EOR implementation.

No.	Net income (*NI*), n.u.	Discount factor (K_d) (Formula 12.19)	NPV^a, n.u. (Formula 12.21)
1	33,424.92	1.0000	33,424.92
2	66,626.38	0.9091	93,994.36
3	73,764.92	0.8264	154,957.1
4	71,385.41	0.7513	208,590
5	73,764.92	0.6830	258,972.5
6	71,385.41	0.6209	303,297.4
7	73,764.92	0.5643	344,925.5
8	73,764.92	0.5131	382,773.1
9	71,385.41	0.4664	416,068.5
10	73,764.92	0.4241	447,351.3
11	71,385.41	0.3855	474,870.8
12	73,764.92	0.3505	500,726.0
Total	**828,182.46**		**500,726.0**

a. *NPV* for months is calculated on an accrual basis.

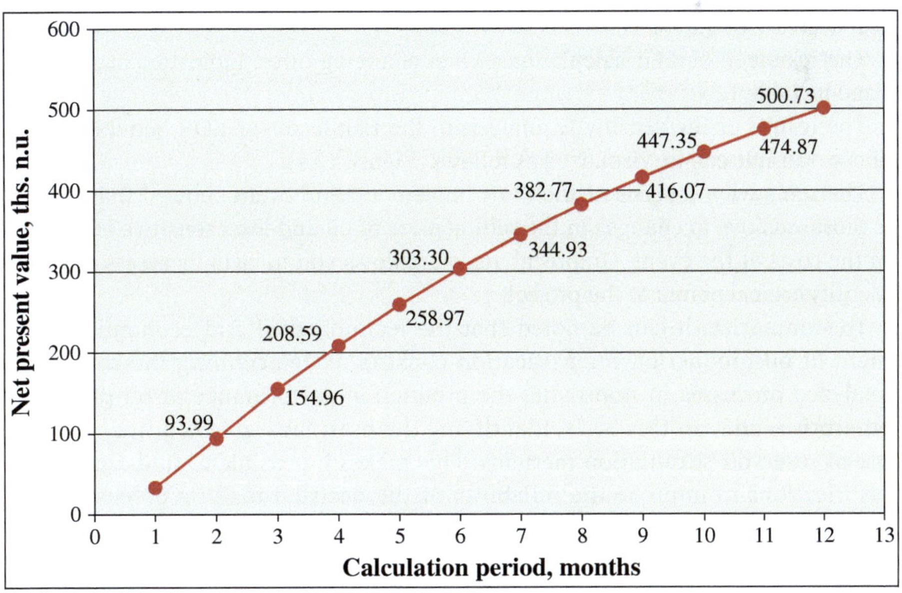

Figure 12.3 Dynamics of *NPV* for the implementation of EOR.

Next, we will analyze the sensitivity of the implementation of EOR to risk factors. The calculation is carried out taking into account changes in each of the indicators that affect the final results *of the NPV*.

Let's consider the impact on *NPV* when each of the following indicators changes, provided the others remain unchanged:

- increase in oil $NPV\,(\Delta Q_t)$,
- oil sales prices $NPV\,(C_i)$,
- variable costs per tonne of oil $NPV\,(Z_{inc})$,
- costs of carrying out measures $NPV\,(Z_i)$.

Let's calculate the *NPV* of EOR measures when the increase in oil $NPV\,(\Delta Q_t)$ decreases by 10% according to Formula (12.21):

$$NPV(\Delta Q_t) = 446{,}619.4 \text{ n.u.}$$

The *NPV* of the base scenario (NPV_{base}) is equal to 500,726.0 n.u. (Table 12.3). Percentage change *in* $NPV\,(\Delta Q_t)$ relative to *the* NPV_{base} the base scenario is determined by Formula (12.24):

$$\Delta NPV = \frac{446{,}619.4 - 500{,}726.0}{500{,}726.0} \times 100 = -10.81\%.$$

Degree of sensitivity to the risk factor NPV_{base} from a decrease in oil growth (ΔQ_t) by 10% is determined by Formula (12.25):

$$F_R = \frac{-10.81\%}{10\%} = -1.08.$$

This means that with a decrease in oil growth (ΔQ_t) by 1% *NPV* of EOR activities will decrease by 1.08%.

The results of similar calculations when changing other indicators are summarized in Table 12.4.

The results of the sensitivity analysis to the risk factor of EOR activities in the above example can be visualized as follows (Figure 12.4).

The sensitivity analysis of the EOR implementation event showed that its *NPV* is most sensitive to changes in the selling price of oil and least sensitive to changes in the costs of the event. Graphical analysis allows you to visually assess risks and identify critical points of the project.

To summarize, it can be noted that the technological and economic assessment of oil production intensification consists of determining the role of the analyzed processes in improving the production performance of oil-producing enterprises and, on this basis, identifying the most favorable conditions for the use of reservoir stimulation methods. This makes it possible to find criteria and justifications to improve the reliability of the decision-making process for the implementation of EOR.

Table 12.4 Summary of calculation results for different indicator variations.

Parameters	Change in parameters, %				
	−10	**−5**	**0**	**+5**	**+10**
Oil production growth					
• $NPV(\Delta Q_t)$, n.u.	446,619.4	473,672.7	500,726.0	527,779.3	554,832.6
• risk factor	−1.08	−0.54	0	0.54	1.08
Oil sales price					
• $NPV(C_i)$, n.u.	250,100.3	375,413.1	500,726.0	626,038.8	751,351.7
• risk factor	−5.01	−2.50	0	2.50	5.01
Variable costs					
• $NPV(Z_{inc})$, n.u.	697,245.1	598,985.5	500,726.0	402,466.4	304,206.9
• risk factor	3.92	1.96	0	−1.96	−3.92
Implementation costs					
• $NPV(Z_i)$, n.u.	504,760	502,743	500,726.0	498,709	496,692
• risk factor	0.08	0.04	0	−0.04	−0.08

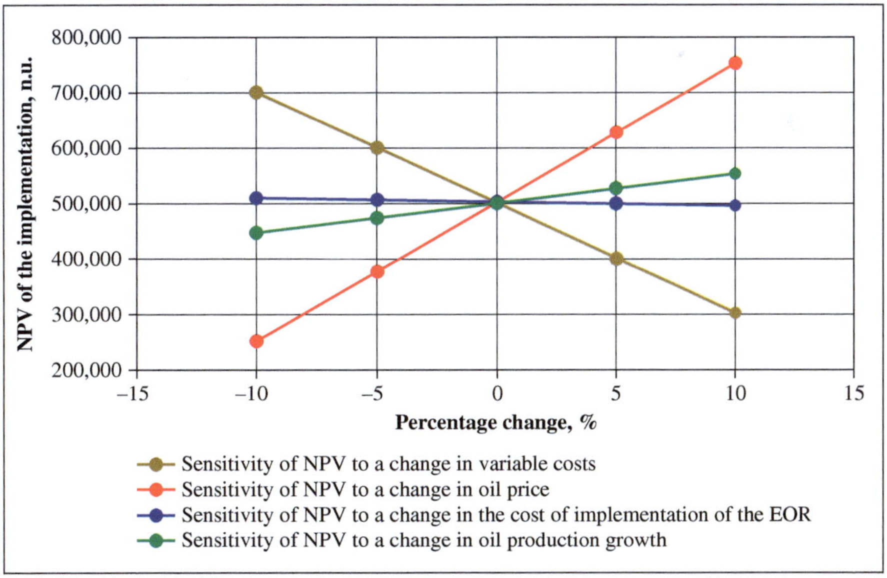

Figure 12.4 Sensitivity of *NPV* to changes in calculated parameters.

Bibliography

Galiullin, F.R. (2005). *Economic Efficiency of Methods for Increasing Oil Recovery at the Late Stage of Field Development*. Ufa branch of the Institute of Economics of the Ural Branch of the Russian Academy of Sciences.

Konstantinova, A. (2016). Increasing oil recovery from oil and gas fields: Methods, economic efficiency, legislative regulation. VMM-15-02.

Ministry of Economy, Ministry of Finance, & Gosstroy of the Russian Federation. (1999, June 21). Methodological recommendations for assessing the effectiveness of investment projects.

Ministry of Fuel and Energy of the Russian Federation. (1996). Methodological guidance for assessing technological efficiency and application of enhanced oil recovery methods (RD-153-39.1-004.96).

Ministry of Fuel and Energy of the Russian Federation. (1996). Regulations for the preparation of design technological documents for the development of oil and gas fields (RD-153-39-007-96).

Scarlett, B. (2009). *Performance Operations: Operational Level*. Elsevier; CIMA Publishing.

Surguchev, M.L. (1985). *Secondary and Tertiary Methods for Increasing Oil Recovery*. Nedra.

Index

Note: Page numbers in *italics* refers to figures and **bold** refers to tables.

Methods for Enhanced Oil Recovery: Fundamentals and Practice, First Edition.
Baghir A. Suleimanov and Elchin F. Veliyev.

r